COMPARATIVE ANIMAL BEHAVIOR

COMPARATIVE ANIMAL BEHAVIOR

Donald A. Dewsbury

Professor of Psychology
University of Florida

McGRAW-HILL BOOK COMPANY

New York St. Louis San Francisco Auckland Bogotá
Düsseldorf Johannesburg London Madrid Mexico Montreal
New Delhi Panama Paris São Paulo Singapore Sydney Tokyo Toronto

COMPARATIVE ANIMAL BEHAVIOR

7 8 9 10 11 KPKP 8 9 8 7 6 5 4 3

This book was set in Times Roman by Black Dot, Inc. The editors were Richard R. Wright and Susan Gamer; the cover was designed by Scott Chelius; the production supervisor was Dominick Petrellese. The drawings were done by J & R Services, Inc. Kingsport Press, Inc. was printer and binder.

Library of Congress Cataloging in Publication Data

Dewsbury, Donald A date
Comparative animal behavior.

Bibliography: p.
Includes index.
1. Animals, Habits and behavior of. 2. Psychology, Comparative. I. Title.
QL751.D452 591.5 77-8885
ISBN 0-07-016673-0

Contents

PART THREE
DEVELOPMENT OF BEHAVIOR

Preface

This book is intended for use with undergraduates taking their first course in the area of comparative psychology and animal behavior. It is written by a comparative psychologist, with a certain emphasis on material from that discipline. However, the area of animal behavior is so truly interdisciplinary that any contemporary presentation must represent some kind of synthesis of material from different related disciplines. Therefore, the book also draws heavily on material from related disciplines. It is hoped that the book may also generate some interest in those disciplines.

This book is short, compared with others covering similar material. That is deliberate: the goal was to create a relatively short book that would provide a solid core for a course in animal behavior, leaving the instructor free to supplement this basic core with related readings which would emphasize material of his or her choosing. Thus, the student would be able to profit from the instructor's expertise in a particular subarea, and, at the same time, to view the area of animal behavior in a somewhat broader contemporary perspective.

The basis for the organization of the book lies in the approach to the study of animal behavior most clearly delineated by Niko Tinbergen. The first task of the animal behaviorist is to observe and describe behavior. However, observation and description are not ends in themselves. Rather, they represent a way to provide a foundation for asking four classes of questions—questions about development, mechanism, evolutionary history, and function. Part One provides introductory material related to the study of animal behavior. Part Two comprises three chapters in which behavior is described. The goal is to familiarize the student with a broad range of behavioral patterns, many of which will be analyzed further in later sections. Parts Three, Four, Five, and Six deal with the four classes of questions to be asked of animal behavior—questions about development, mechanism, evolutionary history, and adaptive significance. The book concludes with a section on learning (Part Seven). The organization of Part Seven is parallel to, and restates, that of Parts One through Six, as it covers description, development, mechanisms, evolutionary history, and adaptive significance.

I have been fortunate to have had many teachers and colleagues who have influenced my approach to animal behavior. While I cannot name them all, I thank them all. The manuscript was typed by Juanita Brown, Donna Gillis, and Martha Hettrick. Finally, I wish to thank Joyce, Bryan, and Laura for their tolerance as they discovered that what they may have heard about the difficulties of having a budding author in the family is all true.

Donald A. Dewsbury

Part One

Introduction to Animal Behavior

This section provides a general introduction to the study of animal behavior and to some of its fundamental concepts. It includes discussions of the activities of animal behaviorists and the questions they ask, a brief history of the field, and a consideration of the various disciplines which together constitute the contemporary study of animal behavior. The section concludes with a survey of animal species and their classification and a summary of the basic principles of evolution.

Chapter 1

The Study of Animal Behavior

The contemporary science of animal behavior is the product of the efforts of a large number of scientists from different disciplines, each working to answer a fundamental set of questions about the behavior of animals. Great diversity stems from the training, questions emphasized, and methods used by different animal behaviorists. Unity comes from a general dedication to the complete understanding of the behavior of animals and a tolerance for consideration of questions and utilization of methods different from one's own so long as research is intelligently and competently conducted. In this chapter we shall begin our treatment of animal behavior with a consideration of the questions asked in the study of animal behavior and a brief sketch of the early history of the field.

THE SCOPE OF THE STUDY OF ANIMAL BEHAVIOR

The noted European ethologist Niko Tinbergen has provided a useful and widely accepted framework within which to discuss the structure of the study of animal behavior. Tinbergen (1963a) pointed out that the first tasks of the study of behavior are observation and description. Once behavior is adequately described, the scientist can proceed to ask four kinds of questions about behavior—questions of immediate causation, development, evolutionary history, and function. A truly comprehensive treatment of behavior entails answers to all four kinds of questions.

Observation and Description

Observation and description provide the foundation upon which the study of behavior is built. It is all too easy for an eager student of behavior to attempt to find answers to complex questions about behavior without an adequate foundation in observation and description. This has been referred to by Lorenz (1973) as "the fashionable fallacy of dispensing with description." An investment of time spent in the often tedious task of simply watching and describing the behavior to be studied may pay considerable dividends when complex questions come under investigation. It is only when one has a firm grasp on the behavioral repertory of the species in question that one will understand the behaviors which may often appear quite unexpectedly. An example may be instructive. Breland and Breland (1961) reported an attempt to condition raccoons to place coins in a piggy bank. They used the principles of operant psychology and had little difficulty in the early stages of training. However, as the raccoons were being trained to drop the coins into the bank, an unexpected behavior developed. The raccoons started rubbing the coins together and dipping them in a container. Had the Brelands had no knowledge of the naturally occurring behavioral patterns of raccoons, they would not have understood this odd behavior. However, as good animal behaviorists, they were able to recognize it as "washing behavior"—a pattern used in food collecting and eating, as, for example, in the removal of the exoskeleton from crayfish. This behavior caused great problems in the training of the raccoons but was recognizable given a solid knowledge of raccoon behavior.

The task of observing and describing behavior is more difficult than it sounds. As soon as the observer begins to watch and describe behavior, some degree of abstraction and bias is inevitable. This is a result of limitations on time and attention. The observer has limited time and cannot watch the animal 24 hours a day the year round. In selecting a time to watch the animals, the observer may miss important behavior patterns. Further, the observer is unable to attend to all aspects of the behavior at once. He or she must make choices concerning which behaviors should be attended to and which should be recorded. When providing verbal descriptions, the observer must force observations into the structure of a familiar language. Many of the problems of attention may be minimized with objective recording devices, such as motion-picture and television cameras. Nevertheless, problems of time limitations remain, since decisions must be made as to when such devices are to be turned on and off. The diligent animal behaviorist must always strive for maximal objectivity and breadth in behavioral description.

Two types of behavioral classifications may be distinguished—*description by motor pattern and description by consequence*. In the former, an attempt is made to describe the actual pattern of muscle movements made by the animal, with minimal reference to the effects of the behavior on the environment. Thus, one might describe a "tail wag" or a "knee jerk." When classifying by consequence, the observer notes the effect of the behavior on the environment. Thus, one may use categories of behavior such as "picking up nesting material" or "depressing a lever." In each case, one specifies the effects of behavior on the environment rather than a sequence of actual movements. Each type of description has its advantages (see Hinde, 1970). With description by motor pattern, one remains close to the actual behavior, imposing minimal interpretation. With description by consequence, one emphasizes the important consequences and directedness of behavior. Often a given consequence can be achieved through any number of different motor patterns. To attend only to particular motor patterns would be to ignore this apparent goal-directedness of much behavior. In practice, the two types of description intergrade; each extreme can be quite useful in the appropriate circumstances.

The Four Questions

With a sound foundation in description, the animal behaviorist can proceed to ask the four classes of questions regarding behavior.

Immediate Causation When asking questions of immediate causation, we study factors influencing behavior over a relatively short time span in the life of an individual organism. The effects of various stimuli, environmental contingencies, sensory and perceptual processes, and physiological mechanisms all are evident in the regulation of behavior. Often we are interested in the mechanisms through which internal and external events affect behavior. Questions of immediate causation can be asked using physiological methods, as, for example, when a physiological psychologist alters hormone levels or removes a portion of the brain. Such questions also can be asked without such intervention, as when an animal behaviorist presents different stimuli to an animal or a psychologist manipulates schedules of reinforcement.

Development When asking questions about development, the animal behaviorist considers a longer time span in the life of an individual organism. The behavioral tendencies of organisms change throughout their lives. In many species, the behavior of the newly hatched or newborn organism is quite rudimentary. Throughout its lifetime, the behavior can be seen to change under the influence of a wide array of factors. The ontogeny of behavior represents the joint interaction of genes and environment in producing a viable organism. Analysis of genetic and environmental factors in the ontogeny of behavior has been an important part of the study of behavior. Behavior during the prenatal period is more complex than is sometimes thought, and events affecting the prenatal organism can have pervasive influences later in life. Many behavioral patterns deteriorate as a part of the aging process. Developmental questions span the full range of the organism's lifetime from conception to death.

Evolution While questions of development and mechanism are concerned with the lifetimes of individual organisms, questions of evolution reach across different organisms and are concerned with longer time spans. Such questions deal with the ways in which behavior can be seen to change as natural selection operates both within species and in the evolution of new species. Many animal behaviorists have conducted studies designed to analyze the ways in which behavior changes as species evolve and the ways in which selection acts to effect such changes.

Function Questions of function are concerned with the role that behavior plays in the adaptations of the organism to its environment. The forces of natural selection appear to have molded organisms so that in many respects they function exquisitely in relation to the particular environment to which they have become adapted. It is this relationship between behavior and the adaptation of an organism to a particular environment that provides the essence of the study of function. There are some remarkable examples of the close relationship between specific behavioral patterns and specific ecological factors in very particular environments. The migratory patterns of birds, food habits of very specialized feeders, and mutual helping behaviors between members of different species provide cases in point.

It is important to remember that, in considering the adaptive significance of behavior, we do not assume that an organism understands the function of its behavior in survival and reproduction. To assume that the organism has a conscious purpose underlying its behavior would be to engage in teleology. Rather than think in teleological terms, the animal behaviorist views the organism's behavior as the result of the action of the forces of natural selection that have acted in the past to produce a genotype that, in the presence of the appropriate environmental factors, produces an organism that acts in a particular way in particular circumstances. The remarkable "goal-directedness" of behavior often seems to occur not because the animal understands what behavior is demanded by a given situation but because evolutionary forces have acted to produce an organism that can do no other in a given situation. The study of adaptive significance has been termed "teleonomy."

The four questions can be grouped into sets of two for purposes of clarification. Thus, Wilson (1975) groups questions of mechanism and development under the rubric of "proximal causation," while questions of evolution and function are considered as "ultimate causation." Alcock (1975) treats the former as "how" questions and the latter as "why" questions.

The four questions overlap and interact in many situations. Nevertheless, it is important that one clearly specify the kind of question one is asking. Often, scientists have been criticized for using a method which is inappropriate for the question of interest to the critic while quite appropriate to the interests of the investigator. Thus, a scientist who provides a perfectly adequate physiological (i.e., immediate causation) analysis of a behavioral pattern should not be criticized for failing to explain its adaptive significance. Such controversies can be avoided if the goals and methods of research are clearly stated and understood.

The Four Questions and the Study of Sleep The distinctions among the four questions might best be clarified through consideration of an example of their application, as in the study of sleep. One must start with careful observation and description of the sleep patterns of the species in question. Sleep then can be studied from the perspectives of all four questions. First, one can ask proximate questions. When studying the immediate causation of sleep, one might study the external and internal factors affecting sleep behavior. One may inquire as to the role of biological clocks, hormones, specific brain structures, drugs, sleep deprivation, and even boring lectures in controlling the onset and regulation of sleep.

As a part of developmental analyses, one might study the ontogenetic changes in sleep patterns that appear, beginning with prenatal sleep and continuing through the entire life-span through to old age.

Genetic factors in sleep can be analyzed with the study of different species and strains within species. The role of critical environmental events in the development of stable sleep patterns can also be studied.

Sleep can also be studied from the standpoint of ultimate causation. One might inquire as to the evolutionary history of sleep. Do all organisms sleep? If not, when did sleep first appear and how have its characteristics changed, both in the broad sweep of evolutionary history and within smaller groups of more closely related species? Allison and VanTwyver (1970) believe that sleep evolved independently among birds and mammals and that paradoxical sleep (see page 58) evolved within the mammalian line after a hypothesized branching of placental from nonplacental mammals.

Finally, the adaptive significance of sleep can be considered. How does sleep contribute to the survival and reproduction of organisms? Is sleep essential for a restoration of some bodily resources which are depleted during the active period? Might sleep function primarily in assuring that the organism is inactive and in a safe place during those portions of the day in which it is susceptible to danger from extreme environmental fluctuations, such as temperature, or from potential predators? It is possible that the reductions in metabolic rate that occur in sleep may conserve energy and prolong life. In inquiring as to the function of sleep, we do not imply that the animal understands these factors. Rather, we consider the possibility that the need to conserve energy and to avoid predators might have acted as forces in natural selection, producing organisms which sleep in a characteristic manner and at characteristic times.

A complete understanding of sleep will require careful observation and description as well as answers to all four classes of questions.

EARLY HISTORY

One can trace interest in the behavior of animals back to the early history of the human species. Early hunters surely were careful students of the behavior of their prey—a fact revealed in numerous cave paintings.

The Study of Animal Behavior before Darwin

The study of animal behavior before Darwin was concentrated in philosophy and natural history. The primary philosophical issue was that of the relationship between humans, other species, and the rest of the universe. If one believes that humans exist completely apart from all other species, a comparative psychology makes little sense. While the study of animal behavior may be interesting and important in and of itself, results would have no implications for understanding human behavior. If, on the other hand, one views humans as a part of nature, studies of animal behavior may be of considerable significance in understanding our own species. These two views are contrasted in the following quotations:

> In animals there are only instincts, but not in man. Instincts in animals seem to operate according to the pattern of physical forces, where the stronger always prevails; for animals are utterly devoid of the freedom which characterizes man (Sheen, 1949, p. 98).
>
> . . . if we remove man from the central point in a comparative science of behavior, this may, in the long run, prove to be the very best way of reaching a better understanding of his place in nature and of the behavioral characteristics which he shares with other animals as well as those which are in him developed to a unique degree (Beach, 1960, p. 17).

Note that the latter position assumes neither that there are no differences between humans and nonhumans nor that results of studies with nonhumans can be generalized to humans in a simple and direct way. Rather, it implies that there are at least some commonalities between humans and nonhumans and that the best way to understand both differences and similarities among all species is to study all species in the same objective way.

Warden (1927), Waters (1951), and Beach (1955) traced the early origins of animal psychology. The dichotomous view of humans and animals can be traced to the Greek philosophers, who proposed that there had been two kinds of creation: humans and gods were the products of rational creation, while the irrational brutes were a separate category of living creatures. They were believed to differ with respect to the number and types of souls they possessed. The history of this view can be traced through the Stoic philosophers, Albertus Magnus, St. Thomas Aquinas, Descartes, and into the present. Aristotle, while accepting the view that humans differed from nonhumans in regard to the nature of their souls, attempted to place all species on a continuous scale, or *scala natura*. Humans were placed at the top of the scale of living organisms.

Aside from philosophy, the major contribution before Darwin stemmed from natural history. The contributions of some of the early naturalists have been reviewed by Gray (1968/1969). Many of the problems studied by the early naturalists are still of scientific interest. For example, Gilbert White (1720–1793) was able to distinguish among three morphologically similar species of birds on the basis of their songs. Ferdinand Pernauer (1660–1731) studied territorial habits, migration, sexual behavior, and the ontogeny of song patterns in birds. The role of falconry in the history of animal behavior has been stressed by Mountjoy, Bos, Duncan, and Verplank (1969).

Darwin

The contributions of Darwin to the development of animal behavior are difficult to overestimate. Perhaps foremost was his formulation of evolutionary theory and its application to humans in *The Origin of Species* (1859) and *The Descent of Man* (1871). These works made a strong case for the continuity of species including humans—a critical prerequisite to a truly comparative psychology, as seen above. However, Darwin's contributions extended even beyond this major influence (Ghiselin, 1973). Darwin himself initiated a systematic study of comparative animal behavior. His best-known contribution in this domain was *The Expression of the Emotions in Man and Animals* (1873). In this book, Darwin reported numerous observations in support of the general view that the expression of the emotions in humans and animals represents a continuum. Indeed, some patterns of expression in humans were regarded as resulting from the inheritance of behaviors that were useful in our ancestors are but now no longer functional.

Behaviorists after Darwin

The relatively brief period between Darwin and the development of twentieth-century ethology and animal psychology was notable for the contributions of a substantial number of animal behaviorists. The study of animal behavior became a truly viable enterprise during this period. While the contributions of these early animal behaviorists were substantial, a full discussion is beyond the scope of this text; thus they will be discussed only briefly. More detail can be found in Boring (1957), Jaynes (1969), Waters (1951), Klopfer (1974), Warden (1927), or other historically oriented works.

George John Romanes In 1882 Romanes, a friend and student of Darwin, wrote *Animal Intelligence,* a classical book in animal psychology. Romanes attempted to apply Darwin's principles further in the development of a comparative science of behavior. He is most famous because of two negative characteristics of his works, his reliance on the anecdotal method and on anthropomorphism. While anecdotes, or unsystematic observations of behavior—often of one-time occurrences by untrained observers—can be of great interest to the animal behaviorist, they must be interpreted with more caution that was typified in Romanes's works. Anthropomorphism is the tendency to attribute excessive human characteristics to nonhuman organisms, another weakness in these works.

C. Lloyd Morgan Morgan was another prominent animal behaviorist, whose outstanding contribution was his *Introduction to Comparative Psychology*, written in 1900. Morgan is best known for opposing unbridled anthropomorphism. According to the often-cited "law of parsimony" or "Lloyd Morgan's canon," "In no case may we interpret an action as the outcome of the exercise of a higher physical faculty, if it can be interpreted as the outcome of the exercise of one which stands lower in the psychological scale" (Morgan, 1894). The admonition that we should strive to accept the simpler of two equal alternative explanations is certainly good advice for many situations. It should not, however, encourage one to avoid complex explanations under circumstances where they indeed seem warranted.

Jacques Loeb Loeb was an extreme advocate of the law of parsimony and attempted to account for nearly all behavior in terms of *tropisms*, which were conceptualized as "forced movements." According to Loeb, animal behavior could be understood as resulting from immediate physical and chemical effects of stimuli upon protoplasm. Thus, stimuli were viewed as affecting animals in much the same rather simple way as that in which they affect plants.

Herbert Spencer Jennings Jennings was an early behaviorist who stressed the descriptive study of the full range of behaviors of species under investigation. His *Behavior of Lower Organisms*, written in 1904, dealt prominently with protozoans. Jennings disagreed with Loeb and believed that the variability and modifiability of the behavior of even the simplest of organisms was incompatible with the tropistic interpretation.

Douglas A. Spalding Spalding was a pioneer animal behaviorist who is best known for a series of empirical investigations of the development of behavior in young chicks. In an attempt to understand the factors controlling behavioral ontogeny, he conducted deprivation experiments in which animals were reared in the absence of certain normal portions of their environment. Spalding also conducted early research on the phenomenon of imprinting (see Chapter 8).

Ivan P. Pavlov Although Pavlov worked in Russia in relative isolation from many of his contemporary animal behaviorists, his work on classical conditioning was to have substantial impact on the development of the study of animal behavior to the twentieth century.

William James James's *Principles of Psychology* (1890) became the standard textbook in psychology for many years and contained much information on animal behavior, including a chapter on instinct and a discussion of imprinting. He did much to facilitate the growth of comparative psychology.

William McDougall William McDougall was a psychologist whose work has had much important influence on the development of contemporary theories of behavior. McDougall developed a "purposive psychology," which assumed that the organism is always striving toward a goal. He is best known in animal behavior for *Social Psychology* (1908). In this book, McDougall attempted to show that all human behavior could be explained as a result of the actions of instincts and the alterations of instincts as a result of experience. His list of instincts included instincts of flight, pugnacity, self-abasement, reproduction, acquisition, and on and on. The list was endless. It was soon recognized that such "explanations" of behavior do not really explain anything but merely name phenomena. The confusion in which one merely names things, but believes that this explains them, has been referred to as the "nominal fallacy." While McDougall made many positive contributions, it was probably he more than anyone else who led psychologists to be highly skeptical of the usefulness of the instinct concept—an orientation that was to be of great importance when psychologists and ethologists began interacting nearly a half century later.

With all these outstanding early behaviorists active, the period around the turn of the century must have been a very exciting time in which to be an animal behaviorist.

SUMMARY

The beginnings of the study of animal behavior lie in thorough observation and description of behavior. Observation and description are followed by consideration of four classes of questions about behavior—questions of immediate causation, development, evolution, and function. A complete understanding of behavior requires consideration of all four classes of questions.

Before Darwin, the study of animal behavior was centered in philosophy and natural history. A group of highly creative scientists of different persuasions was active at the turn of the century and laid the groundwork for the development of animal behavior in the twentieth century.

Chapter 2

Animal Behavior: An Interaction of Disciplines

The contemporary study of animal behavior represents the interaction of scientists trained in a variety of disciplines. Psychologists, zoologists, anthropologists, veterinarians, physiologists, entomologists, and practitioners of various agricultural sciences are among those contributing to the study of animal behavior. There are many differences in emphasis among scientists representing these various disciplines. However, there are so many common characteristics among virtually all of them that a synthesis into a general area of animal behavior is occurring. To understand this contemporary synthesis, one must understand the differences that separated animal behaviorists trained in different disciplines only a few decades ago. We shall consider some of the major features of the different disciplines contributing to the study of animal behavior, with an emphasis on two areas—ethology and animal psychology.

CLASSICAL ETHOLOGY AND ANIMAL PSYCHOLOGY

It is instructive to begin our consideration of classical ethology and animal psychology by contrasting them as they existed at their peaks of divergence as different disciplines. These peaks occurred in the 1940s and 1950s, but the origins of the differences separating the two areas can be traced back to the nineteenth century (Jaynes, 1969).

There is no simple definition of classical ethology. Various writers have defined ethology as "the objective study of behavior," "the biological study of behavior," "the study of instinct," or "the study of behavior by individuals who like their animals." Classical ethology was an approach to the study of animal behavior that developed among a group of innovative European zoologists and which emphasized an evolutionary approach to the study of instinct.

Classical animal psychology is equally difficult to define. It originated, of course, within psychology and with a concern for the evolution of mind and of higher mental processes. Animal psychology developed into a science of animal behavior that was dominated by studies of learning, especially in white rats.

Classical ethology and animal psychology are contrasted in Table 2-1. The differences between the two encompass differences in geographic location, training, typical subjects, emphasis, and methods.

Burghardt (1973) has written of the "ethological attitude," which represented the philosophical basis of ethology and which had an important influence on the study of animal behavior independent of specific theoretical formulations. The ethological attitude was characterized by insistence on (1) studying behaviors meaningful in the natural habitat, (2) beginning analysis with descriptive studies, (3) studying a wide range of species and behaviors, (4) comparing similar behaviors in closely related species, and (5) disparaging the exclusive use of domesticated species. By contrast, most animal psychologists were searching for "general laws of behavior," which they believed transcended particular species and behavior. Partially as a result of this philosophy,

Table 2-1 Some Characteristics of Classical Ethology and Animal Psychology

Feature	Classical ethology	Classical animal psychology
Geographic location	Europe	North America
Training	Zoology	Psychology
Typical subjects	Birds, fish, and insects	Mammals, especially laboratory rats
Emphasis	"Instinct," the study of the evolution of behavior	"Learning," the development of theories of behavior
Methods	Careful observation, field experimentation	Laboratory work, control of variables, statistical analysis

Source: From *Readings in Animal Behavior* edited by Thomas E. McGill. Copyright © 1965 by Holt, Rinehart and Winston, Inc. Adapted by permission of Holt, Rinehart and Winston.

many animal psychologists tended to (1) view the selection of a response as arbitrary, (2) bypass descriptive studies, and (3) concentrate on studies of one domesticated species, laboratory rats.

The practitioners of classical ethology and animal psychology developed their disciplines in relative isolation from each other, and each group came to view its approach as "the objective study of behavior." When ethologists and psychologists began to interact in the 1950s, much controversy ensued. It was out of this controversy that the contemporary synthesis of animal behavior has blossomed.

ETHOLOGY

We shall consider the origins, basic concepts, and recent developments of ethology.

Origins of Classical Ethology

The origins of ethology in the early part of this century can best be seen in the contributions of four men, Charles O. Whitman, Oskar Heinroth, Jacob von Uexküll, and Wallace Craig. Whitman was an American zoologist who studied the behavior of pigeons and doves and is regarded as a founding father of ethology. He is best known for proposing that behavior, like structure, can be studied from an evolutionary perspective. Whitman proposed that we can study the evolution of behavior just as we study the evolution of organs. Heinroth, a German zoologist, was also interested in the study of bird behavior. He was able to document various instances of behavioral evolution in birds of the sort proposed by Whitman.

Wallace Craig, a student of Whitman, made advances in the development of a theoretical model of the control of animal behavior. He termed the stereotyped, species-typical behavioral patterns studied by Whitman and Heinroth "consummatory acts." Craig further noted that not all behavior is so rigidly fixed and invariant. He distinguished "appetitive behavior" from consummatory acts and noted that appetitive behavior is variable, agitated behavior that is terminated by the occurrence of a stimulus releasing the consummatory act. The notion was that under certain drive conditions the animal engages in variable, searching behavior that usually results in the appearance of a particular stimulus, which in turn elicits a consummatory act. Thus, a food-deprived animal engages in variable, "searching" behavior, which is terminated by the appearance of food and the elicitation of relatively stereotyped eating patterns.

Von Uexküll emphasized the fact that only very few stimuli, out of the entire array that continuously bombard the animal, are capable of eliciting responses by the organism. He termed the unique perceptual world of each animal its *Umwelt*. Because of its unique sensory and neural structure, each animal is particularly sensitive to some potential stimuli and insensitive to others. Approach for feeding in the tick, for example, is elicited by the stimulus of butyric acid. Any object emitting butyric acid can elicit approach behavior from the tick, while items with no such odor will be ineffective.

Classical ethology reached its peak in the works of Konrad Lorenz and Niko Tinbergen, two European ethologists who, together with Karl von Frisch, shared the 1973 Nobel Prize in Physiology or Medicine. Lorenz, a student of Heinroth, has become noted for a long series of studies in behavior—especially the social behavior of birds. An innovative pioneer, he has provided a major impetus to the renewed vigor of studies in animal behavior. Tinbergen is an equally creative "curious naturalist" who provided much of the firm empirical foundation for ethology. His book *The Study of Instinct* (1951) represents a high-water mark for classical ethology.

Basic Concepts and Models of Classical Ethology

The fundamentals of classical ethology can be understood through a consideration of some basic concepts and theoretical models which were proposed for the understanding of the proximate causes of behavior. It should be remembered that questions of ultimate causation were also of great importance to the classical ethologists.

Fixed Action Pattern The fixed action pattern, or FAP, lies at the heart of ethological thinking. Fixed action patterns are motor patterns that were the objects of most ethological research and are roughly equivalent to the "consummatory acts" discussed by Craig. Although different authors stress different defining attributes of fixed action patterns, the list generally includes a sample of seven characteristics:

1 FAPs were regarded as *stereotyped*, in that they comprised sequences of motor acts which occurred in rigid, predictable, and highly structured sequences.

2 FAPs were *complex* patterns, a characteristic that distinguished them from simple reflexes.

3 FAPs were considered to be *shown by all members of a species*, or at least by all members of a given sex in the appropriate age range and physiological condition within a species.

4 FAPs were *elicited by simple yet highly specific stimuli*.

5 FAPs were regarded as self-exhausting; that is, the mere occurrence of the FAP reduced the ease with which it could be reelicited. It was not the consequences of the behavior so much as its mere occurrence that made it harder for the behavior to be elicited a second time.

6 FAPs were regarded as *"triggered,"* by which ethologists meant that once elicited, FAPs continued independent of external stimulation. Once initiated, a sequence of motor acts would continue even if the environment changed so that the behavior was no longer appropriate.

7 The occurrence of an FAP was taken to be *independent of experience*; FAPs were essentially complete on their first occurrence.

Ethologists studied a wide array of different fixed action patterns in many different species. The "gaping" response of young nestling birds toward a parent alighting on the nest and about to feed them, the tongue flick with which a frog catches flies, and the various display patterns in the courtship and aggressive displays of birds (see Figure 2-1) are all examples of fixed action patterns. A classic example is provided in the study of egg rolling in greylag geese. These birds build shallow nests in the ground. The goose uses a stereotyped set of behavioral patterns to retrieve eggs that roll out of the nest (Lorenz and Tinbergen, 1938). As it approaches the egg, the goose stretches its

Figure 2-1 Four display patterns of green herons. *(a)* The aggressive full-forward display; *(b)* the snap display of courting males; *(c)* the stretch display; *(d)* the flap-flight display. *(After Meyerriecks, 1960.)*

(a)

(b)

Figure 2-2 The egg-rolling pattern of greylag geese. *(a)* Retrieval of a normal egg; *(b)* attempt to retrieve a large model or "supernormal stimulus." *[From* The Study of Instinct *by N. Tinbergen (1951), published by Oxford University Press.]*

neck toward the egg, touching the ground on the other side (see Figure 2-2*a*). It then engages in a very rigid sequence of movements in a vertical plane that bring the egg back to the nest. Egg rolling in greylag geese fulfills most of the criteria for a fixed action pattern (e.g., it is stereotyped, species-common, elicited by a simple specific stimulus, etc.). Sometimes the egg rolls away as the goose is moving it toward the nest. On some such occasions, the animal continues its retrieval movements through to their completion. This phenomenon demonstrates the extent to which the pattern is triggered—that is, independent of the environment once elicited.

Sign Stimulus The relatively simple yet specific stimuli that release fixed action patterns were termed "sign stimuli." It is sign stimuli which terminate the variable, searching "appetitive behavior," to use Craig's terminology, and release the stereotyped FAP. The red breast feathers of the European robin in another male's territory are a sign stimulus for aggressive behavior, while the bill of the parent is a sign stimulus for pecking by newly hatched gull chicks. (Pecking of the parent's bill elicits feeding by the parent.)

In studying sign stimuli, ethologists use *models,* or artificial representations of specific characteristics of stimuli, in order to determine the critical aspects functioning in the release of the FAP. Sticklebacks are small freshwater fish. In the breeding season, the males develop a red coloration on the undersides of their bodies and become highly territorial. The models shown in Figure 2-3 were used by Tinbergen (1951) to study the sign stimuli which were critical for territorial defense. The highly abstract models with red undersides (series R) were more effective in releasing defense than the more realistic

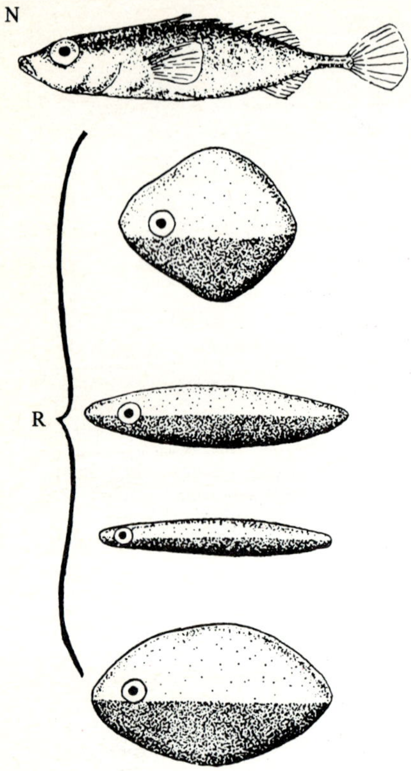

Figure 2-3 Models constructed by ethologists to study sign stimuli releasing attack behavior in sticklebacks. Model N resembles a stickleback but lacks a red belly and is relatively ineffective in releasing attack. Group-R models do not closely resemble sticklebacks but have red undersides and are relatively effective in releasing attack. *[From* The Study of Instinct *by N. Tinbergen (1951), published by Oxford University Press.]*

model lacking the red underside (model N). Apparently, the red belly is the critical sign stimulus responsible for releasing territorial defense in sticklebacks.

Ethologists have been able to construct models which are even more effective in releasing fixed action patterns than are the original stimuli; such stimuli are called "supernormal stimuli." The greylag goose will direct retrieval movements toward very large models of eggs in preference to her own eggs (see Figure 2-2*b*). The gull chick will peck at a very long, thin artificial bill in preference to one more closely resembling that of the parents.

The term "releaser" is often used in the ethological literature. Some authors use the term as an alternative, equal to the term "sign stimulus." Others use "releaser" in a more restrictive way, to refer only to those sign stimuli which are structures, movements, sounds, or odors of a companion animal of the same species as the recipient of the stimulus.

Innate Releasing Mechanism The fixed action pattern was viewed as intrinsically organized and held in check by active inhibitory processes in the central nervous system. The removal of this active inhibition "released" the behavior and thus permitted its occurrence. The innate releasing mechanism (IRM) was the hypothesized locus in the central nervous system on which sign stimuli acted to release fixed action patterns. A lock-and-key analogy is often used to clarify the operation of the sign stimulus and IRM. The sign stimulus was conceptualized as a highly specific key which fits perfectly into the lock, the IRM.

Action-Specific Energy It was noted above the FAPs were thought to be self-exhausting. In order to explain this characteristic, Lorenz proposed an energy model of behavior. Each FAP was thought to have its own reservoir of "action-specific energy" (ASE). The amount of energy in the reservoir increased steadily as the animal refrained from displaying the FAP, and the energy level was depleted by the repeated occurrence of the FAP. The ease with which the FAP was released was a function not only of the characteristics of the sign stimulus but also of the level of action-specific energy at that point in time. As action-specific energy accumulated, stimuli which resembled the ideal sign stimulus to a less and less degree were capable of activating the IRM and releasing the FAP. Finally, the level of action-specific energy could become so great that the pattern could occur in the absence of any stimulus; this was termed a "vacuum activity."

Other Concepts When tendencies to engage in two different acts (e.g., fight or flee) were in conflict or when there was a strong tendency to engage in a particular behavior at the same time there was strong inhibition against such behavior, an unexpected behavioral pattern might occur. Such behavioral patterns were notable in that they were common behaviors except that they appeared out of their usual motivational contexts. These out-of-context behaviors were termed "displacement activities." Two birds might be in conflict at the borders of their territories. Because each was dominant within its own territory, each would be in conflict as to whether to attack or retreat. As a result, the birds might engage in "displacement preening" or "displacement nest building." An alternative resolution to this kind of conflict is that of *redirected activity*. If an animal has both a tendency toward and an inhibition against attacking a conspecific, it might direct its aggressive attack against yet another, less threatening animal. "Redirected aggression" in ethological terminology appears roughly equivalent to what is termed "displaced aggression" in psychoanalytic theory.

Intention movements are the incomplete or preparatory movements that occur at the beginning of an activity (e.g., a bird crouches and engages in various postural adjustments when it is about to fly).

A *taxis*, in the context of ethological theory, is an oriented (directed) movement that occurs along with a fixed action pattern. Once elicited, the FAP was hypothesized to be independent of stimulation, whereas the taxis was oriented. As the greylag goose in Figure 2-2*a* rolls her egg toward the nest, it

further from the rigidity of some classical models toward a rapprochement with other approaches and development of the synthetic approach to animal behavior which is now prevalent. Today, ethology is in a dynamic empirical stage in which it is often not easily distinguished from other branches of the study of animal behavior, with which it interacts to comprise the contemporary study of animal behavior.

ANIMAL PSYCHOLOGY

The contributions of some of the early animal psychologists around the turn of the century were discussed in Chapter 1. At this point, we shall review some of the developments in animal psychology in the twentieth century.

Dominant Trends in Animal Psychology

A pivotal publication in the development of animal psychology was E. L. Thorndike's report (1898) of trial-and-error learning in cats and dogs escaping from puzzle boxes. This doctoral dissertation and its demonstration of the role of reinforcement ("the law of effect") in learning had substantial influence on the development of animal psychology in this century. In the period from 1899 to 1903, laboratories of animal psychology were established at Clark University, Harvard, and the University of Chicago (Boring, 1957). W. S. Small, at Clark, introduced the study of maze learning by white rats. Robert M. Yerkes took charge of comparative psychology at Harvard in 1902. At Chicago, John B. Watson, working under H. H. Donaldson and James R. Angell, completed a thesis on neurological and behavioral maturation in rats. By 1910 there were eight laboratories of animal psychology in the United States and courses in the area offered at another dozen or so schools. The *Journal of Animal Behavior* was founded in 1911; it merged in 1921 with *Psychobiology* to become the *Journal of Comparative Psychology*.

Early animal psychology had a broad comparative base of study. Yerkes studied medusae, earthworms, dancing mice, and primates. Karl Lashley's early bibliography includes studies of paramecia, parrots, monkeys, and terns. Watson and Lashley (1915) published a study of homing in birds. Other animal psychologists studied an equally broad array of species. The shift of dominance from the broad-based animal psychology to one dominated by studies of learning in rats represents an important trend in animal psychology in this century. This trend has been illustrated with a content analysis of the species used as subjects in studies published in the *Journal of Comparative and Physiological Psychology* over this period, as can be seen in Figure 2-5. During this period, the journal became dominated by studies of mammals, especially Norway rats. The early comparative psychology journals also were addressed to a wide range of behavioral patterns (e.g., social behavior, simple action patterns, feeding). Throughout the twentieth century, they came to be dominated by studies of learning (see Beach, 1950). Quality research outside the mainstream often appeared in less traditional journals (see Gottlieb, 1976).

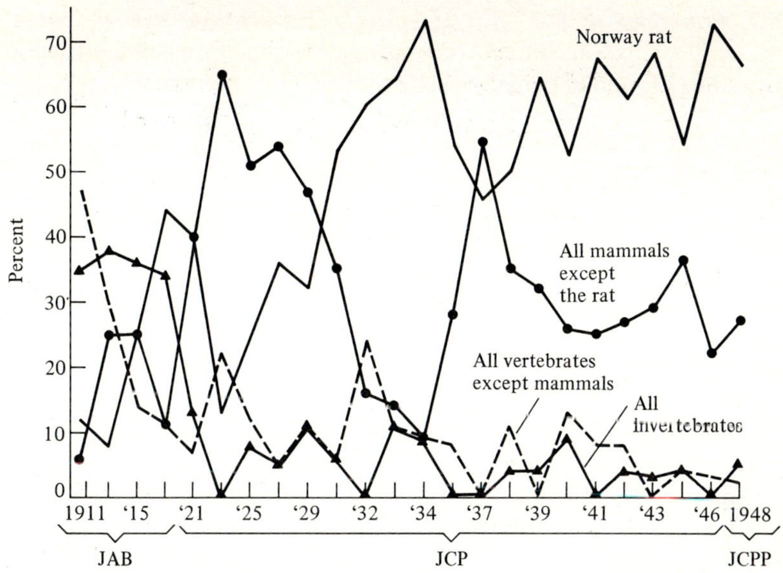

Figure 2-5 Percent of articles appearing in *Journal of Animal Behavior* (JAB), *Journal of Comparative Psychology* (JCP), and *Journal of Comparative and Physiological Psychology* (JCPP) from 1911 to 1948 dealing with different groups of species. *(After Beach, 1950. Copyright 1950 by the American Psychological Association. Reprinted by permission.)*

What happened in psychology to produce these trends away from comparative studies of a variety of behavioral patterns and toward dominance by studies of learning in white rats? One factor was the advent of ***behaviorism***. John B. Watson is given credit for formulating behaviorism in a paper entitled "Psychology as the Behaviorist Views It" (1913). Behaviorists rejected the mentalistic concepts prevalent in much of psychology and concentrated on observable behaviors. Most behaviorists minimized the role of genetically controlled behavior and thus concentrated their attention on the study of learning (e.g., Watson, 1930).

Another important factor in the decline of a broad-based comparative psychology was the conclusion of Thorndike (1911) that the principles governing learning were essentially the same for all species. Thorndike had been a leader in the comparative study of learning. If it was true that the learning process was the same in all species, it made sense to study a convenient species such as laboratory rats. Comparative research was unnecessary, expensive, and inconvenient. The behaviorist-Thorndike tradition in animal psychology has carried through to the present day. Skinner and many operant psychologists have tended to regard the selection of a species for the study of operant principles as a matter of convenience. The Russians, carrying on the tradition stemming originally from Pavlov, displayed a similar view (Voronin, 1962).

By the 1940s and 1950s, animal psychology came to be dominated by grand learning theories and theorists such as Hull, Tolman, Skinner, and Guthrie.

Books entitled *Principles of Behavior* and *The Behavior of Organisms* were based almost exclusively on studies of learning in rats. Emphasis on truly comparative studies of animal behavior had reached a low point in psychology.

Classical Comparative Psychology

While it is difficult to overemphasize the domination of learning theory over animal psychology in the 1930s and 1940s, other important trends can be recognized as well. A smaller number of animal psychologists kept alive the tradition of a broader-based comparative psychology. Robert Yerkes maintained a long-term interest in primate behavior and in 1930 was able to open the Yale Anthropoid Experiment Station at Orange Park, Florida. This facility was to become an important breeding ground for future comparative and physiological psychologists.

Maier and Schneirla (1935) wrote a classical textbook on animal psychology in which they considered the behavior of a full range of species from protozoans to humans. Schneirla's research came to emphasize field studies of the social behavior of ants—a far cry from laboratory studies of learning in rats.

In the 1930s, C. R. Carpenter studied howler monkeys in Panama, spider monkeys in Central America, gibbons in Thailand, orangutans in Sumatra, and rhesus monkeys in Puerto Rico. Karl Lashley kept alive an interest in the study of the physiological bases of a wide array of behaviors, including reproductive behavior. Frank Beach began a classic series of studies on the physiological and sensory determinants of reproductive behavior. Henry Nissen conducted a series of studies of problem solving in apes.

The American Museum of Natural History became the center of a New York group of animal behaviorists who came to emphasize the study of the development of animal behavior. Led by Beach, Schneirla, and Lester Aronson, this group expanded to include such workers as Daniel Lehrman, Jay Rosenblatt, and Howard Moltz. In general, individuals in this "New York epigeneticist group" emphasized the study of the complex interactions of the genes and environment in the ontogeny of behavior. Behavioral epigenesis may be defined as "a continuous developmental process from fertilization through birth to death, involving proliferation, diversification, and modification of behavior patterns both in space and in time, as a result of the continuous dynamic exchange of energy between the developing organism and its environment, endogenous and exogenous" (Kuo, 1967, p. 11). It is the complexity of the dynamic interchange of organism and environment that became the focus of most of the researchers in this group. Many researchers trained in this tradition became important exceptions to the generalization that most animal psychologists studied rats.

Examination of the leading textbooks of comparative psychology written during this period (e.g., Maier and Schneirla, 1935; Moss, 1941; Warden, Jenkins, and Warner, 1934; Stone, 1951) reveals a considerable breadth of approach with respect to the range of species and behavioral patterns studied.

The beginnings of many research areas that were to come into vogue as "new" problem areas in the 1960s and 1970s can be found discussed in these works. Thus, one finds important contributions to the study of the development and evolution of behavior and of its control by stimuli, sensory-perceptual processes, internal secretions, and neural mechanisms. Although outnumbered within animal psychology by students of rat learning, these truly comparative psychologists and their students would later play a major role in developing the contemporary study of animal behavior.

CONTEMPORARY ETHOLOGY AND ANIMAL PSYCHOLOGY: A RAPPROCHEMENT

Beginning in the 1950s, classical ethologists and animal psychologists began interacting on a more regular basis. The first interactions between these two "objective sciences of behavior" were quite bitter.

Criticisms of Ethology

Psychologists attacked the lack of control and experimental rigor which characterized much ethological research. They were appalled at the lack of quantification and statistical analysis. Further, they were skeptical of the energy models of motivation proposed by the ethologists and of the physiological implications of some of the ethological formulations. It appeared difficult to construct a physiological analog of the Lorenzian hydraulic model.

The most substantial basis for criticism of ethology by psychologists lay in the use of the concept of instinct. The early ethologists were somewhat loose in their formulations of the development of behavior. They believed that the instinctive basis of behavior was demonstrated by deprivation or "Kaspar Hauser" experiments, in which animals were reared in isolation. Such animals often displayed many species-typical behavioral patterns. These experiments and the ethologists' reliance on the concept of instinct invoked the wrath of behaviorist and behavioral epigeneticist alike. Early ethological interpretations of deprivation experiments ignored the many external and internal factors that could influence behavior in such experiments. Psychologists pointed out that environmental experience had to be given a broad conception. Many factors interact to produce complex behavior, and many of these factors are not controlled in deprivation experiments. The dynamic interactions that were the essence of behavioral epigenesis were ignored in much early ethological writing.

Criticisms of Animal Psychology

Ethologists could not understand what they perceived to be the narrow interest of many psychologists in the study of learning in rats. Ethologists believed that through domestication and abnormal rearing conditions, many of the important natural behaviors of this species had been lost. The psychologists studied behaviors that were of virtually no obvious importance to the adaptation of real animals surviving in their natural habitats. How could they expect to learn anything important about the evolution and control of behavior from such experiments? The environmentalist bias of most animal psychologists provided a sore spot. Ethologists could see no way in which the complex behaviors of newly hatched insects, for example, could possibly be learned according to reinforcement principles.

Resolutions

Some of the issues that separated classical ethologists and animal psychologists have never been fully resolved. For the most part, however, there is considerable unity despite differences in emphasis in the contemporary study of animal behavior. A general cooling down has taken place with respect to the so-called "nature-nurture issue," the ontogeny of "instinctive behavior," and many other points of initial disagreement. Ethologists have become more sophisticated in their use of experimental controls, statistical analyses, and physiological models. Animal psychologists have come to recognize the potential distortions of natural behavior that can occur in the laboratory, the importance of field studies, and the value of work with a variety of nondomesticated species.

Each group has recognized that they had given emphasis to different kinds of questions. In their concern to answer broad questions of evolution and function, ethologists had tended to write carelessly when dealing with development and mechanism. In their concern with controlled studies of development and mechanism, animal psychologists had tended to ignore evolution and function. Each group had come to criticize the other for not answering the questions in which it was interested (see Chapter 1). Gradually, each group came to recognize the value of the other's research for answering particular kinds of questions. Studies of "unnatural" laboratory behavior can be of great value in understanding the full range of response possibilities of the organism. Minimally controlled field studies may provide the only available valid information regarding the function of behavior in the adaptation of organisms to the natural habitat.

Today, it is sometimes difficult to tell a comparative psychologist from an ethologist "without a scorecard." Psychologists such as Lott, Mason, LeBoeuf, Daly, and others are conducting field studies of social behavior. Ethologists such as Hinde are doing controlled laboratory studies of the development of behavior in monkeys. Psychologists such as Pinsker and Corning are studying invertebrate behavior. Ethologists such as Sevenster and Stevenson-Hinde study learning in the laboratory, while Andrew has become interested in the psychology-like problem of the role of hormones in "persistence." Studies of learning play an important role in ethology, while studies of "instinctive behavior" have increased dramatically in psychology. This is not to say that all differences of emphasis have been lost. Rather, distinctions have become blurred as ethologists and animal psychologists have learned from each other and changed their disciplines as a result of new findings and criticisms from both within and without their respective areas. If the distinction between ethology and comparative psychology is valid today, it must be based on possession of a sample of a large number of defining attributes—a polythetic definition (see Jensen, 1970). Simple definitions have become outmoded.

CONTRIBUTIONS OF OTHER DISCIPLINES

In emphasizing the role of ethology and comparative psychology in the development of contemporary animal behavior, one should not ignore the role of a number of other disciplines.

Physical Anthropology

Anthropologists have long been concerned with human evolution. Analysis of the evolution of human behavior has been a critical part of this activity. In conjunction with psychologists and zoologists, anthropologists have taken the lead in a modern explosion of field research on primate behavior. Anthropologists have limited themselves to primate studies, conducting virtually no research outside that order. However, in their research on primates, they have made a substantial contribution to the development of contemporary animal behavior.

Behavioral Ecology

Many zoologists, for whom the term "ethologist" would be inappropriate, have made substantial contributions to the study of animal behavior. Entomologists, ichthyologists, herpetologists, ornithologists, and mammalogists are zoologists specializing in the study of insects, fishes, amphibians and reptiles, birds, and mammals. Some individuals from each area have conducted some classical studies of animal behavior. The training of these individuals has been in general zoology, and their concern has been with natural history or the adaptation of animals to their environments. They might be grouped under the broad rubric of "behavioral ecology." The common interests of these reasearchers with those of other animal behaviorists and their contribution to animal behavior merit recognition.

Physiology

Developments in physiological sciences have led numerous scientists to begin study of the physiological correlates of animal behavior in ways that are quite compatible with other approaches to animal behavior. Neuroscientists are recognizing the importance of studying naturally occurring behavioral patterns in investigations of the organization and function of the central nervous system. An emerging area of neuroendocrinology has developed and been concerned with the reciprocal interactions of hormones, the nervous system, and behavior—especially reproductive behavior, aggression, and stress.

Other Disciplines

Although somewhat fewer, scientists from other disciplines have made some important contributions to the development of animal behavior. Psychiatrists interested in animal "analogs" for human behavior have become interested in the behavior of nonhuman animals. Social psychologists have initiated studies of the social interactions of nonhuman as well as human animals. Veterinarians and zoo keepers interested in the health and reproduction of their specimens have made valuable contributions. Individuals in various agricultural sciences—such as poultry science, dairy science, and animal science—have all made contributions. "Human ethology" is gaining stature as an independent area, attracting individuals trained in developmental and social psychology and education as well as animal behavior.

ANIMAL BEHAVIOR TODAY

The efforts of scientists of different disciplines have been fused into the contemporary science of animal behavior. This can be seen in the journals, societies, and research activities of different disciplines. The Animal Behavior Society in the United States, the Association for the Study of Animal Behavior in England, and the International Ethology Conference all hold regular meetings which are attended by scientists of a wide range of backgrounds. The *Journal of Comparative and Physiological Psychology* has become more accessible to zoologists, and the mainstay journals of ethology—*Behaviour, Animal Behaviour,* and the *Zeitschrift für Tierpsychologie*—have been opened to a wide range of scientists. There has been an explosion of new scientific journals such as *Applied Animal Ethology, Behavioural Processes, Behavior Genetics, Behavioral Biology, Animal Learning and Behavior, Primates,* and *Biology of Behaviour.*

While differences in emphasis remain, a synthetic study of animal behavior in which scientists of different backgrounds interact harmoniously appears to have come of age.

SUMMARY

Classical ethology developed among a group of European zoologists who came to emphasize the study of instinct and evolution in a wide range of species, especially insects, fish, and birds. A classical ethological model of the proximate causes of behavior was developed and reached a peak in the 1950s.

Animal psychology started out in the twentieth century as a broad-based study of various behavioral patterns in a variety of species, but it came to be dominated by studies of learning in white rats. The comparative tradition was maintained by a smaller number of true comparative psychologists.

In the 1950s, ethologists and animal psychologists began to interact, first by bitterly criticizing each other's formulations. Eventually, many of the issues dividing ethology and animal psychology were resolved, so that ethologists, animal psychologists, and others could interact to develop the modern science of animal behavior.

The contemporary study of animal behavior is in a stage of dynamic growth reflecting the interaction of scientists from many disciplines.

Chapter 3

Animal Species: Their Nature, Classification, and Evolution

In this chapter we complete our coverage of material that is introductory to the study of animal behavior. A comparative animal behavior is concerned with different animal species, and an important prerequisite to an understanding of comparative animal behavior is an understanding of animal species and their evolution. With an elementary background in basic evolutionary principles, we shall be able to allude to the evolution of behavior throughout much of the book. The evolution of behavior will be covered in greater detail in Part Five, whereas the adaptiveness of behavior will be stressed in Part Six. The present coverage of animal species and their evolution deals with the nature of species, their classification, and their evolution, with the latter viewed both as history and as a process.

ANIMAL SPECIES

The problem of establishing a definition of "species" is more complicated than it might seem. Most of us are used to a morphological or structural concept of species. We regard a bird as a blue jay rather than as a cardinal because it looks more like other birds that we have assigned to the class "blue jay" than those of the class "cardinal." However, there are difficulties with the morphological concept. The major problem is that it is often useful to classify individuals that look quite different from one another as belonging to a single species. Often individuals of the two sexes can look quite different; in many instances, early taxonomists were led to classify the male and female of what we now recognize as a single species as belonging to different species. The male and female cardinal, boat-tailed grackle, and mallard duck each show appreciable differences in appearance between the sexes (sexual dimorphism). Individuals can also differ according to their developmental stage. How are we to deal with a frog and a tadpole or with different stages in species that metamorphose if we attempt to apply a morphological species concept?

Because of the difficulties in working with the morphological concept, most biologists now use a "biological" concept. According to this conception, a species may be defined as a group of organisms that are capable of freely interbreeding under natural conditions but which do not freely interbreed with animals of other groups. Essentially, this means that species are defined in terms of their reproductive behavior and reproduction rather than their appearance. Animals belong to the same species if they are capable of breeding freely in nature. There are many cases in which animals of different species can be interbred in captivity but do not interbreed in nature. They are "good" species. Tigers and lions, for example, have been interbred in captivity to produce "tiglons" and "ligers." However, evidence suggests that when the two species coexisted in India, they did not interbreed; tiglons and ligers were not found in nature.

There are problems with the application of the biological species concept too. It is difficult if not impossible to apply this concept in work with fossils and asexually producing forms (see Scudder, 1974). Fortunately, as a result of their breeding behavior, members of the same age and sex groups within species do come to resemble each other, and structural characteristics can be useful in providing an indication of species identification.

Population versus Typological Thinking

One of the fundamental factors in the remaking of modern biology has been the replacement of "typological thinking" with "population thinking." In typological thinking, one views individual events as instances or representations of some ideal. Individual events have meaning only as they represent some fixed, unchangeable idea. Variation is viewed as "noise" or "error variance" in the expression of the ideal in reality.

Population thinking stresses the uniqueness of everything in the organic world and thus is diametrically opposed to typological thinking. Each organism or organic entity is the product of a unique combination of genetic and environmental events. The variability one observes among different organisms which are classified in the same group is as "real," as important, and potentially as interesting as are the similarities that lead one to classify the organisms together. Statistical averages are viewed as abstractions from biological variability rather than as the essence of reality. To the typologists, the type is real and variation is an illusion; whereas to the populationist, the type is an abstraction and variation is real (Mayr, 1963). "No two ways of looking at nature could be more different" (Mayr, 1963, p. 5).

Phrases such as "the rat" and "the dog" when used in preference to phrases such as "rats" or "dogs" are symptomatic of typological rather than population thinking. It is better to use the plural to minimize careless typological thinking. One studies learning in "rats," not "the rat"—unless there is just one rat under study.

The importance of population thinking will become obvious when we consider populations, variation, fitness, and genetics. At this point, it is useful to remind the reader that the species, like the statistical average, represents but an abstraction. There is variation within species just as there is variation among species, although variation tends to have a sharper gradient at species boundaries. To gain comprehensive understanding of species and their behavior, one must consider not only the similarities and differences among species but also the nature of similarities *and individual differences* within species. Individual differences are not "error variance," as often treated in statistical analyses. Each organism is unique; individual differences are real and must be dealt with in the study of behavior.

Reproductive Isolating Mechanisms

As was discussed above, a species is a group of organisms which breed among themselves in nature but not with organisms of other species. The mechanisms that function to keep individuals of different sympatric species (those living in the same general region) from interbreeding are termed "reproductive isolating mechanisms." They may be defined as "biological properties of individuals that prevent the interbreeding of populations that are actually or potentially sympatric" (Mayr, 1963, p. 91). "The mechanisms that isolate one species reproductively from others are perhaps the most important set of attributes a species has, because they are, by definition, the species criteria" (Mayr, 1963, p. 89).

There exists a whole spectrum of devices which function to ensure reproductive isolation among species. The most obvious is a lack of fertility between species; many species are isolated because the sperm of one species cannot fertilize the eggs of the other. However, reproductive isolation is often much more complex than this. Mallard (*Anas platyrhynchos*) and pintail (*Anas acuta*) ducks are among the most common North American species of ducks and are fully interfertile in captivity, even when studied through many types of crosses (F_1, F_2, backcrosses, etc.). Although ducks of the two species may nest side by side in nature, hybrids are uncommon. Many factors, other than mere geographical isolation, can function to keep potentially interfertile species such as these from interbreeding in nature.

Table 3-1 A Classification of Reproductive Isolating Mechanisms

1 Premating mechanisms (mechanisms preventing copulation attempts)
- *a* Habitat isolation (potential mates do not meet because of preferences for different microhabitats).
- *b* Seasonal isolation (potential mates do not mate because they breed in different seasons).
- *c* Temporal isolation (potential mates do not mate because they breed at different times of day).
- *d* Courtship isolation (potential mates do not mate because of selective responsivity to or preference for courtship signals of their own species).

2 Mating-related mechanisms (mechanisms operating around the time of mating)
- *a* Mechanical isolation (mating is attempted but fails because of mechanical structure of genitalia).
- *b* Neuroendocrine isolation (sperm are transferred but critical neuroendocrine responses are not triggered).
- *c* Gametic mortality (sperm transfer occurs, but sperm are not viable in the prefertilization environment).

3 Postzygotic mechanisms (mechanisms reducing the success of interspecific matings)
- *a* Zygote mortality (egg is fertilized, but zygote dies).
- *b* Hybrid inviability (zygote produces a hybrid of reduced viability).
- *c* Hybrid sterility (F_1 hybrid is fully viable but either partially or completely sterile).

A system for classifying reproductive isolating mechanisms is presented in Table 3-1. Three major classes are distinguished—premating mechanisms, mating-related mechanisms, and postzygotic mechanisms. A number of factors can prevent individuals of different species from ever attempting to mate (premating mechanisms). If individuals breed in different years, in different seasons, at different times of day, or in different portions of their habitat, they will never meet at their breeding times and hence will not interbreed even though they are sympatric (seasonal and habitat isolation). In some cases, individuals meet but do not attempt copulation. Often the failure to attempt mating is a function of the inappropriate matching of courtship signals and responsivity to such signals; hence this type of isolation may be termed "courtship isolation."

Mating-related isolating mechanisms include mechanical isolation, neuroendocrine isolation, and gametic mortality. In the former, copulation is attempted but no sperm transfer takes place. Often this is because the design of the reproductive organs prevents sperm transfer; the organs may just not fit together properly (mechanical isolation). In many species, copulation must result not only in sperm transfer but also in the initiation of critical neuroendocrine reflexes in the female. These reflexes result in the induction of ovulation in reflex ovulators or preparation of the uterus for implantation (see Table 12-2). Species may be isolated despite sperm transfer if these neuroendocrine responses are not triggered by the appropriate kind and amount of copulatory stimulation normally provided by the male of the species. This may be termed "neuroendocrine isolation" (see Gray, Kenney, & Dewsbury, 1977). Reproductive isolation stemming from gametic mortality occurs when sperm are transferred but fail to survive in the environment of the heterospecific female.

Postzygotic isolating mechanisms include zygote mortality (death of the zygote after fertilization), hybrid inviability (a reduced chance for survival of the hybrid offspring), and hybrid sterility (complete or partial sterility in the hybrid offspring or deficiencies in the offspring's offspring).

Activities associated with premating activity, copulation, and the carrying of and caring for offspring can entail a considerable expenditure of energy and resources as well as increased risk of predation. The offspring of cross-species matings are almost always at a competitive disadvantage of one kind or other relative to the offspring of within-species matings. For these reasons, evolutionary forces appear to act to produce isolating mechanisms which function as early as possible in a potential reproductive encounter. It is more efficient to break off pairing on the basis of a courtship signal than through the sterility of a hybrid that has been produced through such mating.

TAXONOMY

Taxonomy is concerned with the identification, naming, and classification of animals. Each species has a unique scientific name in a binomial Latin system. The scientific name is always italicized or underlined, with the first word, which specifies the genus, capitalized, and the second word, which specifies the species, in lowercase letters (e.g., *Rattus norvegicus; Homo sapiens*). After its first use, the name of the genus can be abbreviated provided one does not intermix references to two genera with the same first letter (e.g., *R. norvegicus*).

Most species have one or more common names in addition to their scientific names. In some cases the common name can be used almost as reliably as the scientific name (e.g., there is only one species of lion). In many cases, however, reliance on common names can cause great confusion, as common names may differ in different areas and be quite nonspecific. For example, there are hundreds of species of rats and mice; naming a rodent as a "rat" or "mouse" tells the observer more about the size of an animal than about its taxonomic standing (e.g., white-throated wood rats, *Neotoma albigula*, are more closely related to cotton mice, *Peromyscus gossypinus*, than to Norway rats, *Rattus norvegicus*, or cotton rats, *Sigmodon hispidus*). A "gopher" in Florida is a tortoise; in Texas it is a rodent.

Species and genera are but two of the lower taxonomic levels in the complex hierarchical system that is internationally accepted for the classification of animals. Progressively broader major levels in this system are species, genus, family, order, class, phylum, and kingdom. One may refer to the more inclusive taxonomic categories (e.g., phylum) as "higher" levels than the less inclusive (e.g., genus). Classification at these higher taxonomic levels is somewhat more arbitrary than at the species level, and different taxonomists may "lump" or "split" more than others when making these classifications.

A highly selected listing of animal taxa (plural of "taxon"; animal groups of differing taxonomic levels) is presented in Table 3-2. This table was designed to illustrate the classification system and to provide the student with a reference with respect to various animals that will be discussed later in the book. It is highly biased toward mammals and primates. Table 3-2 should not be confused with a complete listing; complete listings can be found in most introductory zoology textbooks.

Estimates of the number of species vary, but there may be about 2 million. At least half of all species are arthropods, listed in a single row in Table 3-2. There are only 5,000 to 6,000 mammalian species. However, much of the research in comparative animal behavior, especially in comparative psychology, has been focused on vertebrates (e.g., Figure 2-5). We view the world through vertebrate eyes.

EVOLUTION AS HISTORY

If a taxonomic system is well constructed, it should correspond with the history of evolutionary diversification through time. It is difficult to reconstruct evolutionary history because much of it occurred before humans began keeping permanent records. Zoologists have used the fossil record, comparative morphology, and zoogeographical evidence together with information from anatomy, physiology, genetics, embryology, ecology, and behavior to reconstruct plausible histories of major segments of the course of evolution.

A representation of the probable evolution of the eutherian (placental) mammals is presented in Figure 3-1. Evolutionary history is best represented as a tree. Animals of base stock, represented as the trunk, radiate and diversify through time to produce a vast array of species. In the case of the placental mammals, for example, the basal stock appears to be found in the insectivore order—the order that includes shrews, moles, and hedgehogs. The radiations from insectivore stock have produced the rodents, carnivores, primates, and other orders of placentals. The lines leading to rodents, carnivores, and primates appear to have been separate radiations from the insectivore trunk. No rodent was an ancestor to any carnivore or primate. An insectivore, perhaps quite different from most living insectivores, was such a progenitor.

A broader representation of animal evolutionary history is presented in Figure 3-2. Once more, the use of the tree analogy should be apparent. Note, for example, that while fishes appear on the evolutionary line that led to mammals, these fishes were not the teleost fishes that are common today but a phylogenetically older group. No teleost fish was an ancestor of human beings.

Understanding of the tree analogy and of the "branching" pattern of evolution is important when comparing species. It can be contrasted with the

Table 3-2 A Partial and Selectively Biased Synopsis of the Animal Kingdom

- Phylum Protozoa—acellular organisms (e.g., *Paramecium, Amoeba*)
- Phylum Porifera—sponges
- Phylum Coelenterata—polyps and medusae
- Phylum Platyhelminthes—flatworms, including planarians
- Phylum Mollusca—molluscs (e.g., snails, clams, octopuses)
- Phylum Annelida—segmented worms (e.g., earthworms)
- Phylum Arthropoda—(e.g., spiders, crustacea, insects)
- Phylum Echinodermata—(e.g., sand dollars, starfishes)
- Phylum Chordata—the chordates, which have a single dorsal nerve cord
  - Subphylum Urochordata—sea squirts
  - Subphylum Cephalochordata—the lancets (e.g., *Amphioxus*)
  - Subphylum Vertebrata—the vertebrates
    - Class Chondrichthyes—fishes with cartilaginous skeleton (e.g., sharks)
    - Class Osteichthyes—bony fishes (e.g., goldfish)
    - Class Amphibia—amphibians (e.g., frogs, toads, salamanders)
    - Class Reptilia—reptiles (e.g., turtles, lizards, snakes, alligators)
    - Class Aves—birds
    - Class Mammalia—the mammals; warm-blooded, generally hairy
      - Subclass Prototheria—egg-laying mammals
        - Order Monotremata—monotremes (e.g., platypus, spiny anteater)
      - Subclass Theria—live-bearing mammals
        - Infraclass Metatheria—pouched mammals
          - Order Marsupialia—marsupials (e.g., opossum, kangaroos)
        - Infraclass Eutheria—placental mammals
          - Order Insectivora—insectivores (e.g., moles, shrews)
          - Order Dermoptera—flying lemurs
          - Order Chiroptera—bats
          - Order Edentata—edentates (e.g., sloths, anteaters, armadillos)
          - Order Pholidata—pangolins
          - Order Rodentia—rodents (e.g., rats, mice, chinchillas)
          - Order Lagomorpha—hares, rabbits, pikas
          - Order Primates—primates
            - Suborder Prosimii—prosimians (e.g., lemurs, tarsiers)
            - Suborder Anthropoidea—the monkeys
              - Superfamily Ceboidea—New World monkeys (e.g., capuchins)
              - Superfamily Cercopithecoidea—Old World monkeys
              - Superfamily Hominoidea—gibbons, apes, and humans
                - Family Pongidae—apes
                - Family Hominidae—hominoids
                  - Genus *Homo*
                    - Species *sapiens*
          - Order Cetacea—cetaceans (e.g., whales, dolphins)
          - Order Carnivora—carnivores (e.g., dogs, cats, bears)
          - Order pinnipedia—seals
          - Order Tubulidentates—aardvarks
          - Order Proboscidae—elephants and mastodons
          - Order Hyracoidea—hydraxes
          - Order Sirenia—sea cows or manatees
          - Order Perissodactyla—odd-toed ungulates (e.g., horses)
          - Order Artiodactyla—even-toed ungulates (e.g., pigs, sheep)

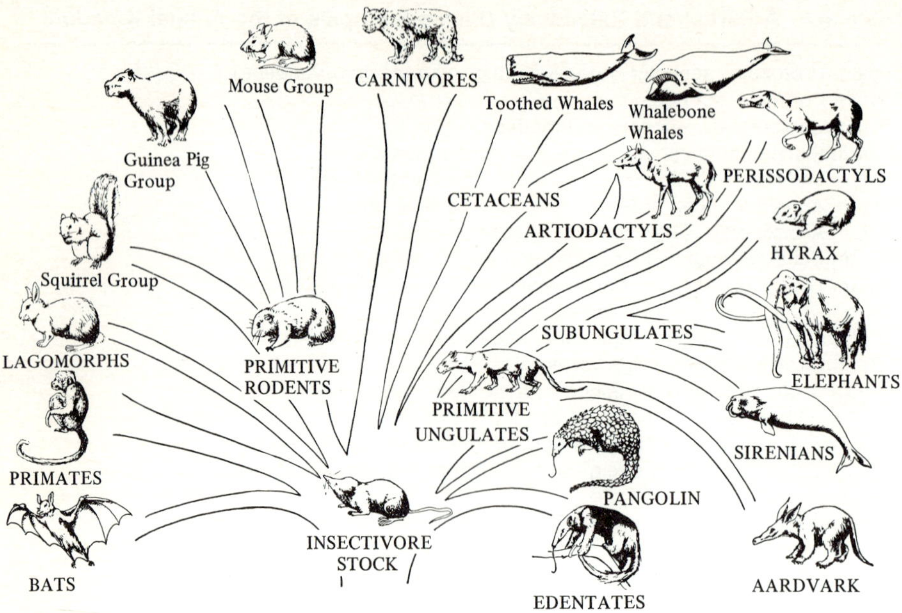

Figure 3-1 Diagrammatic representation of the probable evolution of placental (eutherian) mammals. *(After Romer, 1962.)*

notion of a *scala natura*, or linear concept of species (Hodos and Campbell, 1969). Animal species cannot reasonably be placed on any single scale with some species near the top and others near the bottom. While species can be scaled with respect to size, brain weight, metabolic rate, or any other continuously varying characteristic, no such scale provides a single representation of phylogeny. While fishes are phylogenetically older than primates (Figure 3-2), fishes and primates cannot correctly be said to be "lower" and "higher" species. To speak of "higher" and "lower" implies a single-dimension scale which, as we have just seen, does not accurately represent evolutionary history. Similarly, species other than humans are "nonhuman," not "subhuman," and species other than primates are "nonprimates," not "subprimates." There is no single phylogenetic scale. The vertical dimension in phylogenetic trees represents time, not "height."

EVOLUTION AS A PROCESS

Our understanding of evolutionary processes owes much to the contributions of two outstanding nineteenth-century scientists, Charles Darwin and Gregor Mendel. Darwin, of course, developed the fundamental theory of natural selection and the evolution of species, while Mendel provided a foundation for genetics. For a surprisingly long time after Darwin's and Mendel's contribu-

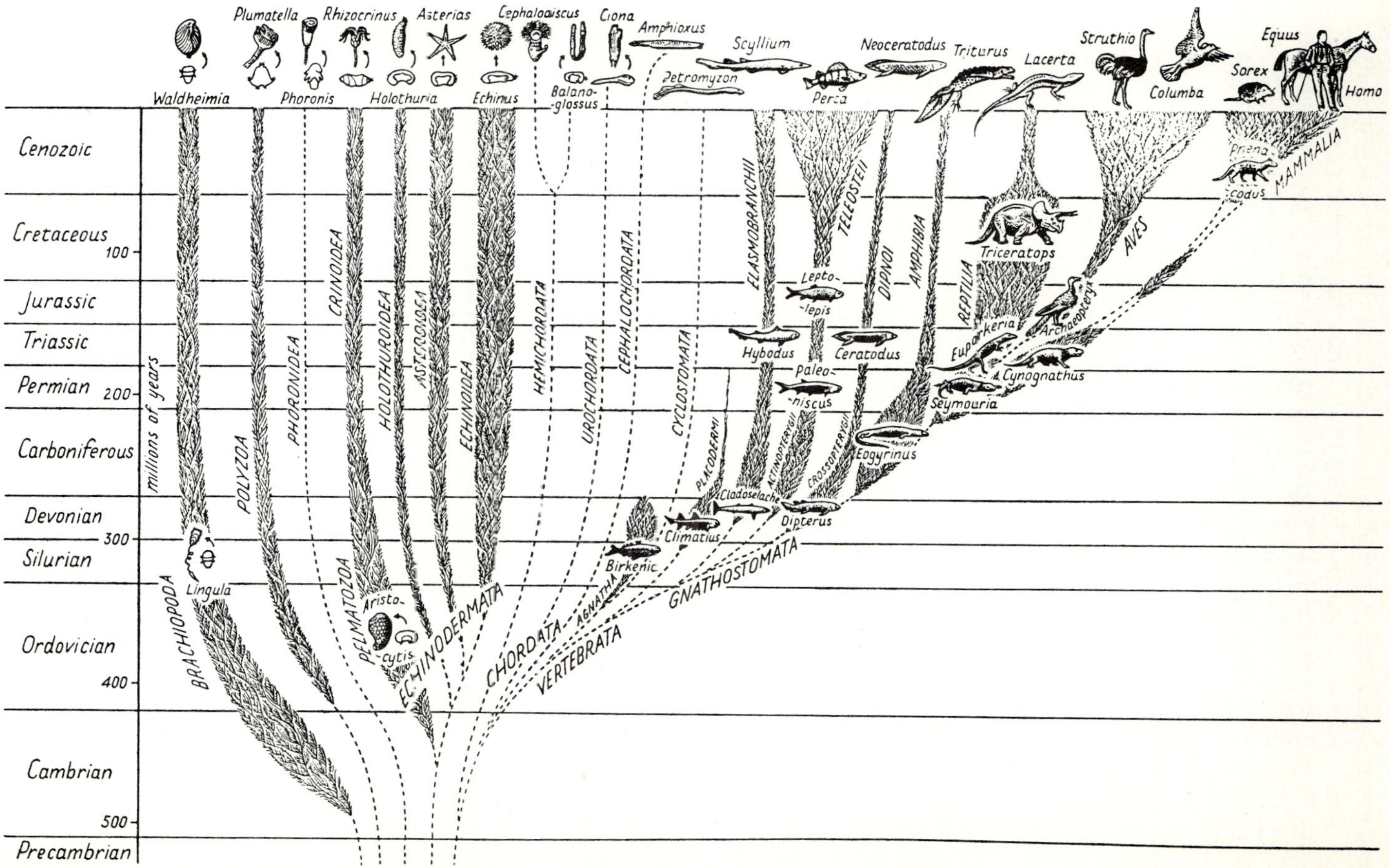

Figure 3-2 Diagrammatic representation of the probable origins of various groups of animals. *[From* The Life of the Vertebrates *by J. Z. Young (1950), published by Oxford University Press.]*

tions, evolutionary theory and genetics developed relatively independently of each other. This was remedied in the 1930s, with the development by R. A. Fisher, Sewall Wright, and J. B. S. Haldane of the so-called "modern synthetic theory of evolution." The major tenets of the modern synthetic theory are accepted by most biologists and will provide the basis for our discussion of evolution.

Basic Genetics

Before proceeding to a discussion of evolutionary processes, it may be useful to review a few concepts from genetics. It is assumed that you have been exposed to this material elsewhere and that this is a review. Should this not be the case, you may wish to consult an elementary textbook in zoology or genetics.

Genes For our purposes, genes are defined as the basic particulate, functional units of heredity. Although this fact is not essential to the conceptualization of genes, we now know that the information in genes is coded as a sequence of bases in large molecules of deoxyribonucleic acid (DNA).

Chromosomes Chromosomes are small structures, found in the nuclei of cells, upon which genes can be said to be located.

Gametes Gametes are specialized reproductive cells—sperm and ova.

Locus The locus is the location or "address" of a gene on a chromosome. Normally a gene always occupies the same location on the same chromosome.

Allele An allele is a particular form of a gene which can be distinguished from other forms or alleles. Alternate alleles of the same gene are capable of occupying the same locus (but not at the same time) and usually have different effects on the development and functioning of the organism.

Diploid Number The diploid number is the number of chromosomes found in each cell of the body except the gametes of sexually reproducing organisms. The diploid number is identical in all normal members of a species (e.g., 46 for humans; 42 for laboratory rats; 8 for fruit flies). With the exception of some sex chromosomes, chromosomes occur in pairs called homologues. One homologous chromosome comes from the mother and one from the father. Members of a pair of chromosomes normally cannot be distinguished from each other microscopically.

Haploid Number The haploid number is one-half the diploid number. In sexually-reproducing organisms, the haploid number of chromosomes is found in each normal gamete. It equals the number of pairs of chromosomes found in somatic cells (23 for humans; 21 for laboratory rats; 4 for fruit flies).

Segregation Segregation is the process that occurs in the formation of gametes, meiotic cell division, in which chromosome pairs separate so that each normal gamete has one complete set of chromosomes, the haploid number.

Assortment Assortment is the process in gamete formation whereby genes which are located on different chromosomes become situated in particular gametes in a manner independent of genes on other chromosomes. Thus, genes which come from the mother and are located on chromosome A have an equal chance of winding up in a gamete together with genes from either the mother or the father that are located on a different chromosome, chromosome B.

Crossing Over Crossing over is the exchange of actual genetic material between homologous chromosomes during gamete formation. Because of crossing over, the composition of chromosomes changes; therefore genes, not chromosomes, must be regarded as the basic units of inheritance.

Mutation A mutation is a change in the actual genetic material of the organism, DNA.

Population A population is a group of organisms of the same species occupying a given space at a given time.

Gene Pool The gene pool is the sum of all of the genes in a population.

Gene Frequency The gene frequency is the proportion of genes of a particular type and hence the relative representation of genes of that type in a gene pool.

Hardy-Weinberg Equilibrium A Hardy-Weinberg equilibrium is in effect when the frequencies of different genes in a gene pool remain constant over a period of generations. Such a situation would prevail when there is random mating, equal viability and reproductive potential across all genotypes, and no mutation.

Natural Selection

Having reviewed some basic concepts from genetics, we are now in a position to discuss the concept that lies at the very heart of evolutionary thinking—natural selection.

The Nature of Natural Selection Natural selection may be defined as the process through which there is a differential production of offspring in the next generation by individuals of different genotypes. The essence of natural selection is that there is a statistical bias in favor of certain genes with respect to their representation in the gene pool of the next generation. A view of the relationships among natural selection, segregation, assortment, and the individual organism has been expressed by Wilson (1975, p. 3):

> Natural selection is a process whereby certain genes gain representation in the following generations superior to that of other genes located at the same chromosome positions. When new sex cells are manufactured in each generation, the winning genes are pulled apart and reassembled to manufacture new organisms that, on the average, contain a higher proportion of the same genes. But the individual organism is only their vehicle, part of an elaborate device to preserve and spread them with the least possible biochemical perturbation. Samuel Butler's famous aphorism, that the chicken is only an egg's way of making another egg, has been modernized: the organism is only DNA's way of making more DNA.

Dawkins (1976) goes so far as to treat organisms as *survival machines* invented and manipulated by genes in their struggle for survival and representation.

For natural selection to work, there must exist in a population: (1) genetic variability, and (2) conditions under which individuals of different genotypes are differentially able to gain representation in the gene pool of the next generation.

Sources of variability The raw material upon which the forces of natural selection work arises ultimately from *mutation*. Mutations, as discussed above, are changes in the actual genetic material, DNA. Normally, mutation is a relatively rare event, with mutation rates being estimated at 10^{-5} to 10^{-6}. A number of factors, such as x-irradiation and certain chemicals, can increase mutation rates substantially. Mutations may be regarded as a nondirected, essentially random events. Mutations alone do not ensure that organisms become better adapted to the environment. Because organisms are the finely tuned products of natural selection and mutations are random, most mutations have detrimental effects.

While variation in actual genetic material must arise through mutation, differences among individuals in a population are in part a function of the combinations of particular genes possessed by different organisms. These arise largely as a result of sexual reproduction. Natural selection works on whole organisms to produce changes in gene frequencies.

Kinds of selection There are many ways through which individuals gain differential representation in the gene pool of the next generation. The first problem for any organism is survival. In most cases, organisms must survive in order to reproduce. Some organisms are better able than others to exploit their habitat and utilize its resources, including potential food and shelter, in order to survive. Differential survival may result from avoidance of predators or from an ability to cope with predators, parasites, or a variety of environmental changes (e.g., extreme temperatures, drought, levels of chemical pollutants). It is this form of survival selection which Darwin viewed as the essence of natural selection and which still may be uppermost in the thoughts of most readers (Mayr, 1972).

For Darwin, sexual selection was a basically different process from natural selection. It was selection based on reproduction rather than survival. Darwin believed sexual selection responsible for some of the more dramatic differences found between the sexes (e.g., bright coloration patterns, large size, specialized song patterns in males). As natural selection has been redefined since Darwin, so that it now is viewed in relation to the differential reproduction of genotypes, sexual selection can now be seen to be but one form of natural selection. Sexual selection is that form of natural selection directly related to differential reproduction, including the finding and acquisition of a mate, copulation, fertilization, and parental care. Following Huxley (1938), two forms of sexual selection are usually distinguished. *Epigamic selection* is based on choices made between males and females. The selection by a female of a particular male as a mating partner as a function of his large rack of antlers or bright coloration would represent epigamic selection. *Intrasexual selection* is selection based on interactions between animals of the same sex, usually between males. An example of intrasexual selection would be competition among males for a territory in a breeding ground. In practice, the forms of selection overlap and distinctions are not clear; e.g., a given territory may contain better resources such as food (survival selection) as well as being more attractive to females (sexual selection).

Other Factors Producing Changes in Gene Frequencies It is worth mentioning that natural selection is not the only process through which changes in gene frequencies can occur. Wilson (1975) lists four other sources of "microevolutionary" changes in gene frequency.

1 *Mutation pressure:* Differential mutation rates in favor of one allele can produce changes in gene frequencies on rare occasions.

2 *Meiotic drive:* Under rare circumstances, one allele can have an advantage over others as a result of differential representation in the formation of gametes.

3 *Genetic drift:* Genetic drift is the production of alterations in gene frequencies through chance factors alone, particularly through sampling error. Genetic drift is of particular importance when dealing with small populations or with new populations founded by a few animals.

4 *Gene flow:* Gene flow is the exchange of genes among different populations of a species, primarily through immigration.

While all four of these factors can create changes in gene frequencies in populations, it is natural selection that is the primary force in evolution.

Teleology and Lamarckism Most biologists agree that differential reproduction correlated with genotype, occurring through the actions of the mechanisms described above, provides the only mechanism through which precise adaptation to the environment can be achieved. Some individuals have proposed that the alterations in genotype occur in anticipation of environmental circumstances. Others have proposed that the environment directs such changes. These are both forms of teleology and are generally rejected. Mutation is essentially a random process; the environment "shapes" the organism only in that environmental contingencies permit differential reproduction correlated with genotype. There is no "direction" or "progress" toward "higher" forms or more "complex" structures. Natural selection works through differential reproduction of genotypes.

While at this point in the book we are stressing the role of genetics so as to facilitate an understanding of evolutionary processes, the role of the environment should not be underestimated. Genes and environment interact in the production of every organism, and the study of environmental factors merits, and later will receive, important emphasis. While the environment is important in the development of the individual organism, Lamarck went a step further to propose that such factors alter genetic material and that therefore acquired traits can be inherited. The notion of the inheritance of acquired traits, or Lamarckism, now is almost universally discredited.

The Level of Action of Natural Selection In principle, natural selection could act at any of a variety of levels (e.g., at the level of the individual, the population, or the species). The notion that selection acts on groups has been used by many scientists but was most systematically developed by V. G. Wynne-Edwards (1962). The most controversial aspect of Wynne-Edwards's proposals was that, under some conditions, individuals curtail reproduction for the good of the group. He proposed that territorial behavior (see Chapter 6) and epideictic displays (behavioral patterns that permit other members of the population to assess population density) have evolved because they permit some individuals to curtail reproduction when overexploitation of environmental resources appears imminent. For example, Wynne-Edwards noted that many species of birds breed in colonies, with each pair occupying a territory of a particular size. Only a portion of the habitat is used for breeding. If a bird is unable to secure a breeding territory in this area, it does not breed in that season, even though it would appear to the observer that nearby areas might be just as good as potential breeding sites. The essence of Wynne-Edwards's proposals is that in situations such as that of the nonbreeding colonial birds, the behavior functions to place a limit on the number of birds that can breed in a given year. When that limit is reached, other birds sacrifice breeding for "the good of the group." Earlier populations that have displayed this pattern of behavior have been successful and their genes have survived. Presumably, other groups failed to curtail reproduction as they approached the limits of the resources—such as food—in their environments, overproduced, and died out. Thus, natural selection would act differentially on the survival of groups. According to principles of *group selection*, behavior not in the best interests of the individual's reproductive activities would be selected for if it was of benefit to the group.

Wynne-Edwards's proposals provoked a barrage of criticism from biologists, most of whom reject his ideas. Perhaps the classic of this genre is the critique by G. C. Williams (1966). As noted above, natural selection is generally believed to act at the level of the individual. Natural selection is the differential reproduction of individual genotypes. Suppose that an individual did refrain from reproducing for the good of the group. How would the behavior be selected for? The individuals which display it do not reproduce. Any genes that would lead to a cessation of reproduction against the best interests of the individual would be selected against, not for. Adaptations appear selected to maximize the reproductive success of the individual over its lifetime. In the case of the territorial bird that refrains from mating, it is likely that attempts to breed in peripheral territories have some severe disadvantage, such as increased exposure to predation. The organism that refrains from reproducing during one year may be better able to reproduce successfully the next year than the organism that tries and fails the first time around. While the tendency to refrain from breeding may appear to be a "sacrifice," it may actually be a part of a long-term strategy to maximize reproductive success. Wilson (1975) has pointed out that there are rare circumstances under which group selection could function. However, these are sufficiently rare for us to dismiss at this point. Selection acts at the level of individuals.

Perhaps at this stage a repetition of an earlier point is in order. Insofar as we can determine, decisions as to whether or not to breed are not conscious choices made by individual birds. Rather, natural selection has acted to favor genes which act in the normal environment to produce organisms that behave in particular ways under particular conditions; they can do no other.

Statements to the effect that behavior can be selected because it is "good for the species," "preserves the species," or "enhances survival of the species" still are common not only in lay language but in some biological and psychological sources as well. It should now be apparent that these statements stem from a misconception. For example, organisms do not reproduce because reproduction preserves the species. They reproduce because reproduction increases representation of their own genes (translated: natural selection has operated to increase the frequencies of genes that produced organisms that reproduced effectively).

Genetic Fitness The concept of "fitness" has undergone several redefinitions. One still occasionally finds reference to "survival of the fittest," in which the term is used to mean that only those organisms of large size and aggressive habit will survive under conditions of "nature red in tooth and claw." The modern conception of fitness is much more subtle than this. Genetic fitness is potentially measurable as the contribution to the gene pool of the next generation by an individual with a particular genotype relative to that of other genotypes. The relation to natural selection should be clear. By definition, natural selection leads to prevalence of genotypes with high fitness.

Adaptation An adaptation is any structure, physiological process, or behavioral pattern that increases genetic fitness. The term may also be used to refer to the process through which natural selection creates changes that increase genetic fitness. As discussed in Chapter 1, determination of the ways in which behavioral patterns are adaptive (i.e., increase fitness) constitutes one of the four major classes of questions in comparative animal behavior.

Altruism Altruistic behavior may be considered to be behavior that acts to increase the fitness of another organism at the cost of a possible decrease in the fitness of the individual engaging in the altruistic act. Curtailment of reproduction under conditions such as those conceptualized by Wynne-Edwards would represent one form of altruistic behavior and has already been considered. However, there are other forms of altruistic behavior that demand consideration. In the social insects, for example, the majority of females in a complex society do not reproduce, acting instead to care for their sisters. On spotting a predator, group-living birds and primates emit a "warning cry," which alerts other group members to the presence of a predator but may make it easier for the predator to locate the signaler. Consideration of such behaviors leads us to three more important concepts: inclusive fitness, kin selection, and reciprocal altruism.

Inclusive fitness W. D. Hamilton (1964) proposed a modification of evolutionary theory that had profound impact on the development of our understanding of evolution in general and the evolution of altruistic acts in particular. Hamilton's thinking began with the recognition that the individual is not the only one to possess certain genes. Among diploid vertebrates, for example, an individual shares half its genes with its mother, father, brother, or sister by common descent. It shares one-fourth of its genes with its uncle and one-eighth of its genes with its first cousin. If an individual can increase the fitness of its close relatives, it can increase the representation of the genes which it shares with them in the gene pool of the next generation. The more distant the relation, the fewer genes will be shared by common descent with another, and hence the less the individual has to gain by increasing the fitness of another.

Hamilton proposed the notion of *inclusive fitness*, which may be defined as "the sum of an individual's own fitness plus its influence on fitness in its relatives other than direct descendants" (Wilson, 1975, p. 586). According to this conceptualization, there will exist circumstances in which individuals can maximize their total, inclusive fitness by performing altruistic acts toward relatives. The likelihood of such acts will vary as a function of (1) the cost of the act in terms of decreased individual fitness, (2) the gain in individual fitness for the relative, and (3) the degree of relatedness of the two individuals. If by sacrificing its own life and breeding potential a vertebrate could ensure that three of its brothers would reproduce at a rate equal to its own where they would not reproduce if the vertebrate should not sacrifice its life, the individual could maximize the representation of its genes in future gene pools by sacrificing and not reproducing. Sacrifice would lead to fitness 1 1/2 times as great as selfish reproduction. Paradoxically, an animal can increase inclusive fitness by not reproducing.

The notion that an organism could maximize its contribution to the gene pool of the next generation by not reproducing at first sounds absurd. However this is exactly what the social insects appear to do. Social insects are haplodiploid (females develop from fertilized eggs and are diploid, whereas males develop from unfertilized eggs and are haploid). Because of this, females share three-quarters rather than one-half their genes with their sisters. Because of these unique genetic relationships, female social insects appear to have been selected for engaging in a strategy of caring for their sisters rather than breeding themselves (Hamilton, 1964; Trivers & Hare, 1976).

Kin selection Selection acting through close relatives can be termed "kin selection." Kin selection can be defined as differential representation of genes resulting from favoring the survival and reproduction of relatives who possess the same genes by common descent.

Reciprocal altruism Altruistic acts are not always directed at relatives. For example, birds and primates emit warning cries which may aid survival in distantly related or unrelated animals, and various aquatic species engage in cleaning symbioses, wherein an organism removes ectoparasites from a member of another species. To explain such behavior, Trivers (1971) proposed the concept of *reciprocal altruism*. Reciprocal altruism involves the trading of altruistic acts by two or more individuals at two different times. In effect, an individual engages in an altruistic act, such as jumping into the water to save a drowning person, in exchange for the "promise" that the other individual will do the same in like circumstances. Warning cries and cleaning symbioses may be explained in this way.

It should be noted that, according to modern evolutionary theory, all acts of altruism can ultimately be traced to selfishness. There are no clear examples of behaviors which are altruistic and which do not, in the long run, appear to increase the inclusive fitness of the individual. Given our discussion of natural selection and fitness, it should be clear why this is so. Selection acts to maximize fitness. "What passes for cooperation turns out to be a mixture of opportunism and exploitation. . . . Scratch an 'altruist' and watch a 'hypocrite' bleed" (Ghiselin, 1974, p. 247).

This presentation of the theory of natural selection has been rather brief. If the material is new to you, you may not fully grasp all the concepts. We shall return to the issues of fitness, altruism, and the evolution of reproductive strategies in Part Six of the book. For the present, it is sufficient that you become acquainted with the general principles. So-called "sociobiological" theory represents one of the most exciting developments of animal behavior in the 1970s, with new theory and tests of theory producing a vigor that is enlivening the discipline. Extensions have been made to the social sciences, as with a controversial presidential address to the American Psychological Association (Campbell, 1975). It is important that you be familiar with these ideas and consider them as the book progresses. The interested student may wish to consult Williams (1966), Ghiselin (1974), Wilson (1975), Dawkins (1976), or LeBoeuf (1977) for more detail.

SUMMARY

Animal species are groups of organisms that are capable of freely interbreeding under natural conditions but are reproductively isolated from other such groups. Population thinking recognizes the importance of interindividual variability and emphasizes the uniqueness of each individual organism. A variety of reproductive isolating mechanisms minimize interspecific breeding in nature.

Each species can be classified in a hierarchical, internationally accepted taxonomic system.

Natural selection is the process whereby there is differential representation in the gene pool of the next generation of certain genotypes. Selection for both survival and reproduction works to alter the frequency and survival of genetic variants produced randomly through mutation.

Natural selection appears to work primarily at the level of the individual. Where altruistic behavior occurs, it generally can be traced to kin selection or reciprocal altruism. The notion of "inclusive fitness" and the idea that individuals can increase their overall fitness through facilitating reproduction by close relatives are of major theoretical significance.

Part Two

Patterns of Behavior

According to the set of guiding principles developed by Tinbergen and chosen for the organization of this book, the study of comparative animal behavior begins with observation and description. It is with behavioral description that we begin consideration of the substantive material in animal behavior. Given an understanding of the kinds of things animals do, we shall later proceed to consider Tinbergen's four classes of questions. This section is divided into three chapters, providing observational and descriptive information on individual behavioral patterns, reproductive behavior, and social behavior.

Chapter 4

Individual Behavioral Patterns

In this chapter we shall consider some of the behavioral patterns displayed by individual organisms in adapting to their environments. The behavioral patterns to be considered are those that permit the animal to acquire food, water, oxygen, and other sources of energy; to maintain its body; to interact with its environment; and to pattern the occurrence of its activities. These are the behavioral patterns critical to survival. The specific behavioral patterns we shall consider are the following:

1. Locomotion
2. Ingestion
3. Thermoregulation
4. Seeking of shelter
5. Avoidance of predators
6. Sleep
7. Body maintenance
8. Elimination
9. Exploration
10. Play
11. Use of tools
12. Biological rhythms

Reproductive and social behaviors will be discussed in Chapters 5 and 6.

Three qualifications are in order. First, the division of material into these three chapters and into sections within chapters is for convenience and clarity in presentation. Many lines of division are somewhat arbitrary; some are ambiguous at best. For example, the seeking of shelter may be important in avoiding predators; exploration may be a part of play; eating affects thermoregulation. Virtually all these behavioral patterns (e.g., eating, seeking shelter, sleep) can be highly social on occasions. Even patterns of elimination can be of great social significance, as in the use of urine in territorial marking. These functional divisions of behavior, then, are for our convenience in presentation and should not be regarded as absolutes.

Second, it should be remembered that "an animal is an adapted organism, and not simply a collection of specialized adaptations" (Bunnell 1973, p. 79). The ways in which an animal adapts to different aspects of its environment interlock and interact to produce a complete functional whole. For example, the feeding pattern an organism adopts affects patterns of locomotion, avoidance of predators, and sleep and, in turn, is affected by each of these. Survival is critical and reproduction is the "bottom line." Selection works to produce total organisms that function in their environments, not to approach perfection in any one domain.

Finally, it is obviously impossible to provide a complete coverage of all these topics in one chapter. We shall try to provide some idea of the diversity of behavior in a given category together with a few interesting examples.

LOCOMOTION

Because it permits the animal to move about in its environment, locomotion is essential to most animal behavior.

Types of Locomotion

There are five fundamental types of locomotion: flagellar movement, changes in body shape, undulation, jet propulsion, and movement of extremities (Maier and Maier, 1970). In *flagellar* movement, the animal moves by beating either a single flagellum or a group of hair-like *cilia*. Such movement is common among protozoans. *Changes in body shape*, such as the extension of pseudopods in *Amoeba*, represent an alternative form of locomotion common among invertebrates. *Undulation*, or writhing, is a pulsating, wave-like locomotor pattern particularly common in snakes, aquatic mammals, and fishes. *Jet propulsion* involves the forceful expulsion of water in an aquatic environment and is found in a number of invertebrate species such as jellyfish and squid. Locomotion by *movement of extremities*—be they arms, legs, wings, or flippers—is common among most vertebrate and some invertebrate species.

Movement of Extremities

Locomotion by movement of extremities can occur in water, in trees, in air, underground, or on land.

In Water Swimming with the use of movement of extremities is found in sea horses and a variety of other vertebrate species. Many mammalian species that are not primarily aquatic—including many rodents, monkeys, and carnivores—have been observed to swim well on occasion.

In Trees Well-developed claws and visual and vestibular systems are characteristic of most arboreal (tree-living) species. Many species move about by clinging to trees and either climbing or walking along the tops of branches. A pattern of *brachiation*, swinging hand over hand while hanging beneath a branch, is found in a variety of species of apes.

In Air While gliding is seen in a number of different taxa, including such species as "flying" fish and "flying" squirrels, true flight is the epitome of locomotion in air, and is found among insects, birds, and bats. The most common flying pattern of birds, flapping flight, entails repeated up-and-down wing movements. The inner parts of the wings function primarily to provide buoyancy against moving air currents, while the outer sections are critical to propulsion (Heinroth and Heinroth, 1958).

Under Ground Fossorial and semifossorial species are those that live all or a large part of their lives underground. Typically such species show numerous adaptations—including reduction in the size of the eyes and the ear pinnae and in the thickness of pelage—that function to minimize resistance to underground passage.

On Land Numerous locomotor patterns may be observed on land. *Bipedal walking*, that performed on two feet, is seen in a variety of mammals, although it is most common in some kangaroos and rodents and in humans. Species such as kangaroos and kangaroo rats typically show a "saltatorial" or leaping pattern of locomotion.

Locomotor patterns in most quadrupedal (four-legged) species tend to be quite stereotyped. When walking, the limbs are moved in the only order of stepping that provides for a steady triangle of support by three of the limbs. As the pace is quickened, as in trotting and galloping, an increase in speed occurs, along with a sacrifice in stability (Gray, 1953). Wildebeest and lions are capable of running at speeds up to 50 miles per hour, while cheetahs have been recorded as running at 70 to 75 miles per hour (Bourliere, 1964).

INGESTION

Feeding, drinking, and breathing are patterns of ingestion critical to the survival of a wide array of animal species.

Feeding

Feeding Strategies Animals can be grouped according to their primary feeding strategies into five basic classes: filter feeders, parasites, herbivores, carnivores, and omnivores (Maier and Maier, 1970). *Filter feeders* all live in water and obtain food by filtering it from the water. Movement of water relative to the animal's body can be provided either by some kind of pumping action or through swimming. *Parasites* typically live entirely off a host animal. While some may live for a long time without killing their host (e.g., tapeworms), others may kill their host rather rapidly. *Herbivores* rely on plants, bark, leaves, seeds, fruits, and roots for their food. The Australian koala is highly specialized and eats only leaves of the eucalyptus tree. Cud-chewing ruminants have four-chambered stomachs to aid digestion. *Carnivores* feed primarily on other animal matter. Animals with diverse carnivorous feeding strategies are found among invertebrates and vertebrates alike. Many species are *omnivorous*, eating a wide range of foods. These species have the least specialized feeding strategies. For example, army ants eat virtually everything they encounter as they move through tropical forests. A wide array of omnivorous patterns can be found among the rodents (Landry, 1970). Even chimpanzees appear omnivorous, showing a surprising degree of organized hunting (Teleki, 1973a, b).

Food-Related Behavioral Patterns From among food-related behavioral patterns, we shall discuss predation, hoarding, manipulation, and regulation.

Predation While some species—such as web-building spiders, praying mantids, and ant-lions—wait passively for prey, most species actively hunt. Once captured, prey may be grasped, suffocated, constricted, poisoned, immobilized, shocked, or eaten alive. The range of adaptations for predation is indeed great.

A mammalian predatory sequence may be divided into the approach to the prey, the kill, and ingestion (Ewer, 1968). Domestic cats stalk their prey crouching low, following each movement, and using a "slink run" to move progressively closer until the ambush occurs. By contrast, cheetahs approach in the open, walking after prey, and use their exceptional speed only when prey start to flee.

The killing bites of mammals typically are delivered near the nape of the neck. The precise pattern used in killing varies with both the species and the circumstances. Prey first may be grasped or pinned using the forepaws. Many species, especially canids, display a ritualized "death shake," in which the prey is repeatedly shaken with lateral movements while in the predator's mouth (Eisenberg & Leyhausen, 1972).

Most carnivores first ingest the head of their prey and then work down the body. Research by Leyhausen has shown the cues on which orientation is based to be primarily tactile; predators respond to the lie of fur or feathers. Ingestion of large prey often is begun in the belly or groin region (Ewer, 1968).

Carnivores generally stop searching and hunting when they are satiated. While surplus killing beyond the amount an animal can eat is rare, it can happen in situations where prey occur in high densities and thus require minimal searching (Kruuk, 1972).

Hoarding Because food is not equally abundant in all seasons, many species, especially rodents, store food or "hoard." Two major forms of hoarding may be distinguished: larder hoarding and scatter hoarding (Ewer, 1968). In larder hoarding, a cache of food is stored in a burrow or hole at the home. Scatter hoarders conceal individual loads of food about a much wider area at sites away from the actual nest. Intermediate strategies also occur. A cache of 8 quarts of well-preserved seeds has been found in the burrows of prairie voles, which are small, field-dwelling rodents (Jameson, 1947). Many hoarding species have large cheek pouches that aid in the transfer of food to the site of storage.

Food hoarding in rodents can be elicited in the laboratory, where the effects of deprivation, early experience, physiological manipulations, and similar variables have been studied.

Manipulation The pattern of manipulating food before ingestion varies greatly with species and food material. Sea otters are known to lie on their backs in the water and use a flat rock placed on the chest as an anvil in opening the shells of mollusks. The forepaws and mouth are used in varying combinations in other species.

Many patterns of food manipulation appear to require minimal experience for their development. When opening nuts, squirrels gnaw one or two furrows in the shell in particular orientations. Eibl-Eibesfeldt (1963) raised red squirrels that had no opportunity to open nuts and found that the gnawing movements were essentially normal, though poorly oriented, on first appearance.

Regulation Animals do not eat continuously. As the amount of time since the last meal increases, individuals of many species increase their activity levels and become particularly responsive to food-associated cues. Ingestion of food is terminated at an appropriate time; obesity is rarely a problem in most animal species. In mammals, the regulation of food intake seems to result in part from the interaction of two regions within the hypothalamus of the brain, the lateral and ventromedial nuclei (Keesey and Powley, 1975). In blowflies, by contrast, the regulation of feeding is much more peripherally controlled (i.e., less dependent on the central nervous system). Feeding is initiated by peripheral receptors and inhibited by input from the distended foregut (Dethier, 1967.)

Drinking

Water is critical to a variety of life processes including digestion, elimination, and thermoregulation. While many species must actively ingest water, others can obtain sufficient supplies through the metabolism of their food.

Different movements may be utilized in drinking. Dogs and wolves adopt a stereotyped posture and scoop water with their tongues, while cattle suck water into the mouth, with the tongue playing only a passive role. Most rodents engage in a relatively stereotyped pattern of licking. Elephants, of course, use their trunks to siphon water and transfer it to their mouths (Bunnell, 1973).

Breathing

The ingestion of oxygen in breathing is so ubiquitous a behavior that it is often overlooked. Virtually all species possess some mechanism whereby fresh air is brought into the body and foul air expelled.

Breathing becomes of major concern with aquatic mammals. These species show remarkable adaptations in their respiratory systems that permit them to dive to greater depths and remain submerged for longer periods than other mammalian species. While maximal submersion times for muskrats and elephant seals have been estimated at 12 minutes, those for bottle-nosed whales are estimated at 120 minutes (Bourliere, 1964).

THERMOREGULATION

Animals are able to survive and thrive only within limited temperature ranges. Regulation of body temperature is a critical aspect of adaptation, particularly in harsh desert and Arctic environments. Birds and mammals are *homoiotherms*; they possess physiological mechanisms enabling them to maintain relatively constant body temperatures despite wide fluctuations in environmental temperatures. By contrast, invertebrates, fish, amphibians, and reptiles are *poikilothermic*, with body temperatures fluctuating in relation to environmental temperatures.

Numerous behavioral patterns function at least in part in thermoregulation. Many species display seasonal migrations that function to minimize exposure to seasons characterized by dangerous temperature extremes. Less dramatic daily movements in and out of shaded areas or burrows function in regulating body temperature on a shorter time scale. In the laboratory, animals will perform arbitrary operant responses with temperature changes as the reinforcement (e.g., Wright & Meyer, 1969). Many species survive harsh winters by *hibernating*. The animal becomes inactive and body temperature, heart rate, and metabolic demands are reduced to a bare minimum. Some frogs, crocodiles, and fishes respond to extreme high temperatures by undergoing *aestivation*, a similar state of inactivity.

SEEKING OF SHELTER

Most species seek some kind of shelter and thus gain protection from temperature extremes, precipitation, and predators. In some cases, the animals may simply move into a cave, crevice, or tree with minimal alteration of the environment. Alternatively, elaborate nests may be constructed that entail a considerable modification of the environment. Nests may be relatively permanent, such as beaver dams, or temporary, such as the sleeping nests constructed by chimpanzees, which are normally occupied for just one night. In many species nesting is particularly associated with reproduction. Nests may first be constructed or may be greatly enlarged as the time approaches for arrival of the young.

Among invertebrates, shelter construction is most prominent among insects. In some species, such as nonsocial wasps, a single female digs and provisions a nest hole. In many other species, however, nests may be quite elaborate and nesting quite social. Termite mounds and honeybee hives provide cases in point.

The structure of honeybee hives varies with the species but always contains a single queen and many workers. Specialization of function is the hallmark of the organization of honeybee workers in building and maintaining a hive. Different individuals build the comb, nurse the brood, clean out cells, make honey, guard the hive entrance, and forage for pollen and nectar. The function of a particular bee changes throughout its life, beginning with work cleaning inside the hive and ending with foraging (Lindauer, 1961).

The selection of a location for new honeybee hives is an interesting process. In the late spring, the queen and about half the workers leave the old hive to the queen's daughters and form a swarm at some distance from the old hive. They remain in this swarm until a new site is selected. Scout bees move out from the swarm to inspect numerous potential nest sites. On returning to the swarm, they perform a dance indicative of the location of the site (see Chapter 5). The intensity of the dance varies with the apparent quality of the potential hive site. Size and protective characteristics of the site appear to be of particular importance. The dancing bees recruit other scouts. The vigor of the dances and responses of new scouts act to enable the swarm to reach a "decision." Eventually, a large proportion of the scouts dances to indicate a particular nest site and the swarm moves off to occupy that location (Lindauer, 1961).

Rodents display a great variety of shelter-seeking behavioral patterns. Beaver dams contain a single-chambered lodge which provides a home for a pair of beavers and their last two litters. Wood rats, also known as pack rats or trade rats, collect large numbers of sticks and twigs from which an extensive hut is built. Nest building in laboratory rats and house mice has been much studied in the laboratory. Both species use cotton, paper, cloth, or similar material to construct bowl- or cup-shaped nests that are sometimes covered.

Chimpanzees, orangutans, and gorillas all construct sleeping nests in trees.

AVOIDANCE OF PREDATORS

As most species are preyed upon by animals of at least some other species, avoidance of predation plays an important role in survival and reproduction. The primary means of protection from predators are concealment, warning of conspecifics, warning of predators, escape, and active resistance.

Concealment

Many prey species conceal themselves from predators by remaining in protected locations—such as burrows, crevices, or huts—when predators are present. Alternatively, concealment can be provided by the actual appearance of the animal itself. Cryptic coloration, which functions to enable animals to blend in with the background, can be found in virtually all animal taxa. Many extreme examples are found in insects, which may resemble leaves, twigs, or even bird droppings. Often, a particular behavioral pattern is associated with protective coloration such that the animal orients itself in a particular relationship to its environment, usually remaining motionless.

Warning of Other Prey

Whatever the species-typical response to predators, one problem is that of locating them. Various patterns of movement, such as periodic scanning and orientation (e.g., in relation to the wind), facilitate detection and location of predators. Often animals forage in mixed-species flocks, such as those of baboons and antelopes. Baboons have highly developed visual systems, while the antelope olfactory systems are particularly sensitive. Animals of both species respond to each other's warning calls and thus are hard to surprise.

Warning calls of birds are responded to by a broad range of species. Typically, warning calls are relatively pure tones with no sharp discontinuities; these are difficult for predators to localize (Figure 4-1*a*; Marler, 1959).

Warning of Predators

Predators find the taste of some prey aversive. Blue jays, for example, typically vomit some time after eating monarch butterflies, which are large and distinctively colored. Such a pattern of coloration serves as "warning" to a predator by providing it with information that this is a noxious prey. Through evolution, numerous nonnoxious species have come to resemble noxious ones, with the obvious advantage of predator avoidance. This phenomenon is known as *Batesian mimicry* (Brower, 1969).

Numerous active behavioral patterns are used to warn predators. The rattling of rattlesnakes and aggressive postures of numerous mammals are examples. Birds will "mob" certain predators, such as stationary hawks or owls, by approaching relatively closely and emitting loud calls and conspicuous displays (Marler, 1959). Sounds used in these circumstances have a wide band of frequencies with sharp onsets and offsets and thus are easily localizable (Figure 4-1*b*). The advantage of conspicuousness in such situations is obvious.

Escape

Rapid and agile locomotion provides the best and probably the most common means of escaping from predators. Various species supplement their locomotor escape patterns with displays that function to distract or startle a potential predator. Still others may adopt a state of tonic immobility ("play possum") as a means of reducing the likelihood of attack (see Gallup, 1974).

Active Resistance

As a last line of defense, the prey may show active resistance to the predator. Prey may strike, grasp, or bite predators while resisting. Skunks and numerous

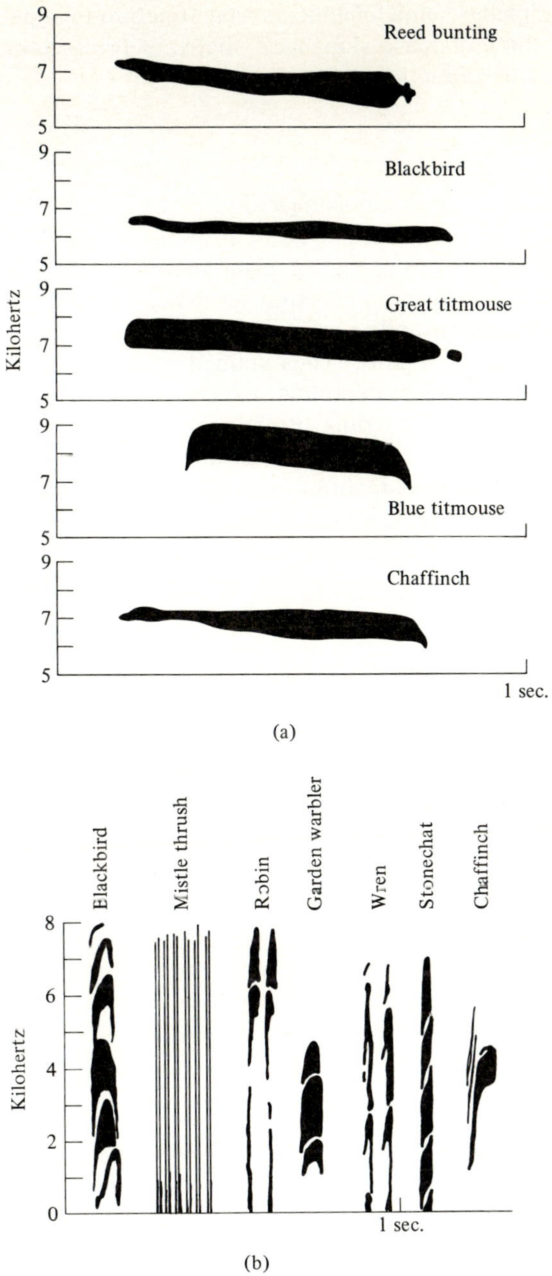

Figure 4-1 Sound spectrograms of various bird calls. A sound spectrogram shows the distribution of sounds of various frequencies (perceived as pitch) on the ordinate against time, plotted on the abscissa. *(a)* The warning calls of five species are difficult to localize as they are relatively pure tones without sharp discontinuities. *(b)* The mobbing calls of seven species are easily localized as they contain a broad spectrum of frequencies and sharp discontinuities. *(After Marler, 1959.)*

arthropod species, such as millipedes, emit chemicals that function to repel predators. Other species have thick or toxic skin, hard shells, or projections such as spines or quills, all of which function to repel predators.

SLEEP

Sleep may be defined behaviorally, electrophysiologically, or by using both kinds of information. Behaviorally, sleep is a period of prolonged inactivity associated with four characteristics: (1) it is often organized in a daily (or tidal) rhythm, (2) an elevation of thresholds for various responses usually occurs, (3) it often occurs in a species-specific site (usually a safe place), and (4) the animal adopts a species-characteristic sleeping posture. Thus defined, sleep has been reported in a wide range of animal species, including insects, mollusks, fish, amphibians, reptiles, birds, and mammals (Meddis, 1975).

Electrophysiological evidence of sleep is provided by an electroencephalogram (EEG). The EEG records patterns of "brain waves" from electrodes either placed externally on the scalp or implanted into the depths of the brain. Electrophysiological evidence of sleep is available only for birds and mammals.

Two major stages of sleep can be distinguished in birds and mammals—slow-wave sleep and paradoxical sleep (see Figure 4-2). *Slow-wave sleep* is named for the characteristic pattern of slow, large-amplitude waves that appears in the EEG record. A state of relaxation can be seen in the loss of muscle tension—as, for example, that recorded electrically from the neck muscles (Figure 4-2).

Periodically during sleep there are dramatic alterations in the EEG and muscular patterns. While the animal remains behaviorally asleep, a "desynchronized" EEG pattern appears and is characterized by low voltage and fast activity. Skeletal muscle tone drops below that characteristic of slow-wave

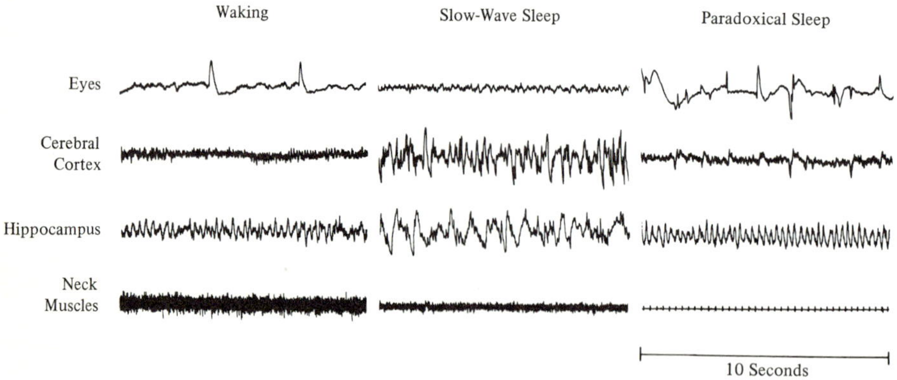

Figure 4-2 Electrophysiological recordings from the brain of a cat showing eye movements, EEG activity in the cerebral cortex and hippocampus of the brain, and muscle tension recorded in the neck muscles during waking, slow-wave sleep, and paradoxical sleep. *(From "The Evolution of Sleep" by T. Allison and H. Van Twyver,* Natural History Magazine, *February, 1970. ©The American Museum of Natural History, 1970.)*

sleep. Because the animal appears asleep while the EEG record resembles that of the waking state, this stage often is referred to as *paradoxical sleep*. One of the most characteristic features of paradoxical sleep is the appearance of *r*apid *e*ye *m*ovements (thus the alternative term for this kind of sleep is "REM" sleep). In humans at least, REM sleep is associated with dreaming. Episodes of paradoxical sleep occur about every 9 minutes in house mice, every 50 minutes in monkeys and human children, and about every 90 minutes in adult humans.

BODY MAINTENANCE

Maintenance of the body surface so that it is clean and free of parasites is important to the health and well-being of many animals. Bouts of self-grooming constitute a portion of the daily routine of most mammalian and some nonmammalian species. Often, grooming occurs right after a meal. Some rodents display a pattern of sandbathing, in which they twist their bodies so as to rub the sides and ventrum on the substrate, thus removing both foreign particles and excess glandular secretions such as lipids (see Borchelt, Griswold, & Branchek, 1976). The preening patterns of birds are readily observed in a number of contexts.

Interspecific maintenance behavioral patterns also occur. Many species of "cleaner fish" are adapted to remove ectoparasites from the skin of much larger fish. Some may even clean parasites from inside the mouths of larger, predatory fish without themselves being eaten.

Among primates, grooming serves not only to keep the pelage clean but also as an important social force in the maintenance of social organization. Grooming is developed to a fine art and may occupy a considerable period each day in some species. Lemurs and lorises have evolved special toothcombs, formed by the lower canines and incisors, that function in grooming. Social grooming may be particularly intense in mother-infant relationships, as a prelude to copulatory activity, or in the assertion of dominance relationships (Jolly, 1972).

Chimpanzees in captivity have been observed to use wooden tools in dental social grooming (McGrew & Tutin, 1973). The "dentist" chimpanzee uses pieces of wood to remove food particles from the teeth of the "patient."

ELIMINATION

Patterns of urination and defecation provide more to interest the animal behaviorist than might at first be apparent. First, many species display characteristic postures or patterns of urination and defecation. In dogs, for example, there is a sexual dimorphism of urinary postures, with the adult female squatting while the male lifts his leg to urinate.

The amount of care used in selecting a location of deposition of urine and feces appears to vary according to the nature and permanence of the nest site. In species of primates in which troops roam about their home range, sleeping in different locations from night to night, little attention is paid to the fate of urine and feces. For species with more permanent nest sites, such as many carnivores, the nest site must be kept clean and control of urination and defecation is more rigid. Burial of feces is commonly observed in domestic cats. The mothers of many mammalian species with protected nest sites ingest the urine and feces of their young, often until the young are old enough to locomote by themselves. This behavior not only keeps the nest clean but minimizes the chance that the young will be located by a predator because of the odor of urine and feces. Whereas ground-nesting gulls go some distance from the nest to defecate, cliff-dwelling kittiwake gulls simply defecate over the rim of the nest.

Patterns of urination and defecation may develop slowly. The mothers of newborn mammals of many species, such as guinea pigs, must stimulate urination and defecation in the young by licking the anogenital region. Some hormonal influences appear to affect the development of different urinary postures in dogs (Beach, 1974).

Urine and feces can be of great social significance, particularly in olfactory communication. Many species mark their territories by depositing urine and feces about them. The familiar cocking of the leg in domestic dogs and backward spray of urine onto objects by tomcats appear to function to spread urine at an appropriate height. Hippopotamuses display a rapid vibration of the tail during defecation, thus flinging feces about in the environment. Chemical signals resulting from elimination are important in regulating reproductive activities as well. Such chemicals can function to block pregnancy in inseminated females, to induce estrus in diestrous females, or to accelerate sexual maturation (Bronson, 1971; Chapter 10).

Enurination, a release of urine by a male in the direction of a female, is often observed in the context of courtship in species such as chinchillas, guinea pigs, porcupines, and rabbits. Such behavior is not a normal part of the courtship sequence and may reflect either "frustration" or rejection (Ewer, 1968).

Numerous comparative psychologists have attempted to assess levels of "emotionality" in animals. As one index, they place the animal in a novel situation and record the amount of defecation.

EXPLORATION

For animals of most species, a knowledge of the surrounding environment can be of great value in facilitating survival and reproduction. By systematically exploring its home range, the organism acquires information regarding the location of resources such as food sources, potential mates, and places of potential safety from predators. Thus, it is not uncommon to find animals exploring their environments even at times when they have recently eaten and drunk and are not in mating condition. While food or water deprivation, an estrous condition, and other factors can facilitate exploratory activity, they are not necessary for its occurrence.

Patterns of exploratory behavior have been the topic of much research by psychologists.

Object Investigation

One approach to studying the investigation of objects is simply to introduce a novel object into an animal's home cage. Glickman and Sroges (1966) studied the reactions of over 300 animals of over 100 different species to wooden blocks, steel chains, wooden dowels, rubber tubing, and crumpled paper balls. A summary of the results is presented in Figure 4-3. Primates and carnivores displayed much more object investigation than did animals with brains of less complexity; reptiles displayed the least object investigation. Characteristically, object investigation was greatest near the beginning of the 6-minute tests and decreased progressively throughout the test.

An alternative strategy in the study of object investigation is to anchor the stimulus in a small alcove and allow the animal to approach it. A photocell ("electric eye") may be used to record the number of approaches the animal makes.

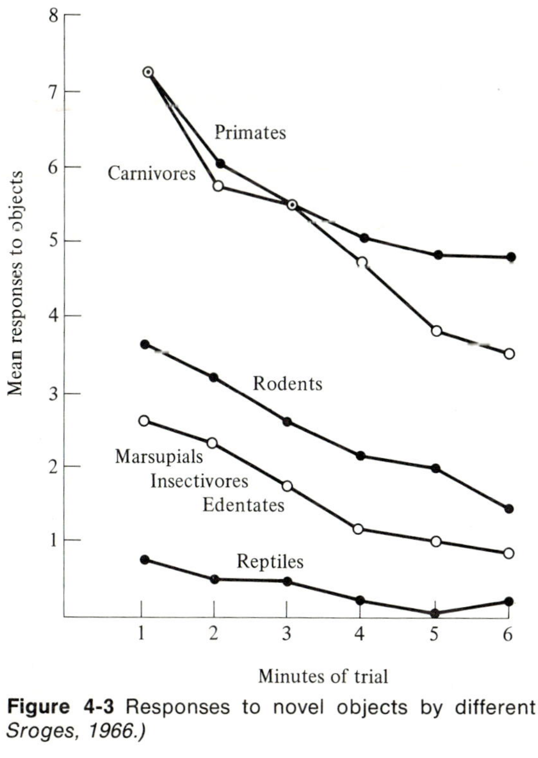

Figure 4-3 Responses to novel objects by different vertebrate taxa. *(From Glickman & Sroges, 1966.)*

Locomotor Exploration

Much interest has also been shown by psychologists in the study of patterns of locomotion displayed in novel environments. Many psychologists have utilized the *open field* in their research (see Walsh and Cummins, 1976). An open field is simply a large open box, often made of wood and painted a flat gray. The floor of the box is marked off into a number of equal-sized squares (Figure 4-4). An animal is introduced into the open field and allowed to move about for a fixed period of time. The observer typically records the number of squares entered and the number of fecal boli deposited. In some tests an attempt is made to provide a categorization of the relative frequency of different behavioral patterns (e.g., sniffing, rearing, grooming).

Some psychologists treat the open-field test as reflecting the "exploratory tendency" of an animal. Others treat behavior in the open field as an index of "emotionality." Animals are considered to be highly emotional when they show much defecation and little locomotion (see Candland & Nagy, 1969; Archer, 1973). As an alternative strategy, one can simply treat the open field as a convenient setting within which to assess the behavioral tendencies of different

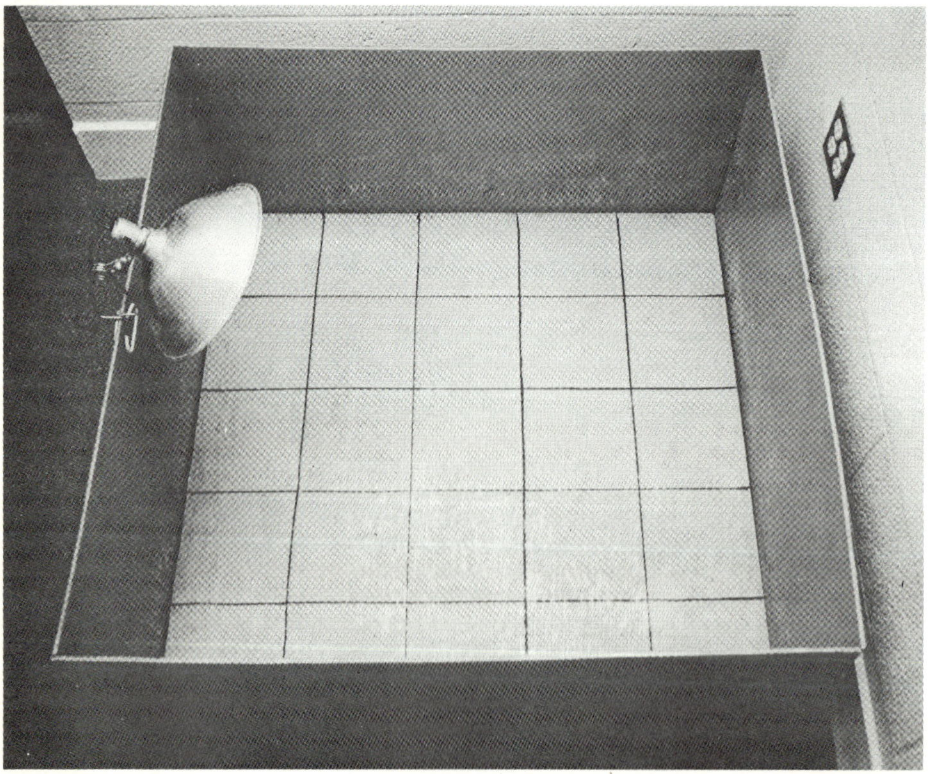

Figure 4-4 An open field. *(Photograph by the author.)*

animals in a minimally structured environment (Wilson, Vacek, Lanier, & Dewsbury, 1976).

While exploration is important to animals, it also entails risks, particularly from predation. To study this, Morrison and Glickman introduced a group of house mice into a cage occupied by a barred owl. The mice that were most active in the open-field test were most likely to be eaten by the barred owl. Brain lesions that increased the tendency to locomote increased the likelihood of predation (Glickman, 1971).

Reinforcement

Animals have been shown to learn a variety of tasks when the only reinforcement is the opportunity to explore. Rats have been shown to learn mazes when the reinforcement is the opportunity to explore a complex, novel environment. Rhesus monkeys will depress a lever when the reinforcement is the opening of a window facing an interesting environment. Monkeys preferred to view other monkeys, electric trains, food, and an empty chamber, in descending order (Butler, 1960). Stimulus complexity appears to be one critical determinant of exploration in mammals (Walker, 1964).

Neophobia

Under some conditions, animals will avoid novel objects, a pattern termed *neophobia* (Barnett, 1963). This response is particularly common where there is a sudden change in a familiar situation. If wild Norway rats, reared in the laboratory and accustomed to feeding from a wire basket at the back of their cage, are required to eat from a basket at the front of the cage, they may refuse food for several days. This is an example of neophobia. Tame laboratory rats begin to explore the basket sooner and daily food intake is less affected. It appears that the neophobic response has been reduced through the domestication of rats (Barnett, 1963; Mitchell, 1976; Barnett & Cowan, 1976).

PLAY

A variety of animal species, especially mammals, engage in behavioral patterns that we can readily label as play. Young carnivores, such as domestic cats, will repeatedly manipulate an object such as a ball, setting it in motion, pouncing on it, chasing it, rolling it, and capturing it. Young wolves and foxes will engage in frantic activity chasing their tails. Otters have been observed playing by sliding down a snowy slope (Bourliere, 1964). In a social context, young carnivores and primates are often observed in repeated bouts of chasing, mock fighting, and running.

While most observers would agree that these are playful behaviors, provision of a definition of play has been a difficult task. All play behaviors appear to occur at times when no other behavior critical to survival—such as feeding or avoiding predators—is essential. Somehow, play behavior lacks the "seriousness" of much other behavior. Play appears "pleasureable" to the participants. In the social context, there may be a solicitation of play by one animal of another, and a signal that "what follows is play" (Bekoff, 1974a). Play is more characteristic of young mammals than of older ones.

Various students of play have treated it from either a functional or a structural point of view (Fagen, 1974). An attempt to provide a structural characterization of play was made by Loizos (1966). She noted that most play entails the restructuring of sequences of adult motor patterns, such as predatory or sexual patterns, in a limited number of ways:

1 The sequence of acts may be *reordered*.

2 The individual movement patterns that make up the sequence may be *exaggerated*.

3 Certain movements from within the sequence may be *repeated* over and over again.

4 A normal sequence of movements may be *fragmented*, in the sense that it is terminated earlier than would usually be the case, as a result of the introduction of irrelevant activities.

5 Some movements may be both *exaggerated and repeated*.

6 Individual movements within the total sequence may *never be completed*.

In addition, acts normally functioning in quite different motivational contexts may be intermingled.

The function of play in the lives of animals has been the subject of much speculation. Because play is most characteristic of young mammals, most behaviorists believe it plays some role in the development of normal adult behavior. Play may allow the young animal to practice and perfect the motor patterns and social skills that will be important in adult life. It certainly provides exercise and information regarding the surrounding environment. Research related to the functions of play is very difficult but badly needed.

USE OF TOOLS

A diverse array of species use tools in a variety of contexts, particularly in getting food and in defensive situations. Alcock (1972) has provided an excellent review of tool use in feeding and proposed the following definition: "Tool-using involves the manipulation of an inanimate object, not internally manufactured, with the effect of improving the animal's efficiency in altering the position or form of some separate object" (Alcock, 1972, p. 464). With such definition, behavioral patterns such as social interactions and web construction by spiders would not be included as instances of using tools.

Examples of use of tools in animals are legion. The use of stones by sea otters in opening mollusk shells and of wooden devices by chimpanzees in delivering social dental care have been cited above. Archerfish shoot droplets of water at terrestrial prey, knocking them into the water where they can be captured. Various species of birds use twigs or cactus spines to dislodge insects from inaccessible places such as tree bark. Other birds have been reported to direct rocks and stones at the eggs of other species and thus to crack them open for eating.

Van Lawick-Goodall (1967) described two interesting instances of tool making (the modification of objects that will function as tools) by chimpanzees. In one case, twigs or grass stems are first cleared of their obstructing side branches and then inserted into termite nests. Termites that cling to the stick are ingested. In the other instance, chimpanzees modify leaves by chewing and crumpling them, thus increasing their efficiency for use as sponges in gathering drinking water from inaccessible hollows in trees.

It is clear that tool-using behaviors have evolved independently many times and that they often represent but minor modifications in the existing behavioral repertoires of the animals displaying them.

BIOLOGICAL RHYTHMS

One of the most striking characteristics of behavior is its organization in time. Because virtually all animal species display orderly behavioral cycles, it is possible to predict the time at which a given behavioral pattern will occur. The time spans of different biological rhythms can be quite diverse. Periodic cicadas may emerge and innundate an area once every 13 or 17 years. Cyclical fluctuations in the responsiveness of nerve cells occur on the order of milliseconds. There are annual rhythms, tidal rhythms, feeding rhythms, estrous and menstrual cycles, etc. Indeed, one can view virtually all animal behavior as the study of rhythms and the factors that control rhythms underlying different behavioral patterns at different times.

Much research has been directed at the study of rhythms that last about 24 hours. These are termed *circadian rhythms* (from Latin *circa*, "about," and *diem* "day"; "about a day"). The behavior of many species of birds and mammals appears organized about such circadian rhythms. *Diurnal* species tend to sleep at night and to be active in daylight, while *nocturnal* species are active at night and sleep in daylight. *Crepuscular* species are active at dawn, dusk, or both.

Wheel running is an easily measured and often studied response in research on circadian rhythms. An animal is placed inside a small wheel that is mounted in a vertical position on an axle. A small cage is provided so that the animal can enter and leave the wheel. A series of levers and counters is arranged so that it is possible to record the number of times an animal turns the wheel in a given period of time. If a nocturnal rodent, such as a hamster, is placed in such an apparatus with alternating periods of light and dark every 12 hours, one finds an orderly pattern of running. The hamster begins running at about the time the lights go out and ceases when they go on. Such a rhythm is said to be *entrained*—it is coupled to the occurrence of a rhythmic environmental event, in this case the light-dark cycle. The periodic environmental stimulus to which a rhythm is entrained is called a *Zeitgeber* or "time giver" (see Rusak & Zucker, 1975).

A major question in the study of circadian rhythms has been whether all rhythms are exogenous (dependent upon the periodic occurrence of some *Zeitgeber*) or endogenous (requiring no regular external stimulus for their maintenance). In testing for endogeneity, animals typically are kept in constant light or constant darkness while their activity or other rhythms are recorded. Under conditions of constant light or constant darkness, it is common for many animals to continue regular periodic fluctuations of activity, although the cycles may be slightly longer or slightly shorter than 24 hours (see Figures 4-5 and 4-6). According to "Aschoff's rule," the direction and rate of drift from a precise 24-hour cycle is a function of the light intensity and of whether the animal is normally diurnal or nocturnal (Marler and Hamilton, 1966). Although there are known exceptions to this rule, it has proven generally useful.

The continuation of rhythms in constant light or darkness suggests that the rhythm may be endogenous, but it is still possible that some subtle, undetected cue is acting as a *Zeitgeber*. Several lines of evidence suggest that rhythms can be truly endogeous (Marler & Hamilton, 1966):

1 *Translocation experiments:* If an apparently endogenous rhythm were actually entrained to some uncontrolled stimulus that was related to the earth's

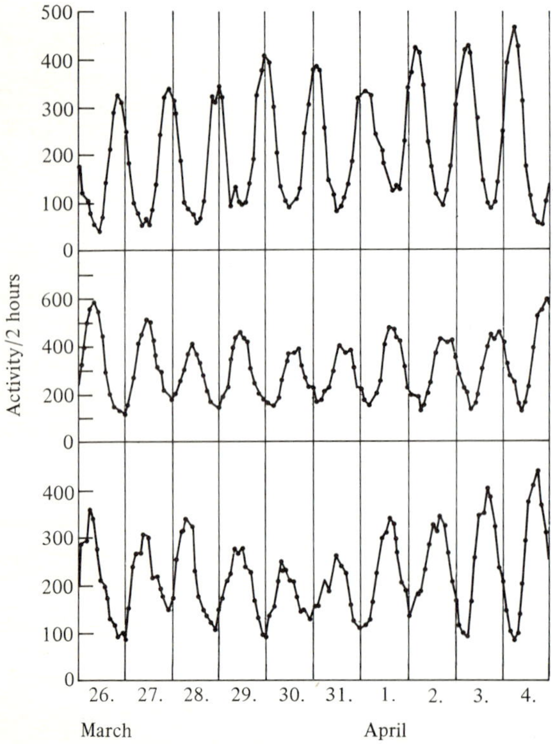

Figure 4-5 Activity of three different mice as summed over successive 2-hour periods. Animals were raised and tested in constant light, yet they display a circadian rhythm of activity. *(From Aschoff, 1960.)*

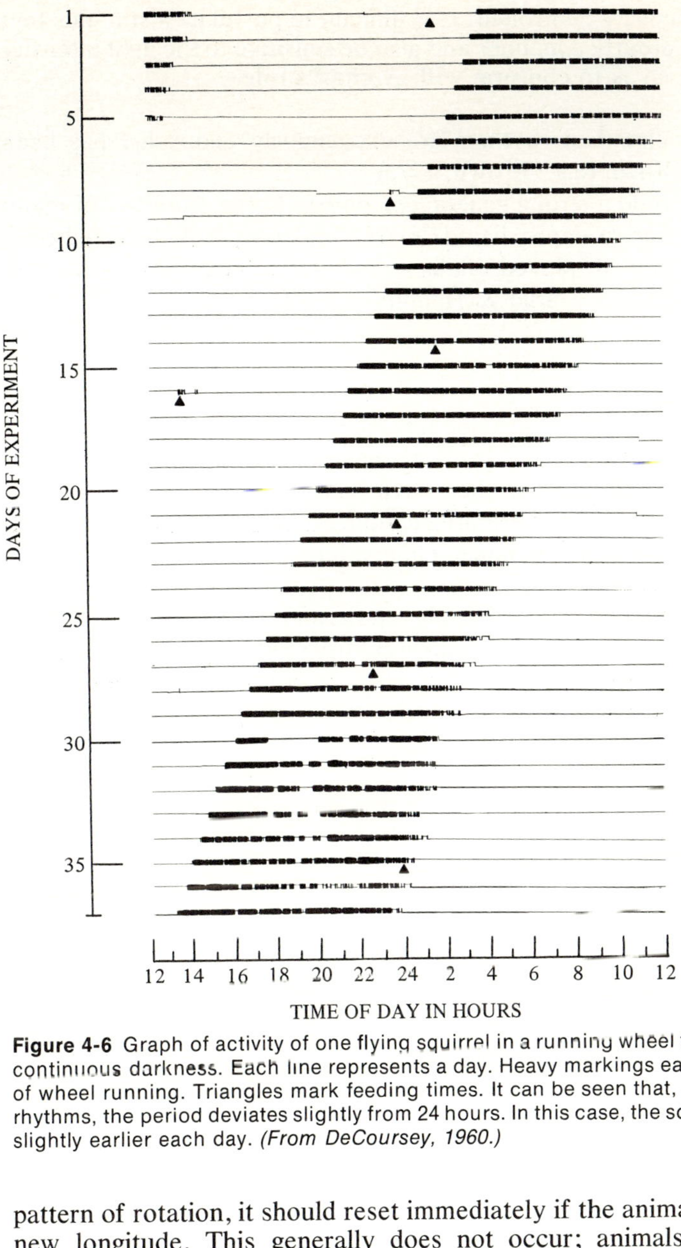

Figure 4-6 Graph of activity of one flying squirrel in a running wheel for 37 days while kept in continuous darkness. Each line represents a day. Heavy markings each day indicate periods of wheel running. Triangles mark feeding times. It can be seen that, as in most free-running rhythms, the period deviates slightly from 24 hours. In this case, the squirrel begins its activity slightly earlier each day. *(From DeCoursey, 1960.)*

pattern of rotation, it should reset immediately if the animal is transported to a new longitude. This generally does not occur; animals transported under constant conditions maintain their original rhythms. (The difficulty of adjusting to a new rhythm is all too familiar to travelers suffering from "jet lag.")

2 *Stopping the clock:* Cooling animals to near freezing can virtually stop the hypothetical "biological clock" of an organism. On warming, the animal resumes its cycle out of phase with any presumed subtle environmental cue.

3 *Deviation from 24 hours:* The very fact that most "free-running" rhythmic cycles are slightly longer or shorter than 24 hours renders it unlikely

that they are exogenously controlled. It is difficult to postulate a stimulus that would change in so orderly a manner and also be sensitive to the light intensity within the chamber so as to conform with Aschoff's rule.

The view that circadian rhythms are exogenously controlled has been defended by F. A. Brown (e.g., Brown, 1972).

There appears to be a strong genetic component to the ontogeny of many circadian rhythms. For example, lizards reared under various distorted regimens of light and dark all displayed similar circadian rhythms when transferred to constant conditions (see Marler & Hamilton, 1966).

SUMMARY

The goal of this chapter was to provide introductory, descriptive information regarding a variety of behavioral patterns that function to allow individual organisms to survive and adapt to their environments. Included were discussions of locomotion, ingestion, thermoregulation, seeking of shelter, avoidance of predators, sleep, body maintenance, elimination, exploration, play, use of tools, and biological rhythms.

The diversity of individual behavioral patterns and the ways in which these enable organisms to survive should be apparent from even this cursory survey. No behavioral pattern exists in isolation; all behavioral patterns interact to produce an organism that is able to survive and reproduce in a particular environment.

Chapter 5

Reproductive Behavior

Biological fitness is a function not only of an animal's ability to survive but also of its ability to contribute to the gene pool of the next generation and subsequent generations—to reproduce. An organism that copes successfully with its problems of obtaining food and water, finding shelter, and avoiding predators, yet fails to reproduce, still may have a fitness of zero. Reproductive behavior is of obvious importance to fitness and lies very close to the very definition of species. Because of its obvious importance, and because animal behaviorists generally treat reproductive behavior as neither individual nor social behavior, we shall consider reproductive behavior in a separate chapter. We shall consider asexual versus sexual reproduction, the components of a complete cycle of sexual reproduction, and three examples of specific reproductive cycles.

SEXUAL VERSUS ASEXUAL REPRODUCTION

The essence of sexual reproduction is the production of new genetic combinations. In the most typical cases, a male and female mate and produce individuals with genotypes that are not identical to the genotype of either parent. In some species new genotypes can be created by processes other than the mating of a male and female. Some protozoans, such as paramecia, undergo a process of *autogamy*, in which a new homozygous genotype is produced from a single animal. Some species, including some flatworms and snails, are *hermaphroditic*, possessing both functional male (sperm producing) and female (egg producing) organs. Some hermaphroditic species are capable of self-fertilization (Bermant & Davidson, 1974).

Not all reproduction is sexual (i.e., produces new genotypes). Paramecia, for example, are able to undergo binary fission wherein a single individual splits in half to create two new daughter organisms that are genetically identical to the original animal. Hydra, a group of coelenterates, are capable of producing new individuals identical to themselves by a process of *budding*. Several new organisms may be produced from a single budding region. Still other species, such as many insects and some fishes, are able to reproduce by a process of *parthenogenesis*, the development of offspring from unfertilized ova (Bermant & Davidson, 1974).

The vast majority of species, particularly more recently evolved species, reproduce sexually by a union of gametes from a male and a female. Evolutionary theorists are not in agreement concerning the reasons for the prevalence of sexual reproduction. Because sexual reproduction has certain costs, it must confer some rather substantial benefits. The primary reasons proposed for the prevalence of sexual reproduction are (1) a long-term advantage to populations that are able to evolve faster than others because they reproduce sexually, (2) a long-term advantage in that sexual reproduction facilitates speciation (formation of new species), and (3) a short-term advantage in that sexual reproduction permits individual parents to diversify their own offspring to cope with unpredictable changes in the environment (Stanley, 1975; Williams, 1966; Wilson, 1975).

BEHAVIORAL PATTERNS ASSOCIATED WITH SEXUAL REPRODUCTION

A first characteristic of reproductive cycles in many species is that they occur seasonally. A complete reproductive cycle can be carried through courtship, mating, sequelae to mating, and care of eggs and young.

Seasonality

Many species, particularly those living in temperate zones, breed during only part of the year. The spring peak of reproductive activity in birds has been a source of inspiration in much literature. For other species, however, breeding behavior occurs at other times of the year. Many species of deer and elk breed in the fall; wolves and coyotes breed in midwinter; many species of seals and sea lions breed in the late spring and early summer. Within species, breeding seasons often vary as a function of latitude. The common denominator in these diverse breeding seasons lies in the time of birth. Although different species have different gestation periods, most give birth in late spring and early summer. It appears that the major advantage to seasonality lies in timing the birth of offspring so that they appear when there is good weather and a peak in available food resources. Mating seasons and gestation periods appear to interact to result in the production of young in late spring and early summer for most species (see Sadleir, 1969).

The nature of the proximate stimuli responsible for the initiation of breeding activity varies from species to species. Temperature, rainfall, growth of plant life, and length of day all vary with season and can all influence reproductive activity in some species. As it is the most reliable correlate of seasonal changes in many habitats, length of day seems to control the onset of breeding in many species. In both male and female Japanese quail, for example, maintenance on days of short length (8 hours light and 16 hours darkness) results in a reduction in the size of various reproductive organs and a loss of sexual behavior. A change in the photoperiod to long days (16 hours light and 8 hours darkness) results in a resurgence of both morphology and behavior (Sachs, 1969; Adkins, 1973).

Courtship

Courtship behavior functions to bring together two animals of different sexes of the same species under conditions where mating is likely to occur and to be successful. The first problem is that of simply locating a potential mate. The importance of mating with a conspecific and the role of courtship behavior in reproductive isolation are discussed in Chapters 2 and 13. Obviously, the conspecific must be of the opposite sex if reproduction is to be successful. Courtship often entails a complex sequence of interacting signals, such as that shown in Figure 5-1, that function to ensure that an animal mates with an appropriate partner. Timing is an important part of successful reproduction; both male and female must be in appropriate physiological condition for mating. This is ensured by the synchronization of cycles that results from the interactions of environmental stimuli and of the two animals themselves. The classical ethologists pointed out that many courtship patterns contain elements of conflict, often because the initial response of an individual to a stranger in its vicinity may be aggressive.

Mating systems vary greatly from species to species. Some species, such as some swans and geese, are truly monogamous and pair for life. Many migrant birds form pair bonds that last for a single season. Some primates are serially polygamous, in that they pair with several individual partners, each for a definable period of time. In simultaneous polygamy, an individual maintains simultaneous pair bonds with several individuals of the opposite sex. Many mammalian species appear to be completely promiscuous, with many copulations and no pair bonds (see Brown, 1975).

Patterns of courtship vary greatly among different species. A few examples follow.

Arthropods Initial detection often occurs via olfaction. For example, silk moths, *Bombyx mori*, are noted for the extreme sensitivity of the male to a sex attractant, bombykol, produced by the female. Males are attracted from great

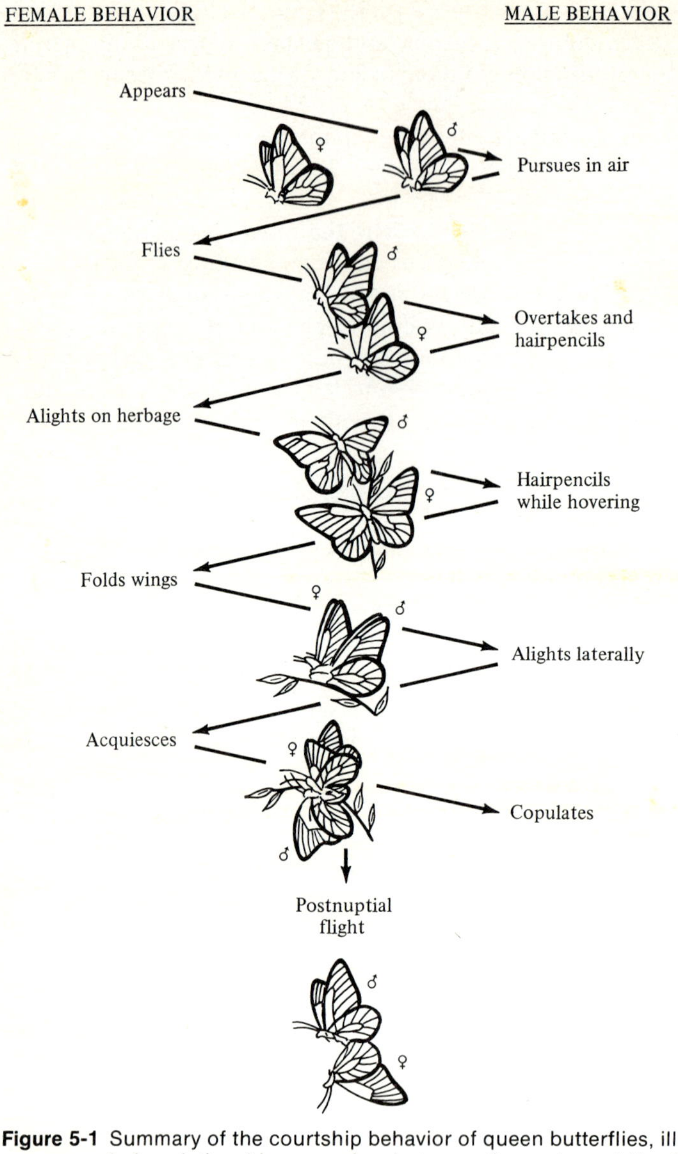

Figure 5-1 Summary of the courtship behavior of queen butterflies, illustrating the stimulus-response chain relationship occurring between the male and the female. *(After Brower, Brower, & Cranston, 1965.)*

distances; a single molecule of attractant can evoke a nerve impulse in a receptor cell (Schneider, 1974). Crickets produce a variety of sounds that function in attracting females as well as in other aspects of courtship behavior.

Visual signals are also common. Male fiddler crabs of the genus *Uca* display species-characteristic, courtship-related displays with their enlarged chela (large claw). Salticid spiders show prominent visual display patterns in courtship.

Figure 5-2 Courtship signals of male fireflies of different species. This figure shows the patterns of light flashes occurring as males fly along different, species-characteristic flight paths. *(After Lloyd, 1966.)*

Among the most spectacular visual courtship patterns are those of fireflies (see Figure 5-2). As the males fly over fields or woods, they display species-characteristic flash patterns as shown in the figure. Females respond to the signal of a male of their own species with a brief flash. The latencies and signals are both species-characteristic and temperature-dependent. The male responds to the female's signal and moves closer and closer as the pair continues to exchange signals, until the male alights and mates (Lloyd, 1966). An experienced observer can attract a male by using a penlight to mimic the female response to the male's flash patterns.

Much research has been done on the courtship of fruit flies of the genus

Drosophila (see Spieth, 1974). Various combinations of behavioral patterns—such as wing vibration, leg vibration, wing semaphoring, circling, and licking—make up the courtship patterns of different species.

Fishes Different species of fishes display diverse courtship patterns. The behavior of tropical aquarium fishes has been most carefully described. Blackchin mouthbreeders, *Tilapia melanotheron,* display four "pure" courtship patterns—patterns that occur in virtually no other contexts. These are the nod (rapid tilting down and forward), quiver (a pattern of head shaking), nip (biting the substrate), and skim (remaining motionless near the nest) (Barlow & Green, 1970). The most prominent display in the courtship of guppies is the adoption of a "sigmoid" or S-shaped posture by the male. Male platys approach females with a sidling movement and also display patterns of backing toward the female, quivering of the body, and swinging of the gonopodium (Clark, Aronson, & Gordon, 1954).

Amphibians and Reptiles Male bullfrogs adopt territories, from which they emit their familiar loud choruses. Females appear attracted by such choruses. Many crocodilian species emit loud roars.

The courtship behavior of anoles (American chameleons), *Anolis carolinensis*, is familiar to many Americans with backyards. The male displays a rhythmical up-and-down bobbing of his body coordinated with the exposure of his dewlap, a bright red flap of skin beneath the chin (Crews, 1975). Courtship in most species of snakes is based on tactile stimulation of the female and olfactory stimulation of the male. The dramatic so-called "courtship dances" in which two snakes become closely intertwined have generally proved to be male-male interactions, presumably aggressive in nature (Porter, 1972).

Birds Birds have provided some of the most dramatic examples of courtship behavior. The complex displays of grebes, gulls, ducks, herons, and other birds have been favored subjects of study by ethologists. The attention paid to birdsongs and the legends surrounding them is indeed substantial.

Lorenz studied various courtship patterns of male ducks (see Figure 5-3). These can be seen in many duck ponds. In the *grunt-whistle*, the male lowers its bill to the water, arching the body upward, flicking the bill, and emitting a loud whistle followed by a grunting sound. The *head-up–tail-down* display is accompanied by a loud whistle. In the *down-up* display, the breast is dipped into the water and the bill is jerked upward and outward, flipping a column of water as it goes (see McKinney, 1969).

Four displays of green herons are depicted in Figure 2-1. Males set up territories. The full-forward display functions to drive away other males from a resident male's territory. The calls of the males attract females. Females are initially repelled with full-forward displays, but females persist and courtship is eventually initiated with the appearance of the snap display and the stretch display. After pairing, the two birds fly about the territory, occasionally

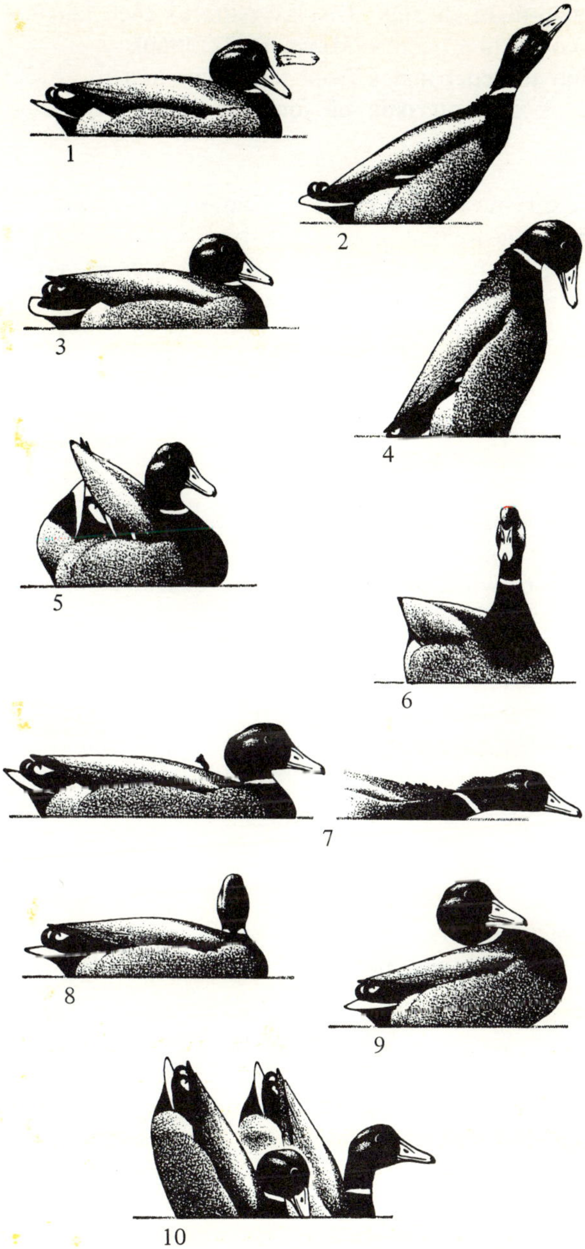

Figure 5-3 Ten courtship poses found in mallard and other surface-feeding ducks: (1) initial bill-shake, (2) head-flick, (3) tail-shake, (4) grunt-whistle, (5) head-up–tail-up, (6) turn toward female, (7) nod-swimming, (8) turning the back of the head, (9) bridling, (10) down-up. *(From K. Z. Lorenz, "The Evolution of Behavior." Copyright © 1958 by Scientific American, Inc. All rights reserved.)*

showing the more intense flap-flight display. Display patterns then become more contact-oriented until copulation occurs (Meyerriecks, 1960).

Many birdsongs function to repel males from territories and to attract females. Song patterns are as characteristic of some species as are their appearances.

Mammals Ewer (1968) surveyed some courtship patterns of mammals. Olfaction plays a large role in the regulation of courtship in many species. Patterns of anogenital investigation and urine testing are common. After smelling a female, the males of many species display a *Flehmen* response, in which the neck is extended and the upper lip is curled. This pattern appears to function more in facilitating perception of the odor than as a display.

Female mammals often solicit mountings, sometimes by approaching a male, nuzzling or licking him, and often by darting from him. Much of the running from the male done by females appears more as solicitation behavior than escape.

Courtship in bottle-nosed dolphins entails vocalization, mouthing of the partner, nuzzling of the partner's genitalia, rubbing of bodies, stroking with flukes or flippers, displaying of the white underside, leaping, chasing, and head butting (see Puente & Dewsbury, 1976).

Mating

The consummation of courtship activity is the occurrence of actual mating, wherein fertilization of ova is accomplished. Patterns of mating are almost as diverse as are courtship patterns, including both internal and external fertilization via a great variety of processes. A select sample of mating patterns will be described.

Arthropods In many species, sperm are "packaged" in a *spermatophore*, a kind of bag or sack containing the sperm. Perhaps the heights of impersonality are reached in some species of mites, pseudoscorpions, millipedes, and springtails, in which the male deposits a spermatophore on the substrate and the female comes along later to pick up the package. The pair may never meet. Many aquatic species have external fertilization (e.g., the male horseshoe crab, *Limulus*, hooks himself on behind the female and remains there until eggs are released). Most terrestrial forms have some kind of internal fertilization, wherein the spermatophore is placed directly into the female using some kind of appendage (see Alexander, 1964).

Fishes Fertilization in fishes can be either internal or external. In anabantid fish, such as gouramis and bettas, a bubble nest is built and spawning generally takes place beneath it. Eggs and sperm are released simultaneously and eggs are fertilized as they float up to the nest. In other species—such as guppies, platys, and *Gambusia*—a gonopodium (modified anal fin) is used to transfer sperm to the female. Male nurse sharks grasp the hind edge of one of the female's pectoral fins, turning her on her back, and insert their specialized claspers (Budker, 1971).

Amphibians and Reptiles Most species of salamanders accomplish internal fertilization through use of a spermatophore, although external fertilization and parthenogenesis occur in some species. Most frogs and toads have external fertilization, with eggs fertilized as they are laid. Exceptions include tailed frogs, which have an intromittent organ, and African toads, in which cloacal glands of the male and female are brought together. Typically, a ripe female frog orients toward a calling male and is clasped by him in a state termed "amplexus." Eggs are released and fertilized in several bouts. If a male clasps another male or is clasping a female that has finished spawning, a release call by the recipient results in release by the clasping male (Rabb, 1973).

Reptiles display internal fertilization through use of intromittent organs. Among anoles, the male delivers a neck grip, twists his tail around that of the female, and inserts his hemipenis.

Birds In contrast to their courtship activities, the copulations of birds generally show little variability. In most species there is no penis, and thus sperm are transferred from the cloacal gland of the male to that of the female. In copulation, the two cloacal glands are brought together in the "cloacal kiss." In Japanese quail, for example, the male grasps the feathers of the female's neck in his beak, mounts her by standing on her back, and orients until the cloacal glands can be brought together (Sachs, 1969).

Mammals Much of the research on mammalian copulatory behavior has been done using laboratory rats, *Rattus norvegicus*, as subjects. Their copulatory behavior will be described in some detail. If a receptive female is introduced into a cage containing a vigorous male, a predictable sequence of events ensues. After initial courtship behavior (sniffing, chasing, soliciting by the female), the male pursues the running female. As the male mounts the female from behind, the female adopts a stereotyped posture, termed "lordosis." The head and hindquarters are elevated to produce a concave arching of the back, while the tail is deflected to the side. Two kinds of copulations typically occur. In some, called "intromissions," the male's penis is inserted into the female's vagina for a period of about $^1/_4$ second before the male displays a rapid dismount. Although there may be extravaginal pelvic thrusting by the male prior to insertion, there is no intravaginal thrusting. After about ten such brief intromissions spaced about a minute apart from each other, the male displays a different pattern, called "ejaculation." The male shows a kind of convulsive thrusting during ejaculation and clasps the female for a few seconds before he dismounts (in a stereotyped vertical posture). A complete group of intromissions ending with an ejaculation is termed a "series." Ejaculation and the associated sperm transfer never occur without prior intromissions. Ejaculation is followed by a temporary cessation of sexual activity. After about 5 minutes, activity resumes and a second series occurs. Normally, rats will complete about seven ejaculatory series before they may be considered satiated, according to a criterion of 1/2 hour with no intromissions (Beach & Jordan, 1956).

Male or female rats will learn an arbitrary operant response when the reinforcement is the opportunity to engage in copulatory activity with a partner of the opposite sex. If a satiated male is exposed to a novel female rather than his original partner, he is likely to resume copulatory activity.

Not all mammals copulate in the same way as laboratory rats. A system for the classification of mammalian copulatory patterns is presented in Figure 5-4 (Dewsbury, 1972). According to this schema, patterns of copulatory behavior are classified with respect to four criteria: (1) Is there a "lock" or mechanical tie between the penis and vagina during copulation? (2) Does intravaginal thrusting occur? (3) Are multiple intromissions prerequisite to ejaculation? and (4) Do multiple ejaculations occur in a single episode? In this system there are 2^4 or 16 possible copulatory patterns. Each species can be classified. Rats have no lock, no intravaginal thrusting, multiple intromissions prerequisite to ejaculation, and multiple ejaculations. If you follow these answers—no, no, yes, yes—in Figure 5-4, you will see that rats display pattern 13.

The copulatory behavior of dogs is quite different from that of rats (Hart, 1967). The most prominent feature is a lock or tie between the male and female.

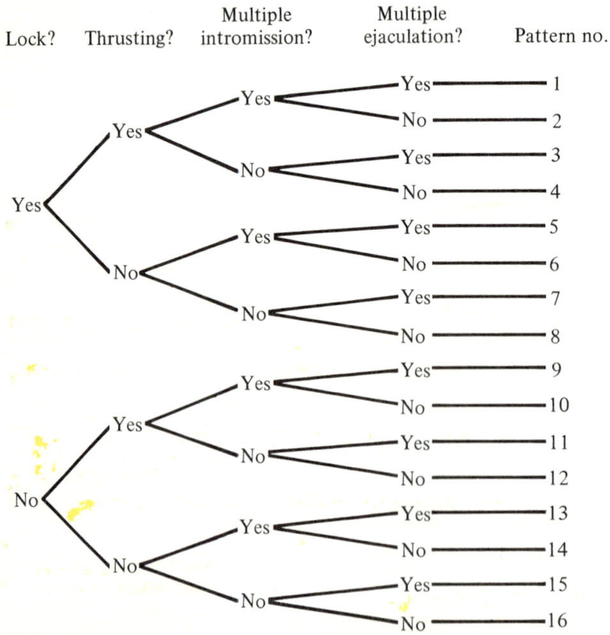

Figure 5-4 Patterns of copulatory behavior of male mammals. Each species can be classified according to whether the males (1) lock, (2) display intravaginal thrusting, (3) require multiple intromissions before ejaculating, or (4) display multiple ejaculations in an episode. There are 2^4 or 16 possible patterns. *(After Dewsbury, 1972.)*

Table 5-1 Some Patterns of Copulatory Behavior in Male Mammals

Common name	Latin name	Lock?	Thrust?	Multiple intro-missions?	Multiple ejacu-lations?	Pattern
Dog	*Canis familiaris*	yes	yes	no	yes	3
Wolf	*Canis lupus*	yes	yes	no	yes	3
Golden mouse	*Ochrotomys nuttalli*	yes	no	no	yes	7
House mouse	*Mus musculus*	no	yes	yes	yes	9
Montane vole	*Microtus montanus*	no	yes	yes	yes	9
Rhesus macaque	*Macaca mulatta*	no	yes	no	yes	11
Bonnet macaque	*Macaca rudiata*	no	yes	no	yes	11
Meadow vole	*Microtus pennsylvanicus*	no	yes	no	yes	11
Norway rat	*Rattus novegicus*	no	no	yes	yes	13
Mongolian gerbil	*Meriones unguiculatus*	no	no	yes	yes	13
Bison	*Bison bison*	no	no	no	yes	15
Black-tailed deer	*Odocoileus hemionus*	no	no	no	no	16

Source: After Dewsbury, 1972.

The male's penis becomes engorged inside the vagina, and it is then very difficult for the pair to separate, even if they struggle. Typically, the male dismounts part way through the copulation and the pair stand back to back. Copulations often last about 20 minutes, in contrast to the very brief insertions of laboratory rats. Dogs have a lock, display intravaginal thrusting, do not require multiple intromissions prior to ejaculation, and attain multiple ejaculations. They display pattern 3. Copulatory patterns of several species of mammals are shown in Table 5-1.

It should be remembered that copulatory patterns vary in many ways, and this is just one convenient system for classification. Within any one behavioral category, different species may show quite different postures and quantitative measures of copulatory activity.

Sequelae to Mating

The primary aftereffect of copulatory activity is the development of fertilized ova. In *oviparous* (egg-laying) species, fertilized eggs develop outside the bodies of adult animals. Nutrients must be provided in the egg. In *viviparous* (live-bearing) species, the embryo develops in the reproductive tract of its mother and obtains nutrients from her. *Ovoviviparous* species bear live young, but they develop from eggs that develop inside the parent. Nourishment comes from within the egg. Some species of fishes and snakes are ovoviviparous.

In some mammalian species, mere transfer of sperm from male to female is not sufficient for the initiation of pregnancy. In addition, critical neuroendocrine reflexes must be initiated in the female. In some species, females ovulate only after mating. If males display little copulatory activity, it is possible for viable sperm to be transferred to the female's reproductive tract, but she will not become pregnant because ovulation was not induced. In laboratory rats and many other rodent species, ovulation is spontaneous, but preparation of the uterus for implantation of the fertilized ovum occurs only after sufficient amounts of copulatory activity (see Conaway, 1971, and Table 12-2, page 235). Fertilized ova will not implant and reproduction will not be successful unless a sufficient amount of stimulation is derived from copulatory activity. In species in which ovulation or preparation of the uterus for implantation is contingent upon copulation, a whole chain of neural and hormonal responses must be triggered by copulation if pregnancy is to be successful (see Chapter 12).

If males are to maximize their fitness, they should ensure that it is *their* sperm, rather than those of another male, that fertilize the female's ova. In many species, males have evolved mechanisms to neutralize the effects of previous matings by other males and to reduce the chances that other males will mate with a female after they do (Parker, 1970). Males of different species may engage in prolonged copulations, deliver a plug that acts as a stopper in the female's reproductive tract, or guard the female; all these behavioral patterns function to reduce the chances that another male will mate with the female. In some insects, specific chemicals occur in the seminal fluid and induce nonreceptivity in the female, thus making it unlikely that she will mate with another male.

Care of Eggs and Young

Many species display some kind of care of their eggs, young, or both.

Arthropods Some species of crustaceans retain their eggs in some kind of brood sac or chamber. Little active care is shown by the parent. Some species of wasps provide food for their developing young but provision their nests only prior to depositing eggs.

Parental care of offspring through to the adult stage is common only in the social insects. The eggs of honeybees are deposited in hexagonal-shaped cells in the hive. Adults periodically visit the cells, inspect them, clean them, and feed the developing larvae. After the adults seal the cells with wax, the larvae pupate; they emerge as adults about 10 days later.

Army ants, *Eciton hamatum*, live a cyclical existence, alternating between a nomadic state and a stationary phase. The fluctuations between these two stages are correlated with and appear controlled by changes in the developing broods. In the stationary phase, a brood in the pupal condition is present and a new brood in the egg stage is produced; whereas a brood in the larval stage is present during the nomadic phase (Schneirla, 1957). When in the nomadic phase, larvae require much food and great daily raids and nightly emigrations occur. During the stationary phase, the developing larvae mature and enter cocoons; the activity level of the colony is then markedly reduced.

Fish Among egg-laying species of fishes, deposition of eggs in an appropriate place is one means of ensuring their development. Some species build nests, such as the bubble nest of bettas, the circular depressions in the substrate of sunfishes, or the nest tunnels of channel catfish. Other species scatter eggs on plants (carp) or over shoals, sand, gravel, or boulders (lake trout); or they deposit them in rope-like masses (yellow perch). A number of species, such as some catfishes, brood eggs in their mouths. In sea horses and pipefishes, the female places eggs in a special brood pouch on the male. Other fishes are viviparous or ovoviviparous.

Many species, such as sunfish, actively defend the hatchlings. Active herding into protected places sometimes occurs. In the mouthbreeding *Tilapia*, a familiar aquarium fish, young escape into the oral cavity of the female in times of danger (Lagler, Bardach, & Miller, 1962).

Amphibians and Reptiles Some frogs and toads brood eggs on their backs. In Surinam toads, eggs are hatched in specialized folds of skin on the female's back. Among midwife toads, it is the male that carries the eggs. Some snakes coil about their eggs, thus warming them. The nests of crocodilians range from a simple hole excavated for deposition of the eggs, as in most reptiles, to mounds of vegetation and other material that are constructed for this purpose.

Although true parental behavior appears to be somewhat rare among amphibians and reptiles, some species display such behavior. For example, alligators are known to protect both nests and young.

Birds Most birds incubate their eggs in a nest by sitting on them. One parent may do all the sitting, or a mated pair may take turns. Parental care is highly developed in many birds. Many species, including gulls, bring food to the nest for their young. Defense of the young against predators, as in colonial nesting marine birds, can be quite intense.

Parasitic bird species, such as cuckoos and cowbirds, do not show parental behavior. Instead, they lay their eggs in the nests of other species. A variety of adaptations, including large size and specific appearance of the eggs, function to maximize the chances that the other birds will care for the eggs.

Among the most elaborate incubation patterns are those of Australian mallee fowl. They build large mounds of rotting vegetation. An elaborate sequence of behaviors, including opening the nest and covering it with sand, function to maintain it at the appropriate temperature (Frith, 1962).

Mammals Parental care is highly developed in mammals.

Parturition The time of day at which birth occurs varies from species to species and tends to be highly species-characteristic (see Jolly, 1973). Typically, the female gives birth in the nest or in another secluded place.

Schneirla, Rosenblatt, and Tobach (1963) divided parturition in cats into four stages: the contraction interval, the emergence interval, the delivery interval, and the placental interval. Most gross body movements occur during the contraction and emergence intervals. The female squats, strains, or crouches and may display a rubbing or rolling pattern. Licking is directed at her own body, previously born kittens, or the surroundings. Licking becomes more intense during the delivery interval. In addition to herself and previous kittens, the cat licks the umbilical cord and placenta and may ingest fetal membranes or the cord. In the placental phase, the female licks and eats the placenta and continues to lick herself, her kittens, and her surroundings. This general sequence of behavioral patterns is typical of many mammalian species.

Nursing The presence of mammary glands and associated nursing behavior constitutes one of the defining attributes of mammals. Nursing occurs in all mammalian females, usually beginning quite soon after birth. The three stages of mother-young nursing interactions distinguished for cats and rats (Schneirla, Rosenblatt & Tobach, 1963; Rosenblatt & Lehrman, 1963) are typical of many species. Stage 1, *female approach*, is typified by the female's initiating most nursing bouts as she approaches the kittens. In some species, the female adopts a characteristic nursing posture. Most ruminants nurse their young while standing; manatees nurse their young under water (Bourliere, 1964). Among cats, the female lies near the kittens and arches her body around them. As in many species, kittens tend to develop specific nipple positions, so that the same kitten nurses from the same nipple time after time. Stage 2 is a period of *mutual approach*. The young become more active and either mother or young may initiate the nursing bout. In stage 3, *young approach*, the young initiate most nursing bouts. Given an opportunity, the female will often escape to a secluded shelf or other shelter. The gradual transition from female-initiated care to mutual initiation of care to greater independence of the young is characteristic of many mammalian species.

Rabbits often visit and nurse their litters just once a day, in contrast to many other species. Deer mice, on the other hand, may spend over 20 hours a day in contact with their young (King, 1963).

Grooming Grooming of the young occurs in most species. The pelage is kept clean, often by licking, although primates use their hands extensively. Licking of the anogenital area appears necessary for stimulation of urination and defecation in some species.

Nest building Many species build nests primarily around the time of parturition, while many others build larger and more elaborate nests at this time. The preparturitional nests of laboratory rats are not elaborate, resembling mats. After the birth of her litter, however, the female rat constructs a dome-shaped nest, which completely covers the litter and contains a tunnel through which she may enter and exit. Similar nests are built by deer mice. Rabbit nests are even more elaborate (Ross, Sawin, Zarrow, & Denenberg, 1963). The mother digs or excavates a burrow and gathers hay, straw, or other material. At this time, her normally fast hair begins to loosen. She plucks this hair and incorporates it into the nest. Well-constructed nests are well packed and hollow in the center, completely surrounded by hay and hair. Among primates, only some prosimians and humans construct birth nests (Jolly, 1972).

Transport of young Transport of young usually is accomplished with the young either in the mother's mouth or on her stomach, chest, or back. In many rodents, canids, and felids the mother grasps the young in her mouth. The young may be grasped by the belly, often curving their bodies toward the mother's head; by the nape of the neck; or by skin on the back. Many arboreal species, such as sloths, carry their young on their stomachs. Sea otters commonly swim on their backs, carrying the young out of the water on their stomachs (Bourliere, 1964). While a baboon mother carries her newborn infant on her ventral surface, a gradual transition begins at about 5 weeks, so that the older infant typically rides jockey-style on the mother's back (Hall and DeVore, 1965). In some species, the young cling to the nipples so tightly that retrieval is rarely necessary. The author has chased escaped female wood rats about a colony room for 20 minutes as they scaled walls and cages, and as the young wood rats never lost contact with the mother's nipples.

Defense of young Defense of the young is common in most mammalian species. Parents tend to be much more irascible when in the vicinity of their young than at other times. Many instances of attacks by large wild mammals on humans can be traced to interference with parent-young interactions. King (1963) noted that he could check for the appearance of litters among captive deer mice using 10-inch forceps. The females were easily chased from their nests when the forceps were introduced. After the birth of their litters, however, the mothers would savagely attack the forceps. The mother moose keeps all potential intruders, including those of her own species, away from her calf, maintaining a "ring of sliding territoriality." By contrast, elk mothers hide their calves in a safe place and often protect their young by distracting predators away from the hiding place toward themselves (Altmann, 1963).

Play and tuition While much play occurs among the young, mother-young play interactions also are common. Parents of many species tutor their young with respect to many of the skills necessary for adult life. For example, many species of cats teach their kittens to hunt (Schneirla, Rosenblatt, & Tobach, 1963).

Paternal behavior While most research has focused on mother-young interactions, the role of the adult male must also be considered. Paternal behavior is displayed in a number of species of nonhuman primates (Mitchell, 1969). In the laboratory, the males of many species of rodents display virtually all the characteristic behavioral patterns of the mothers except lactation. Whether these behavioral patterns occur in nature is yet to be determined. Dudley (1974) found improved survival, greater body weights, and higher body temperatures under various conditions among captive young California mice, *Peromyscus californicus*, when the male was present with the female during development.

THREE REPRODUCTIVE CYCLES

Because of the great diversity of patterns of reproductive behavior, coverage of different species in this chapter has of necessity been somewhat patchy. Yet reproductive cycles represent unified wholes, adapted to the successful production of viable offspring. Three examples of diverse reproductive cycles will be treated in somewhat more detail.

Sticklebacks

"The three-spined stickleback, *Gasterosteus aculeatus*, is unquestionably the best-studied fish in the world. It may fairly be called the white rat of European ethology" (Bermant & Davidson, 1974). The reproductive behavior of sticklebacks represents a complex interaction of male and female based on some specific sign stimuli (see Tinbergen, 1952a; 1965a).

Sticklebacks are small fish (about 50 millimeters or 2 inches long). Out of the breeding season, they are undistinguished grayish-green fish living in schools. In the spring, increases in length of day produce increases in the secretion of sex hormones and a migration from deep freshwater or the coastal saltwater sea to shallow, freshwater spawning grounds. Here, stimulated by temperature changes, the male undergoes dramatic changes in both appearance and behavior. His underside becomes a bright red and he leaves the school to set up a territory. The territory is defended against other males and even females. In one common threat gesture, the male adopts a vertical posture with the head down, thus exposing the red belly to the intruder. The red belly appears to intimidate other sticklebacks (Figure 2-3).

The male next builds a nest, first digging a shallow pit and then filling it with plant material, such as algae and bits of weeds. He then secretes a sticky kidney fluid over the loose pile. This material functions as a glue and the nest is given a tunnel shape as the male swims through it to complete it.

With the nest complete, the male goes through another color change, the upper portion of his body turning a bluish-white. This color pattern attracts females to the territory. The egg-plump form of a female stimulates the male to display a stereotyped "zigzag dance" (Figure 5-5). The male "dances" back and forth in a zigzag pattern that Tinbergen (1951) has interpreted as a ritualized combination of movements toward attack and swimming to the nest. The male may brush the female with his spines. The female signals acceptance of the male with a head-up posture. The male shows the female to the nest and indicates its entrance by adopting a posture in which he turns on his side and points toward the inside. The female enters the nest. The male hovers above the nest and nudges the base of the female's tail with his snout (Figure 5-5). This stimulates the female to spawn. After the female leaves the nest, the male enters and fertilizes the eggs.

After mating with about two or three females, the male begins his parental duties and resumes his protective coloration. He aerates the eggs by fanning water over them, using a stereotyped motor pattern. When the brood hatches, about a week after fertilization, the male guards the young. If one strays too far from the nest, the male takes the hatchling into his mouth and spits it back into the nest.

Reproductive behavior in sticklebacks represents a complex interaction between some sign stimuli derived from the male and others from the female to produce a complete, integrated, and successful reproductive pattern.

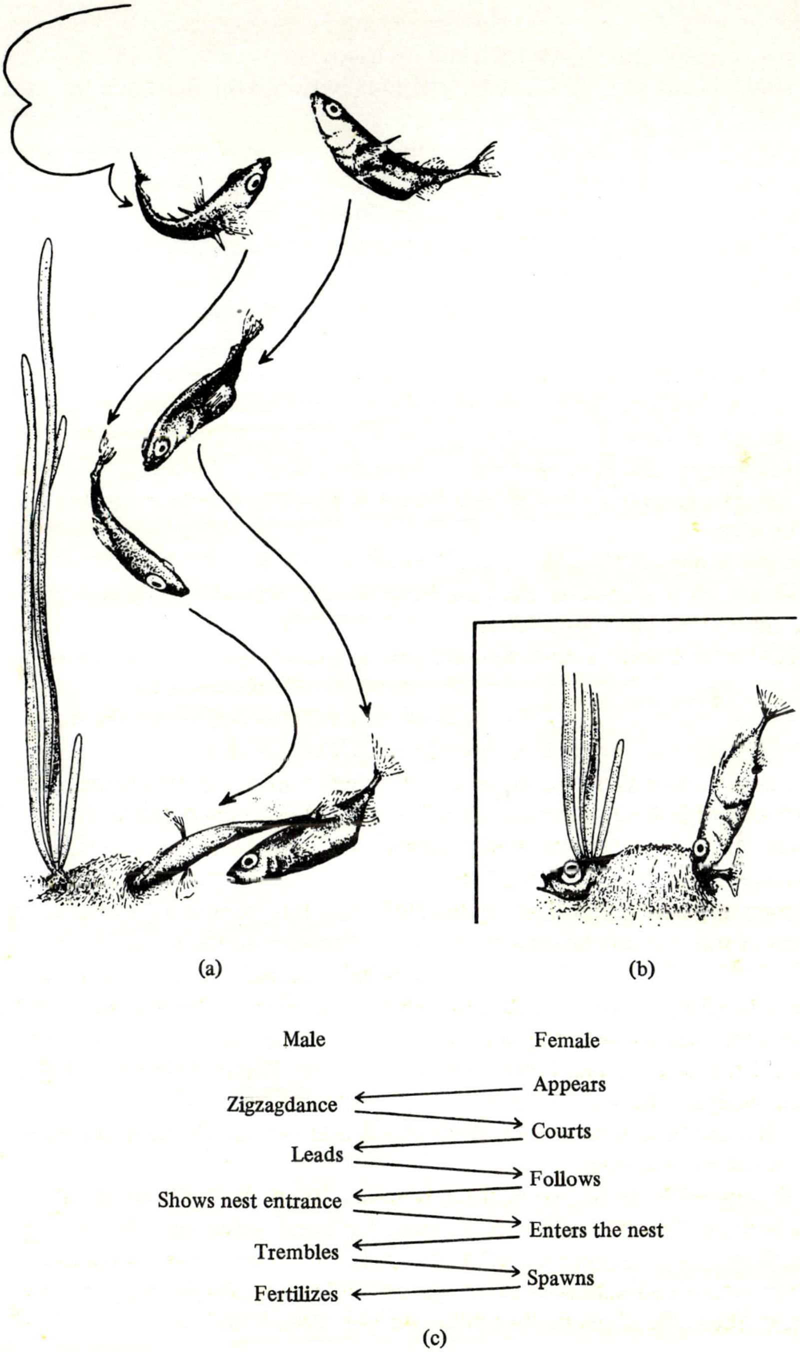

Figure 5-5 The mating behavior of sticklebacks. Included are *(a)* the zigzag dance, *(b)* stimulation of the female in the nest by the male, and *(c)* a schematic representation of successive steps in the courtship interaction. *[From* The Study of Instinct *by N. Tinbergen (1951), published by Oxford University Press.]*

Ring Doves

As the classical ethologists Tinbergen and his associates analyzed reproductive cycles in sticklebacks into complex interactions between the male and female, so the comparative psychologists Daniel S. Lehrman and his associates analyzed the reproductive cycles of ring doves. Ring doves, *Streptopelia risoria*, are small relatives of domestic pigeons which breed well and display complex behavioral patterns in captivity (Lehrman, 1964; 1965). They have a black, semicircular "ring" on the back of the neck which gives them their name.

In Lehrman's test situation, a pair of ring doves is introduced into a breeding cage containing a nest bowl and a supply of nesting material. At this point, the female's oviduct weighs about 800 milligrams, and the crops of both male and female weigh about 900 milligrams. The primary activity observed on the first day is courtship behavior, in which the male struts about the cage and emits a distinctive bow-coo display, a pattern unforgettably imitated by the late Dr. Lehrman. The male bows toward the female in a stereotyped manner and emits a distinctive cooing sound. After several hours, the birds announce the selection of a nest site, a concave place such as the inside of the nest bowl, with a characteristic nest-coo display.

The pair then spend about a week engaged in nest building and copulating. The male gathers most of the nesting material (straw, pine needles, etc.) and carries it to the female, who takes the material from the male and incorporates it into the nest. While there is some role reversal, the male gathers and the female molds most of the nesting material. During this period the pair will not sit on eggs, even if they are provided by the experimenter.

After about a week of nest building and copulating, the female becomes noticeably attached to the nest. She spends most of her time at the nest, and it is even possible to pick her up and find the nest still clutched in her claws. This indicates that she is about ready to lay an egg. The female's oviduct has increased five fold, to about 4,000 milligrams. She lays the first of her two eggs at about 4:00 P.M. on a given day, and the second egg follows at about 9:00 A.M. the next day.

The behavior of the birds changes dramatically at the time of egg laying. Whereas the birds would not sit on eggs earlier in the cycle, sitting becomes a prominent activity during this phase. The female typically begins sitting on the eggs some time between the laying of the first and second eggs. The male begins to sit the day the second egg is laid. The birds alternate in sitting on the eggs, with the male typically taking a 6-hour shift in the middle of the day, while the female sits continuously for the remaining 18 hours. The eggs hatch after about 14 days, but the pair continue to sit on the young for a few days in about the same pattern used in sitting on the eggs.

At the time of hatching, the parents' crops have increased in size, so that they now weigh about 3,000 milligrams. The parents feed the young with "crop milk," a regurgitated substance produced by the epithelial lining of their crops.

The young leave the nest at about 10 to 12 days of age but continue to beg for food. At about 14 to 15 days, the parents become somewhat less responsive to begging, and the young begin to peck at the ground for their own food. When the young are about 20 days of age, the bow-coo display reappears and a new reproductive cycle begins. The complete cycle—courtship, copulation, nest building, egg laying, incubation, and care of young—lasts about 6 to 7 weeks.

Lehrman and his associates have conducted a long series of experiments regarding the stimulus and hormonal determinants of reproductive behavior in ring doves (see Chapter 12). It is worth emphasizing that throughout the cycle there are dramatic changes in the behavior of both male and female. At one stage, they copulate and build a nest; at another stage, they sit on eggs. These dramatic shifts in the readiness to perform certain acts result from the joint interaction of the two birds and from the stimulation, primarily auditory and visual, they derive from each other.

Rhesus Monkeys

Primate behavior tends to be somewhat more variable than the behavior of other groups of animals, and the type of behavior observed can vary considerably as a function of the location of the study. Reproductive behavior in rhesus monkeys, *Macaca mulatta*, has been studied in the laboratory (Kuehn and Young, 1965; Michael and Saayman, 1967; Harlow, Harlow, & Hansen, 1963), in their natural habitat in India (Southwick, Beg, & Siddiqi, 1965), and in an introduced population on Cayo Santiago, Puerto Rico (Carpenter, 1942; Kaufmann, 1965; Loy, 1971).

Free-ranging rhesus live in groups of adults and subadults, with an average of from 10 to over 50 animals per group, depending on the circumstances. Although copulatory behavior can be elicited at all seasons in the laboratory, copulation in the field is most frequent in the fall. Females display menstrual cycles of about 28 days. They typically can become receptive at two stages of the menstrual cycle, intermenstrually, presumably at the time of ovulation, and perimenstrually, at about the time of menstruation. The duration of the female's period of receptivity has been reported to be 6 to 11 days. As the female becomes receptive, there is a noticeable swelling and reddening of her "sex skin" in the genital region.

Receptive females form consort relationships with individual males. During these periods the male and female remain in close association with each other; they move together, feed together, and engage in frequent grooming of each other. Consort relationships can last from a few hours to a few days. Typically, a female may form consort relationships with about three different males within an estrous period. A number of behavioral patterns appear associated with copulatory activity. Females "present" to the males by approaching them and adopting a stereotyped posture in which the swollen red sex skin is oriented toward them. This behavior appears to function as an "invitation" to copulation. A "sporadic arm reflex," rapid flexion and extension of the dominant hand and arm, has been reported in rhesus females in an apparent high state of sexual excitement. Increases in a "yawning" facial expression and in aggressive behaviors also have been reported.

Little or no displaying may immediately precede copulation in both India and the laboratory. The male mounts the female by placing his hands on her back and grasping her hind legs with his feet. Rhesus monkeys display copulatory pattern 11 (Table 5-1) with no lock, intravaginal thrusting, multiple intromissions not prerequisite to ejaculation but usually occurring, and multiple ejaculations in a session. In laboratory tests, the interval from first intromission to ejaculation can average from 2 to 14 minutes, depending on the animals and testing conditions. An average of about five mounts without intromission and four mounts with intromission precede ejaculation. Approximately 4 to 10 intravaginal thrusts occur on intromissions. After ejaculation, there is a distinct postejaculatory refractory period, followed by a resumption of copulatory activity for several additional ejaculatory series.

Females have been observed to have estrous periods during their pregnancies.

The female rhesus withdraws somewhat from the normal activities of the group for a few days before and after the birth of an infant. Her maternal behavior may be divided into three stages: attachment and protection, ambivalence, and rejection. During the first stage, the infant spends much of its life in actual contact with its mother. The newborn does little but nurse and sleep. In the first week or so the infant develops the ability to crawl about somewhat. By about 3 months of age, the young rhesus leaves the female for periods of play, exploration, and feeding. As the young rhesus develops, the frequency of positive responses by the mother—such as grooming, manipulation, cradling, submitting, presenting, clasping, restraining, and retrieving—decreases, while the incidence of negative responses, such as threatening and punishment, increases. Weaning begins gradually, but nursing reaches its final termination sharply with the birth of a new infant, about a year after that of the first. Juveniles tend to associate with their mothers for a period of several years, although they are not permitted to nurse after the birth of the new infant.

Male rhesus play little role in caring for the young in the natural habitat, although they can show some care in captivity (e.g., Redican and Mitchell, 1973).

The progressive change in the mother-young relationship is a function of changes in both the mother and the young, which are integrated into a complete, functional reproductive cycle.

SUMMARY

Different animal species have adopted a wide range of reproductive patterns, each functioning in the particular environment of the species in question. While asexual reproduction occurs in some species, sexual reproduction appears more common.

Behavioral patterns associated with sexual reproduction focus about seasonality, courtship, mating, sequelae to mating, and care of eggs and young.

Many animals breed seasonally, with births occurring in times of optimal environmental resources.

Courtship patterns function in the location and identification of potential mates of the appropriate sex and species and in the synchronization of their reproductive activities. Mating may entail either internal or external fertilization and is associated with complex, species-characteristic behavioral patterns. Sequelae to mating include pregnancy and can include both the initiation of critical neuroendocrine reflexes in the female and a termination of female receptivity.

Care of eggs, young, or both is displayed in many species.

In order to provide an integration of the diverse aspects of complete reproductive cycles, the cycles of sticklebacks, ring doves, and rhesus monkeys are described.

Chapter 6

Social Organization and Social Behavior

Darwin's phrase "survival of the fittest" was misinterpreted by nineteenth-century social philosophers to imply that animal life entailed a direct and continuous competitive struggle—"nature red in tooth and claw." This, in turn, was used to justify "hands off" policies in human social and economic affairs. In Chapter 3 it was noted that, according to current evolutionary thinking, individuals should act to maximize their own fitness—even if this is not in the "best interests of the species." Does this not also mean that life should be a never-ending competitive struggle? The answer, of course, is no. Natural selection works through differential reproduction and in ways far more subtle than would be expected on such a simplistic formulation. The individuals of many species live in highly structured, orderly societies. Rather than being distributed at random throughout their environment and in competition with all they meet, the animals of many species live in compact groups, within which most social interactions are friendly. Paradoxically perhaps, social living has proven an effective way in which to maximize levels of individual fitness.

In its broadest sense, social behavior is the behavior of organisms when they are interacting with one another. Because of its obvious importance to both animals and people, social behavior has been the focus of much research in animal behavior. But social behavior and social relationships are extraordinarily complex; how are we to approach their study so that the analysis of social systems becomes manageable? As in the analysis of many phenomena, scientists have tended to start from one end or the other—the whole or the part (see Crook, 1970). Animal societies are structured. Some scientists have attempted a holistic approach, attempting to study complete organizational patterns of animal societies and to understand their determination. This is a reasonable strategy. However, it is also true that structured societies represent the sum total of a large number of complex individual social interactions. Some scientists have attempted to start with analysis of these individual social interactions and to work upward toward the complete society. Both approaches have their merits, and both will be considered. Analyzing sociality is somewhat like analyzing a football play—one must study both the total flow of players on the field and the effectiveness of individual blocks and tackles. We shall begin with a treatment of social organization and conclude with an analysis of individual patterns of social interaction.

SOCIAL ORGANIZATION

We begin with the holistic approach to the study of sociality—the study of social organization.

Definitions

Not all groupings of animals are organized societies. Eisenberg (1965) proposed five defining attributes of organized societies:

1 *A complex system of communication:* All organized societies have some form of complex communication system.

2 *A division of labor based on specialization:* In organized societies, animals of different castes, sexes, or age groups have different functions in maintaining the society.

3 *Cohesion, a tendency for members to remain together:* The individuals constituting a society tend to remain in close proximity to one another.

4 *A permanence of individual composition:* The individuals making up a society tend to be the same from day to day; there is little migration from the group.

5 *A tendency to be impermeable to conspecifics that are not members of the group:* Most organized societies resist immigration by "outsiders."

Organized societies can be contrasted with other kinds of social groupings. All organized societies appear to be formed on the basis of a social response—a response to other organisms. Yet not all groupings formed by social responses are organized societies. A male and female may come together to breed, as do male and female sticklebacks. Yet such mateships lack the degree of permanence characteristic of organized societies. Flocks, herds, and schools may be similarly contrasted with organized societies. True *aggregations* may be contrasted with organized societies in that they are formed as a result of nonsocial responses, usually responses to environmental factors. The high density of flying insects about a bright outdoor light attracts many tree frogs. Yet neither the insects nor the tree frogs fulfill any of the five criteria for organized societies; both are aggregations rather than societies.

McBride (1976) presented a theoretical model for the study of social systems. Based on communications with a considerable number of animal behaviorists, McBride developed a set of questions to be considered in the study of any pattern of animal social organization. Included are questions related to social phases, castes, specialization in groups, patterns of dispersal, reproductive behavior, environmental factors, and dynamic aspects of social organization. Such a system should provide an excellent foundation for systematic studies of social organization.

Bases of Organization

Dominance and territoriality represent the two major organizational bases in many societies.

Dominance Organization by dominance is characteristic of species from a diverse range of taxa.

Definitions A dramatic characteristic of many dominance interactions is the appearance of overt hostility. This has led some authors to utilize the occurrence of such hostile behavior as a defining attribute of dominance. Thus, Klopfer (1974) states: "*Dominance* is inferred whenever one individual is able to chastise another with impunity" (p. 154).

However, not all dominance interactions involve overt "chastisement"; for example, they may entail only spatial displacement. A subordinate male baboon may be sitting in a given location. On the approach of a dominant male, the subordinate gives way and the position is occupied by the more dominant male. Most animal behaviorists would recognize this as an instance of a dominance interaction, even though there was no overt fighting or even threatening.

A more useful definition of dominance may be given with respect to differential privileges accruing to different members of a group. Van Kreveld (1970) defines dominance in general terms as "a priority of access to an approach situation or of leaving an avoidance situation that one individual has over another" (p. 146). Approach situations include access to food, mating partners, opportunities for aggression, and space. Avoidance situations include remaining in undesirable areas as well as having to endure threats, attacks, and punishments. The essential feature of dominance is that there is some resource that is of limited availability and is contested. One animal has gained privileged access to the limited resource relative to another animal. Once a dominance relationship is established, priority is rarely contested. Note that dominance is not absolute; it exists relative to a specific other animal. Individual recognition of the members of a group is a necessary prerequisite to dominance relationships.

The "peck order" of chickens The classical examples of dominance hierarchies are from the so-called "peck orders" of chickens. The organization of an established flock of chickens is based on a set of dominance relationships. If food is given in a restricted location, it becomes apparent that one animal has priority of access to the food. It will exclude other chickens from the area of the food, perhaps by delivering a peck at them. The other animals never return the peck. Such pecking is particularly prominent in newly established flocks; hence Schjelderup-Ebbe, an early student of dominance hierarchies, referred to dominance hierarchies as "peck orders." Typically, in a group of a dozen or so chickens, there is but a single despot or alpha animal. The alpha animal has priority of access over all other animals in the flock. A second animal, the beta individual, has priority over all animals in the flock except the alpha animal. In an ideal *linear hierarchy*, there is a perfect straight-line ordering of all animals in the flock, so that each is dominant to all animals below it in the hierarchy and subordinate to all animals above it in the hierarchy (Guhl, 1956). The hierarchy gives dominant animals priority of access to food, water, roosting places, mates, and any other appropriate resource.

Not all hierarchies are perfectly linear. For example, triangular relationships may occur, in which animal A dominates B, B dominates C, but C dominates A. Such triangular relationships tend eventually to break down in chickens but not in all species.

Dominance hierarchies in other species Among arthropods, dominance hierarchies have been reported in certain species of crabs, crayfish, and cockroaches. Platys and lizards have been reported to display dominance hierarchies. Among mammals, dominance hierarchies have been reported in animals as diverse as various species of rodents, canids, dairy cattle, reindeer, and a great variety of nonhuman primates. While this survey is by no means complete, it should be apparent that such a mode of organization is common in a diverse array of species.

The dominance hierarchies of some species are more fluid than those of chickens. In addition, the ordering of animals may differ when different resources are considered. In some studies of both rats and nonhuman primates, the male that would be considered the alpha male on the basis of aggressive interactions is not the one that does the most mating. If access to females were the defining attribute, he would not be the dominant male at all. We should not expect all instances of dominance hierarchies to be as neat and reliable as the classical descriptions.

Territoriality Territoriality is related to dominance and represents a means of social organization that is also quite widespread.

Definitions As is the case with dominance interactions, the most dramatic aspect of territorial behavior is overt aggressive behavior. This is reflected in the most widely accepted definitions of territory. Most animal behaviorists adhere to the simple definition proposed by Noble (1939): a territory is "any defended area." The male stickleback provides an excellent example. During the breeding season, the male leaves the school and stakes out a distinct territory, defending its borders by displaying to intruders. Examples of such territorial defense can be found in a wide variety of insects, fishes, amphibians, reptiles, birds, and mammals.

As was the case with the definition of dominance, the simplest definition in terms of aggressive interactions may not be the most useful one. Kaufmann (1971) noted some interesting instances of social organization which would not be considered as territorial using traditional definitions. Coatis, Central American carnivores related to raccoons, live in male-female bands. The band occupies a "core area" of its environment about 80 percent of the time. The core areas of different bands do not overlap. However, these core areas are not overtly defended against other bands. Even when one troop enters the core area of another, there is but fleeting hostility. A similar lack of hostility was observed among whiptail wallabies, close relatives of kangaroos. Again, groups of wallabies had privileged use of a particular area without overt defense. Kaufmann further cited some results of studies with two species of birds, Steller's jays and bicolored ant birds. In these two species there is again no overt defense of territory, nor is there exclusion of other animals from it. However, the resident is dominant in feeding situations in its area and subordinate to other animals in feeding situations in other areas. None of these examples fit the traditional definition of territory, yet all have something in common with territoriality.

Kaufmann (1971) proposed that the problem be resolved with a redefinition of territoriality. He proposed that we define territoriality in terms of priority of access to resources. Following Kaufmann, we might define a territory as *an area in which the resident enjoys priority of access to limited resources that he or she does not enjoy in other areas*. Thus defined, the concepts of dominance and territoriality are seen to have much in common. Both may be given a behavioral definition in terms of overt aggressive interactions. However, it appears that fundamental to each is an arrangement whereby there is a relatively stable system for the distribution of resources, without the necessity of repeated overt fighting whenever a resource might be contested. In a dominance hierarchy, an individual has privileged access to resources regardless of location. With territorial organization, different individuals have priority of access to limited resources depending on their location.

Dominance hierarchies and territories can be seen to grade into each other rather than to exist as absolute alternatives. In some species of lizards, a dominant male has privileged access to resources wherever they might be, while the remaining individuals apportion resources by means of territoriality. House mice have been observed to adopt a territorial organization at low population densities, but they shift to a hierarchical structure based on dominance when population densities increase (Davis, 1958).

While the definition of territory in terms of access to resources may never replace that in terms of overt defense in the use of most animal behaviorists, it does focus discussion on some of the important and interesting aspects of territorial organization and their relationship to dominance.

Forms of territoriality Patterns of territorial organization are extremely diverse. Various authors (e.g., Nice, 1941; Fisler, 1969) have attempted to delineate and describe contrasting forms of territorial organization. Wilson (1975) lists just five forms:

Type A: A large, defended area within which sheltering, courtship, mating, nesting, and most food gathering occur (e.g., benthic fishes, arboreal lizards, insectivorous birds, some small mammals).

Type B: A large defended area within which all breeding activities take place but which is not the site for most food gathering (e.g., nightjars and reed warblers).

Type C: A small defended area around a nest (e.g., many colonial birds, such as ibises and herons, and sphecid wasps).

Type D: Pairing and/or mating territories (e.g., insects such as damselflies and dragonflies, birds such as sage grouse, and ungulates such as Uganda kob).

Type E: Roosting positions and shelters (e.g., many species of bats, starlings, domestic pigeons).

Such classification systems represent useful first approximations but fail to grasp the full range of variability that exists with respect to different patterns of territorial organization. Inevitably, some forms fall between categories. Some of the dimensions along which territorial forms of organization can be said to vary are the following:

1 *Individual versus group residency:* In some species, such as sticklebacks, a single individual stakes out a territory and defends it against conspecifics. In many diurnal songbirds, a mated pair defends a territory. In other species, a whole group of individuals, such as a troop of vervet monkeys, may occupy and defend a territory.

2 *Defended versus nondefended territories:* If one accepts the traditional definition of "territory," all territories are defended (by definition). With the revised definition, one may distinguish territories in which there is overt defense (e.g., sticklebacks, Uganda kob) from those in which there is not (e.g., whiptail wallabies, coatis).

3 *Exclusive versus nonexclusive occupancy:* Exclusion may be practiced against all or some conspecifics. In some systems, the territory may be defended against all conspecifics except during a brief period of mating. In the "classical form," the area is defended only against conspecifics of the same sex. Alternatively, as in the lizards cited above, an animal may defend a territory against all but a single alpha male. In the examples of Steller's jays and bicolored ant birds, there is no exclusive occupancy, only privileged access to resources within the territory.

An interesting form of territoriality in cats was described by Leyhausen (1965). Leyhausen reports that the cats residing in a given area learn each others' daily pattern of activity and apportion territories accordingly. A male may thus occupy a given region during one part of the day but avoid the area, deferring to another male, during other periods.

4 *Feeding versus nonfeeding:* Some territories, such as those of some diurnal insectivorous songbirds, are large enough so that food can be found within the territory. By contrast, most shorebirds, such as gulls and ibises, nest colonially and must leave the colony in search of food.

5 *Mating versus nonmating:* Black-headed gulls may engage in courtship and copulation in a "pre-territory," an area outside of the breeding colony. By contrast, species with *lek* systems use their territories for virtually nothing but breeding. A lek is a communal mating ground. In classical examples, males return to a traditional breeding area and stake out and defend relatively small breeding territories. Females enter and breed with particular males. Typically, males with "choice" territories engage in a disproportionate share of the mating. Among birds, grouse and ruffs display leks, while among mammals, Uganda kob display a similar breeding pattern.

6 *Rearing young versus nonrearing:* Colonial shorebirds and sticklebacks rear their young on their territories; animals with a lek pattern do not.

7 *Fixed versus floating:* Most territories must be related to a fixed unit of space in order to be classified as a true territory. However, bitterlings (small fish) lay eggs inside the mantle cavities of certain mussels and defend the area around the mussel even if the mussel moves. A case can be made for including this as a form of territoriality (Wilson, 1975).

Related Concepts Several concepts related to dominance and territoriality need to be distinguished from them.

Leadership The term "leadership" generally is used to refer to the ability of a given individual to influence the movement pattern of the group as it goes from place to place. Thus, the leader determines the time, rate, and direction of the group's movements. The most effective "leader," in this sense, is not necessarily the most dominant individual. Experience often counts more than physical prowess in determining leadership. In many primate troops, it is the females that play an important role in controlling group movements (Baldwin, 1971; Hinde, 1974).

Home range The home range of an individual is the area that it habitually travels in the course of its normal activities. Many species, such as most rodents, occupy relatively small home ranges for their entire lives. Home ranges may be completely undefended and convey no privileged access to resources; they should not be confused with territories.

Individual distance Many species act so as to keep a certain distance between themselves and other conspecifics. Such a distance is referred to as "individual distance" and should not be confused with a territory. Similarly, defense of young would not normally be regarded as an instance of territoriality. Mere proximity in space has been found to increase the likelihood of aggressive behavior in species as diverse as birds and primates (Marler, 1956; Chalmers, 1968).

Flight distance The concept of flight distance refers to the distance to which an individual will permit a predator to approach before it flees. Flight distance should not be confused with territoriality, home range, or individual distance.

Complications While it is convenient to classify various aspects of animal behavior into orderly categories, it must be remembered that animal behavior is quite complex and often does not fit into the neat pigeonholes created by animal behaviorists. This does not completely destroy the validity of the concepts, but it does necessitate occasional words of caution.

Patterns of social organization tend to be labile. While it is possible to study individual social interactions in the laboratory, the social organization of at least the larger species tends to break down in captivity, both in laboratories and in zoos. Further, social organization in nature can vary from habitat to habitat within a species. For example, in areas where seasonal changes in vegetation make a nomadic life necessary, wildebeest males herd females and defend them, mating with them as they migrate. In more stable areas, males defend areas as females move in and out of them (Estes, 1966). As indicated above, organization can shift from dominance hierarchies to territory as a function of population density. Wilson (1975) suggests that this "behavioral scaling" may have a genetic basis and may permit animals to adopt different functional "strategies" adaptive in different situations.

As in sticklebacks, patterns of social organization often vary with the season. Animals may be organized into schools in the nonbreeding season and become highly territorial during the breeding season. Organizational patterns may change within a day. While breeding on Seahorse Key in Florida, white ibises are territorial. However, during the day, one member of the mated pair goes off to feed, joining a flock to fly to the mainland.

Finally, while all the interactions considered thus far are intraspecific (within a species), interspecific interactions also occur. For example, interspecific territorial defense has been observed. Among reef fishes, the area defended varies according to the species of the intruder; defense occurs early against conspecifics and later against fishes of certain other species (Myrberg & Thresher, 1974).

Examples of Organized Societies

Many examples of organizational patterns have been studied and space will not permit an exhaustive treatment of any. However, the patterns of at least three groups merit some brief consideration.

Social Insects Insects show a great diversity of patterns of social organization. The most complex form of insect social organization, *eusociality,* has been reported only in ants, bees, wasps, and termites (see Wilson, 1971). Eusociality is defined in terms of three attributes: (1) individuals of the same species cooperate in the care of the young; (2) there is a reproductive division of labor, with some more or less sterile individuals working in behalf of fertile nest-mates; and (3) there is an overlap of at least two generations in life stages, contributing to colony labor (Wilson, 1975).

The social insects tend to live in large colonies often containing thousands of individuals. Generally, these individuals are closely related genetically. Division of labor within such colonies typically is so strong that groups with different functions are regarded as "castes."

Birds Like insects, the forms of social organization found in birds are quite variable. Etkin (1964) has described three general patterns that are representative of a substantial number of species.

Diurnal songbirds tend to form stable pair bonds for a breeding season and to occupy territories within which they can feed on material such as insects or plant matter. There is usually little sexual dimorphism, little courtship, sharing of parental duties, and a disruption of the pair bond on departure of the young.

Marine birds, such as gulls and ibises, breed in small territories located within breeding colonies. They travel well outside the breeding colony in search of food. The main threat from predation is on the eggs and young. These species tend to have little sexual dimorphism, elaborate courtship, sharing of parental duties, well-developed nest-relief ceremonies, and well-developed powers of individual recognition.

Flock-living, surface-feeding birds, such as some species of ducks, tend to move about in flocks with females that are protectively colored, while male coloration may be quite bright. There are no reproductive territories, although the female often defends the nest site. Parental care is left to the female, courtship may be quite elaborate, and there is a premium on individual recognition, as in "imprinting."

Nonhuman Primates Primates live in a bewildering array of organizational patterns, the essential features of which are just beginning to be understood. Southwick and Siddiqi (1974) distinguished six grades of primate social organization.

1 Solitary, except for mother-infant pairs (orangutans).

2 Monogamous adult pairs with recent offspring (gibbons, marmosets).

3 Single-male groups with bonded females and offspring (patas monkeys).

4 Aggregate single-male groups with bonded females and offspring (hamadryas baboons). (These are generally assemblies of groups of type 3.)

5 Multiple-male or age-graded male groups (rhesus macaques).

6 Diffuse social parties without consistent or stable groups (chimpanzees).

It is apparent that organizational patterns in nonhuman primates run the full gamut from solitary forms and pair-bonded species to "diffuse social parties." The multiple-male or age-graded male organization is the most common primate pattern. Determinants of these organizational systems will be considered in Chapter 15.

INDIVIDUAL SOCIAL INTERACTIONS

Overall patterns of social organization are the sum product of individual social interactions and at the same time constrain the kinds of individual interactions that can and must occur. Much research has been devoted to the study of individual social interactions.

Animal Communication

Communication is the essence of all social behavior. It is difficult to imagine any social behavior without communication or communication that is not, in some sense, social.

Communication may be said to have occurred when an animal performs an act that alters the behavior of another organism. Such interactions are often quite specialized and usually are adaptive for one or both organisms.

Communication Systems A communication system can be seen to involve seven essential components (adapted from Sebeok, 1965, and Klopfer & Hatch, 1968):

1 *Sender:* an individual which emits a signal
2 *Receiver:* an individual whose behavior is altered by the signal
3 *Channel:* a pathway through which the signal travels (e.g., vocal-auditory channel)
4 *Noise:* background activity in the channel that is unrelated to the signal
5 *Context:* the setting in which a signal is emitted and received
6 *Signal:* the behavior emitted by the sender
7 *Code:* the complete set of possible signals and contexts

The way in which the behavior of the receiver is altered depends on the nature of all these factors and can be affected by the identity of the sender, the identity of the receiver, the channel used, the extent of background noise, the context within which the signal is emitted and received, and, of course, the nature of the signal itself.

Methods in the Study of Communication The three primary types of methods used in the study of animal communication are correlation, experiment, and training.

Correlational methods The most common method of studying communication systems is simply to observe animals and to make inferences about communication systems on the basis of reliable correlations with both antecedent and consequent events. After extensive observation, it becomes apparent that a particular signal occurs under a given set of conditions and has a reliable effect on another organism.

In the case of courtship behavior, for example, acts are considered to be courtship signals on the basis of the situations in which they occur and the effects they have on another organism. In the case of courtship, the signal occurs in the period before copulation begins and has the effect of increasing the probability that copulation will occur. The courtship displays of green herons (Figure 2-1), queen butterflies (Figure 5-1), and fireflies (Figure 5-2) were all analyzed using observational-correlational methods. The differentiation between mobbing calls and warning calls in various species of birds (Figure 4-1) was based on the nature of the antecedent events (i.e., the kind of predator) as well as the effects on other animals (typically vigorous mobbing, immobility, or flight).

Perhaps the most famous example of animal communication is that of the dances of honeybees (von Frisch, 1971). Forager bees scout the area surrounding a hive in search of nectar and pollen. When a rich source of food is found, its location is signalled to other foragers (Figures 6-1 and 6-2). The returning bee performs a ritualized "dance" on the vertical comb of the beehive. The dance is attended to by other foragers, who are able to locate the food source on the basis of information provided by the signaller. If the food source is less than 100 meters from the hive, the *round dance* is performed. The signaller dances in a circular pattern (Figure 6-1*a*) and other bees leave the hive and locate the food. If the food source is more than 100 meters from the hive, the bee performs a dance that contains information about both the distance and direction of the food source, the *waggle dance* (Figure 6-1*b*). Distance is

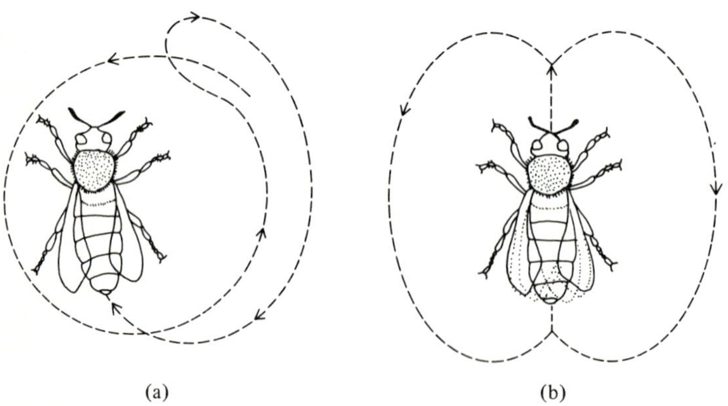

Figure 6-1 *(a)* The "round dance" of honeybees. The round dance is given when a food source is located less than 100 meters from the hive. *(b)* The "waggle dance" of honeybees. The waggle dance is performed when food is located more than 100 meters from the hive. During the straight run in the dance, the bee's abdomen moves from side to side and produces a buzzing sound. *(Reprinted from Karl von Frisch:* Bees: Their vision, chemical senses, and language. © *1950 by Cornell University. Used by permission of Cornell University Press.)*

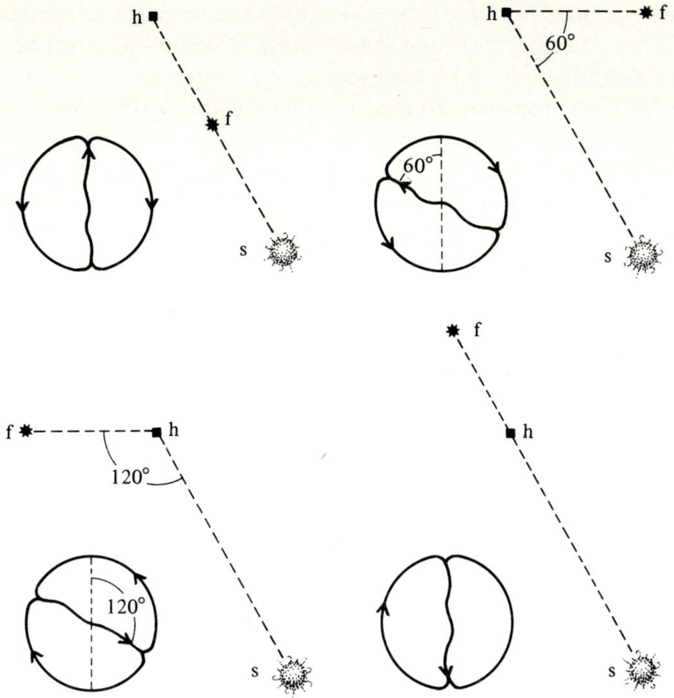

Figure 6-2 A diagram portraying the way in which directional information is encoded in the waggle dance of honeybees. When a food source is located in the direction of the sun, the straight run of the waggle dance is oriented directly up. The angle of displacement from the vertical of the straight run encodes the angle relative to the sun at which the food source is located. *Key: h*—location of hive; *f*—location of food; s—sun. *(Reprinted from Karl von Frisch:* Bees: Their vision, chemical senses, and language. *Copyright 1950, 1971 by Cornell University. Used by permission of Cornell University Press.)*

indicated by the duration of the straight run of the dance and by the buzzing sound emitted during the dance (Wenner, 1962). Direction is indicated by transforming the direction to the food source relative to the location of the sun into an angle of the straight run of the dance relative to gravity (see Figure 6-2). The other foragers leave the hive and locate the food source.

The correlational method is an extremely valuable tool in the study of communication systems. However, it is not without its perils, as can be illustrated from research on honeybee communication. The first thing one learns about the correlational method in most good statistics courses is that *correlation does not imply causation.* We cannot infer that one event causes another simply because the two events tend to occur together. In the present situation, the returning bee dances; the other foragers leave the hive and locate food. Is information about the food location necessarily derived from the dance?

There can be no doubt that the honeybee dances contain the information necessary for location of the food source. An experienced human observer can watch the dance and determine which of several sites is the referent. Lindauer (1961) observed swarming bees performing these identical dances in reference to potential hive sites and was able to locate the sites by decoding the message of the dance. However, a group of American investigators proposed that honeybees do not actually use the information contained in the dance when locating food sources (e.g., Wells and Wenner, 1973; Wenner, 1974). Essentially, they proposed that experienced foragers are able to locate food sources on the basis of the smell of the food on the dancing bee, together with their knowledge of the location of the food sources of the area. Much controversy has ensued regarding the extent to which dance information is actually used in the transmission of information regarding the location of food sources (see Gould, 1975; 1976). Differing experimental procedures used by researchers in the different groups contributed to the occurrence of divergent results. The resolution appears to be that information from the dance, odor, or both can be important, depending on the circumstances. Wenner and his associates have made a positive contribution in pointing to the fact that just because a bee does a dance and another locates food does not necessarily mean that the information regarding location was derived from the dance. However, von Frisch made a critical pioneering contribution in describing and analyzing this fascinating communication system.

Experimental methods An alternative approach to the study of communication is the experimental approach, that of manipulating an independent variable and observing its effects on a dependent variable. The ethologists' usage of models in the study of communication systems is an example of this approach. In the study of the signal initiating aggressive interactions in male sticklebacks (Figure 2-3), for example, the characteristics of models, including their shape and the color of their undersides, were varied. Different models proved differentially effective in eliciting attack by the resident males (the dependent variable). Similar methods have been used in the study of territorial defense in robins, courtship in butterflies, and the escape behavior of birds (Tinbergen, 1951).

Training methods Most attention in the study of communication systems has been devoted to study of the behavioral patterns normally displayed by animals in their typical interactions in the natural habitat. An alternative approach is to attempt to teach an animal to use a new communication system, one not normally utilized in nature. With such an approach, one may learn little about the behavior of the animal in nature but much about the nature and capacities of the organism under study.

The most famous examples of the study of communication using training methods are those in which humans have attempted to teach various sorts of "language" to chimpanzees. Early attempts, such as those of Hayes and Hayes and Kellogg and Kellogg, consisted of efforts to teach chimpanzees to respond to and emit the sounds characteristic of human speech. Although chimpanzees exhibited an ability to respond to a diverse array of human speech sounds, little progress was made in teaching chimpanzees to speak (Kellogg, 1968).

More recently, investigators have recognized that much chimpanzee communication is gestural and that alternative communication systems might more effectively be used in the study of language in chimpanzees. Gardner and Gardner (1969) succeeded in teaching Washoe, a chimpanzee, the complex system of gestures that constitutes the "American Sign Language" used by deaf humans in North America. Premack (1971) succeeded in teaching Sarah, a chimpanzee, a vocabulary of 130 words, which she could use in a complex manner, in which the animal manipulated plastic symbols that adhered to a magnetic board. Rumbaugh and Gill (1976) used a computer-controlled system to develop an automated grammatical system, "Yerkish," with their chimpanzee, Lana. These studies have revealed that chimpanzees are able to learn to use communication systems, the complexity of which formerly was thought to be well beyond the capacities of any nonhuman organism.

A Message Set It appears that most species have relatively uniform-sized sets of available signals, typically ranging from about 15 to 45. This may be because there is a limited number of kinds of information that must be transferred from one organism to another. Smith (1969) proposed a set of 12 message categories that was designed to encompass virtually all kinds of messages used in animal communication:

1 *Identification:* Identification messages function to identify the category or class of the sender.

2 *Probability:* The relative probability of a given course of action by the sender may be communicated.

3 *General set:* Some messages are used in such diverse situations that it is difficult to attribute a single function to them.

4 *Locomotion:* Some messages occur only during or prior to locomotion.

5 *Attack:* Some messages indicate that attack is probable.

6 *Escape:* Some messages indicate that escape is likely.

7 *Nonagonistic subset:* A set of messages indicates that agonistic acts are unlikely.

8 *Association:* Some messages occur when one organism approaches or remains near another.

9 *Bond-limited subset:* Bond-limited messages occur only among individuals with established bonds, such as mates, and among parents and offspring.

10 *Play:* Some messages are uniquely associated with play.

11 *Copulation:* Some messages occur before and during copulation.

12 *Frustration:* Some displays occur only when a given probable behavior pattern is somehow thwarted.

Agonistic Behavior

The term "agonistic behavior"—a broad one—is used to encompass a wide range of behavioral patterns related to interanimal conflict, including fighting, defensive behavior, fleeing, and "freezing." Such conflict situations arise repeatedly in the lives of animals and are associated with a large number of specialized behavioral patterns, including those of overt fighting, fleeing, etc., and those of signaling the probabilities of such behaviors (e.g., threat).

Aggression Much controversy has surrounded the utilization of the concept of "aggression" in the study of animal behavior. The major problem lies in the apparent impossibility of providing an adequate definition of aggression that is usable in studies of animal behavior. We all "know" what we mean when we use the term. However, in working with animal behavior, we strive to define all concepts in terms of objectively observable and measurable phenomena. Thus, Buss (1961) proposed that aggression be defined as "a response that delivers noxious stimuli to another organism" (p. 3). Wilson (1975) defined aggression as "a physical act or threat of action by one individual that reduces the freedom or genetic fitness of another" (p. 577). By these definitions, the following acts would be classified as aggressive: (1) a wolf kills and devours a sheep, (2) a farmer beheads a chicken for his Sunday dinner, (3) a firing squad executes a prisoner, and (4) a Boy Scout attempts to help an old lady but trips her instead (see Johnson, 1972). By Buss's definition, an assassin who fires a shot at the President and who misses, a tennis player who smashes his racquet after missing a shot, and a small boy who daydreams of beating up the neighborhood bully all would be considered as engaging in nonaggressive acts. The problem, of course, is one of motivation or intent. However, motivations and intentions are difficult to measure in animals. We are then left with the dilemma that behavioral definitions are inadequate and those relying upon inferences concerning motivation are impossible.

The solution to the problem of "aggression" is simply to treat the word as a convenient, loosely defined aid to communication and organization, recognizing that we cannot provide an adequate definition and that we are probably lumping together a number of diverse phenomena. How then are we to study aggressive behavior? Rather than attempt to redefine a common-sense concept from everyday language, we may attempt to develop new concepts on the basis of objectively collected data (see Lewis, 1929). In the case of aggression, we might begin studying a broad range of behavioral patterns that would be loosely labeled as "aggressive" and sharpen our definitions as we progress. We may find that there were really 5, 10, or 100 different phenomena that should be differentiated out of the mass of behavioral patterns with which we started.

An attempt to begin work toward an empirical catalogue of aggressive behaviors was begun by Moyer (1968; 1976). Moyer proposed that, as a first approximation, we might consider eight different classes of aggressive behavior. Each may be defined in terms of the conditions under which it occurs and may be differentiated from all other forms. While the list surely will be modified, it has the advantage of focusing attention on the diverse array of aggressive patterns that sometimes may be lumped together. Moyer's list of kinds of aggression included the following:

1 *Predatory aggression:* This form is provoked by the presence of a natural object of prey.

2 *Intermale aggression:* This form of aggression is most effectively stimulated by the proximity of two males not familiar with each other.

3 *Fear-induced aggression:* Fear-induced aggression is characteristic of confined or cornered animals and is preceded by movements toward escape.

4 *Irritable aggression:* Irritable aggression differs from other forms in that it is elicited by a broad range of diverse stimuli, many of which appear aversive to the organism.

5 *Territorial defense:* Territorial defense occurs only in an area in which an animal has established itself and which it defends against intruders.

6 *Maternal aggression:* This form is elicited by the proximity of some agent that is threatening to a female's young.

7 *Instrumental aggression:* Instrumental aggression is a learned response, the occurrence of which is reinforced by some extrinsic reinforcer.

8 *Sex-related aggression:* These aggressive responses are elicited by the same stimuli that elicit sexual responses.

It is easy to be critical of this first approximation to a list of categories. What of female-female aggression? of paternal aggression? Many animal behaviorists would treat predatory behavior as a completely separate behavioral category not included within the rubric of aggression. The important point is to view this as a self-correcting system for the development of a viable "taxonomy" of aggressive behavior. Such a multifactorial approach to aggression stands in contrast to the view that aggression is a "unitary drive."

Social Facilitation

"Social facilitation" is a term applied to the phenomenon in which the mere presence or behavior of another organism produces an increase in the probability, rate, or frequency of a behavioral pattern in another organism. Zajonc (1965) divided studies of social facilitation into two classes; those dealing with *audience effects,* wherein the facilitating organisms are passive spectators, and those dealing with *coaction effects,* wherein both individuals are engaged in the same behavior. Zajonc (1965; 1969) has surveyed studies of social facilitation in animals. In both rats and domestic chicks, the presence of another organism that is eating food results in a facilitation of eating by another organism (Harlow, 1932; Tolman & Wilson, 1965). Larsson (1965) found a facilitation of copulatory behavior in rats resulting from the presence of another copulating pair. Similar results have been found for locomotor exploration in rats and for speed of learning to peck at a stimulus in young chicks (Hughes, 1969; Frank & Meyer, 1974).

The presence of a second organism can result in an interference with performance rather than a facilitation of it. Thus, both Gates and Allee (1933) and Zajonc, Heingartner, and Herman (1969) found that cockroaches took longer to learn to negotiate a complex maze when they learned in the presence of another organism rather than alone. Zajonc (1965) interpreted these results in the context of classical learning theory, that of Spence (1956). Zajonc contended that the presence of another organism results in increased "arousal" in an animal. Increased arousal is hypothesized to facilitate dominant, high-probability responses and to interfere with less dominant, low-probability responses. Whatever the merits of this explanation, the phenomenon whereby the mere presence of a second organism can facilitate the behavior of a first is an interesting one.

Imitation

Imitation is concerned with the learning of a response of low initial probability as a result of observing another animal exhibit the response. In imitation situations, the presence of a second organism has a directive effect in addition to any facilitating effect on the behavior of another organism. For example, Darby and Riopelle (1959) demonstrated that rhesus monkeys could observe and learn from the responses of other monkeys in a discrimination learning situation. Both monkeys were rewarded for selecting one of two stimuli that were presented. Performance by the observing monkey was better when the demonstrating animal made a mistake than when it was correct. Hall and Goswell (1964) found that young patas monkeys avoided a box after their mothers showed a startle response on opening it. Imitation plays a role of great importance in behavioral development, especially among mammals such as primates and carnivores (e.g., Baldwin, 1969).

Cooperation

Cooperation entails the joint and mutual interaction of two or more animals in performing a task. A classical study of cooperation was conducted by Crawford (1941). Chimpanzees were first trained individually to perform a learned task in order to obtain food. They had to push four panels in a fixed order: yellow, red, green, blue. If panels were pressed out of sequence, no reward was given. Next, two chimpanzees were placed in cages so that each had access to two of the four panels. In this situation, cooperation was necessary if either animal was to be rewarded with food. Cooperative behavior occurred in that (1) animals watched their partners and coordinated their own behavior in conjunction with that of the partner and (2) animals solicited the partner by using manual gestures which functioned to direct the partner to press a panel. One chimpanzee would pull, push, or turn its partner in the direction of the panels.

Competition

In most competitive situations, two organisms compete for a limited resource. If one animal obtains access to the resource, the other is excluded. Thus, this situation differs from the cooperative situation in that successful performance by one animal decreases rather than increases the probability that the other animal will obtain a reward. Bayroff (1940) studied competition in laboratory rats. Pairs of rats were forced to swim under water against a current to reach an exit. While the first animal to reach the exit was allowed to escape, the second rat was trapped under water for a short time. Pairs were equally matched with respect to swimming speed. Rats generally swam faster when in competition than when swimming alone. A problem with this and other studies of competition is that of demonstrating that it is the contingencies of competition rather than the mere presence of a coactor (i.e., social facilitation) that improve performance. In a study in which dogs had to run for a food reward, Scott and McCray (1967) found running speed to be greater when dogs were tested in pairs rather than alone. However, when competition was added to the paired (i.e., social facilitation) situation, running speeds tended to deteriorate.

Affiliation

Animals of many species tend to approach and remain near conspecifics. Indeed, it is possible to have an animal learn an arbitrary task when the only reinforcement is the opportunity to be with a conspecific. Feral monkeys were more likely to respond than were animals reared in a restricted laboratory environment. Much research has been conducted on the determinants of similar affiliative responses in laboratory rats (e.g., Sloan & Latane, 1974; Latane, Joy, Meltzer, Lubell & Cappell, 1972).

SUMMARY

The animals of many species live in organized societies with complex communication systems, specialization of function, cohesion, permanence of individual composition, and impermeability. Sociality can be studied either by attempting to understand the total structure of the society or by assessing its component individual social interactions.

Many societies are organized with dominance hierarchies or territorial structure as their basis. Both entail a privileged access to resources on the part of some individuals. With the territorial structure, privileged access is limited to a particular location. There are many forms of territoriality.

Communication may be said to have taken place when an animal performs an act that alters the behavior of another individual. Correlational, experimental, and training methods have been used in the study of communication.

It appears that aggression is not a single form of behavior reflecting the operation of a unitary drive; rather, the term "aggression" represents a convenient if poorly defined, catch-all for considering diverse forms of behavior.

Social facilitation, imitation, cooperation, competition, and affiliation are forms of social behavior that have been studied in the laboratory.

Part Three

Development of Behavior

Parts Three, Four, Five, and Six are concerned with the four sets of questions to be asked about behavior—questions of development, mechanism, evolution, and function. The first to be considered are questions of development—the ontogeny of behavior in the life-span of the individual. The development of behavior represents the product of the joint, continuous, dynamic interaction between genes and environmental factors to produce the organism and its behavior. This section, dealing with development, is divided into three chapters: the first deals with the role of the genes in the development of behavior, the second with experiential factors in development, and the third with the interaction of genetic and environmental factors—the "nature-nurture problem."

Chapter 7

Behavior Genetics

Students have long been interested in both speculation and research regarding the role of inherited factors in behavior. Such interests can be traced from the Greeks through Leonardo, Montaigne, Darwin, Galton, Yerkes, and Tryon (McClearn & DeFries, 1973). However, the discipline of behavior genetics is quite young. Credit for establishing behavioral genetics as a well-defined research area is given to John Fuller and W. R. Thompson, with the publication of their *Behavior Genetics* in 1960. In the last 20 years, research on the genetic influences on animal behavior has progressed at a rapid pace. While different behavior geneticists have divergent interests, it is convenient to organize research in behavioral genetics about attempts to find answers to six kinds of questions (modified from Thiessen, 1972):

1 Is the observed behavior affected by variations in genotype?
2 Assuming a genetic influence is found, how many genes are operating?
3 What proportion of the measured variability in behavior is attributable to genetic and to environmental factors respectively?
4 What developmental and physiological processes intervene between the genes and their expression in overt behavior?
5 What can behavior-genetic analysis tell us regarding probable adaptive significance of a behavioral pattern?
6 How do the genes and the environment interact to produce behavior?

This chapter is organized around these six fundamental questions, methods that can be used in answering them, and some of the kinds of answers that have been found. Discussion of material in behavioral genetics requires the use of a few basic concepts of genetics. Some of these were defined in Chapter 3 (pages 38–39). It may be worthwhile for you to go back to review those basic concepts before proceeding.

IS OBSERVED BEHAVIOR AFFECTED BY VARIATIONS IN GENOTYPE?

A rigid environmentalist position would be that genes play virtually no role in the control of behavior. Such a position is generally regarded as untenable today. The evidence for a role of the genes in controlling behavior, while always somewhat indirect, is overwhelming. Virtually all the research to be discussed in this chapter points to the same conclusion—variations in genotype profoundly influence observed behavior. The data should convince even the most hard-nosed skeptic. While the whole chapter provides an answer to the question of genetic influence on behavior, at this point we shall consider three methods of answering it—with the study of strain differences, selection, and interspecies hybrids.

Strain Differences

The basic strategy for using strain differences to determine whether genes affect behavior is simple. There are two major factors in the development of behavior—genes and environment. The strategy is to locate organisms of differing genotypes, hold environmental factors constant, and determine whether the genotypic factors influence behavior.

The major difficulty in genetic research is that genes are not directly observable. Indirect methods must be used to make inferences about genotype. How, then, are we to find organisms that differ in genotype? Many behavior geneticists have used *inbred strains*. Individuals within an inbred strain are virtually isogenic—all genetically identical and homozygous at all loci. Inbred strains of house mice *(Mus musculus)*, the most popular animals in research in mammalian behavioral genetics, are the result of at least 20 generations of brother-sister matings. A brother and sister from a genetically heterogeneous population are bred to produce offspring. A brother and sister from those offspring are bred to produce more offspring. With each successive generation, some alleles are irrecoverably lost from the line. After about 20 generations, the strain approaches complete isogenic status. Note that animals are chosen for breeding on the basis of genetic relationship, brother and sister, in contrast to the selection based on anatomical, physiological, or behavioral traits which characterizes the artificial selection studies to be discussed below. Largely as a result of the breeding program at the Jackson Laboratory in Maine, a large number of inbred strains is available for study. Each is named according to international convention (e.g., C57BL/6J).

When animals of different inbred strains are reared in identical environments and display consistent differences, it is likely that these differences have a genetic basis. Such results have been found for a variety of behavior patterns, including open-field activity (Thompson, 1953), aggressive behavior (Southwick & Clark, 1968), sexual behavior (McGill, 1962), nest building (Lynch & Hegmann, 1972), and learning (Wahlsten, 1972). Indeed, it is more difficult to find strains that do not differ than strains that do.

A comprehensive bibliography of 1,222 behavior studies using genetically defined mice has been prepared by Sprott and Staats (1975).

Cross Fostering Although strain differences are indicative of a genetic effect, it remains possible that some subtle environmental factor might be acting to produce the observed differences. One method of controlling for such factors is the *cross-fostering* experiment. Newborn litters of mouse pups from two different strains differing in adult behavior are switched at birth, so that the mothers of one strain rear young of the other strain and vice versa. Control animals are reared by their own mothers or by foster mothers of their own strain (in-fostering). An experiment using a cross-fostering procedure was reported by Southwick (1968). The A/J strain and the CFW strain differed greatly with respect to aggressive behavior. The results of Southwick's cross-fostering experiment are presented in Figure 7-1. Rearing A/J mice with a

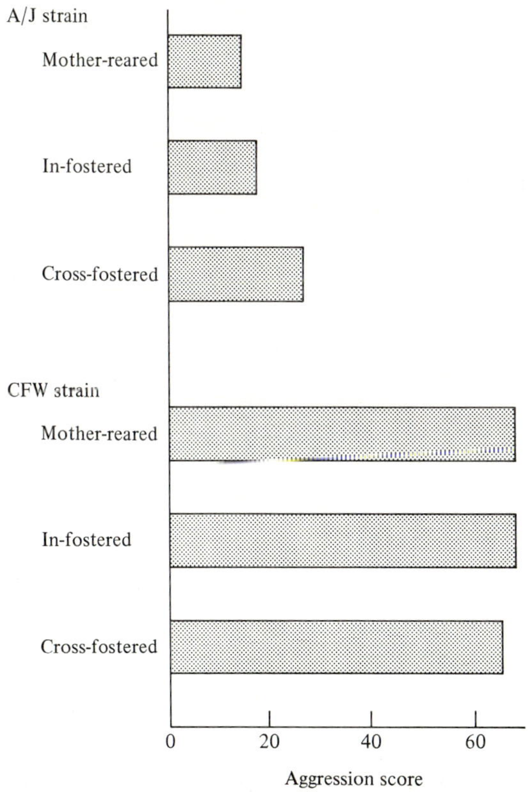

Figure 7-1 Results of a cross-fostering experiment on two inbred strains of mice: the A/J and CFW strains. Mean scores on an aggression index (the sum of the number of chases, attacks, and fights) are given for mice of each strain reared by their mothers, in-fostered, and cross-fostered. *(Based on data from Southwick, 1968.)*

CFW mother significantly increased their aggression scores. However, the change is small compared to the size of the strain difference. Rearing CFW mice with an A/J mother had little effect on their aggressive behavior. Thus, the difference in aggressive behavior between these strains appears to be primarily of genetic origin.

Ovarian Transplantation With the cross-fostering experiment, one eliminates uncontrolled postnatal maternal factors but does not alter the prenatal environment. It is possible to separate certain genetic effects from effects mediated by the mother's prenatal internal environment by using ovarian transplantation techniques. Animals of inbred strains reject tissue grafts from other strains. However, two inbred strains can be crossed to produce a first filial generation (F_1). The F_1 mice accept transplants of ovaries from either parental strain. By mating these females with appropriate males, one can then compare the behavior of mice of two different strains developing in inbred or F_1 mothers. Using this method, DeFries, Thomas, Hegmann, and Weir (1967) found genetic factors predominant in determining open-field behavior and maternal environment more important with respect to body weight.

Selection Studies

Studies of artificial selection differ from attempts to create inbred strains in that animals are chosen for breeding on the basis of their phenotype rather than their genotype. A *phenotype* is any measurable morphological, physiological, or behavior trait of an organism. Body weight, metabolic rate, and open-field activity are phenotypes. *Genotype* refers to all or a part of the genetic material of an organism. To begin a study using artificial selection, a group of males and females from a genetically heterogeneous population is tested on some characteristic. The high-scoring males are then mated with high-scoring females and low-scoring males are mated with low-scoring females. The offspring of these matings are tested for the phenotype and mates are again selected on the basis of their scores. Over successive generations, one thus creates two different lines—one selected for high scores on the phenotype and one for low scores. As a control for changes in the environment, a third strain is random-bred. If there is a genetic component to the phenotype under study, the scores in the high and low lines should diverge over successive generations. The scores of the random-bred strain should not change systematically.

The selection method was used by DeFries, Hegmann, and Halcomb (1974) to study open-field behavior. Starting with relatively heterogeneous stock, they selectively bred two high lines, two low lines, and two control lines for 20 generations. The results of their experiment are presented in Figure 7-2. It can be seen that over the 20 generations, the scores of the different lines diverged—those of the animals in the high lines increased, those in the low lines decreased, and those in the control lines drifted but slightly. These data

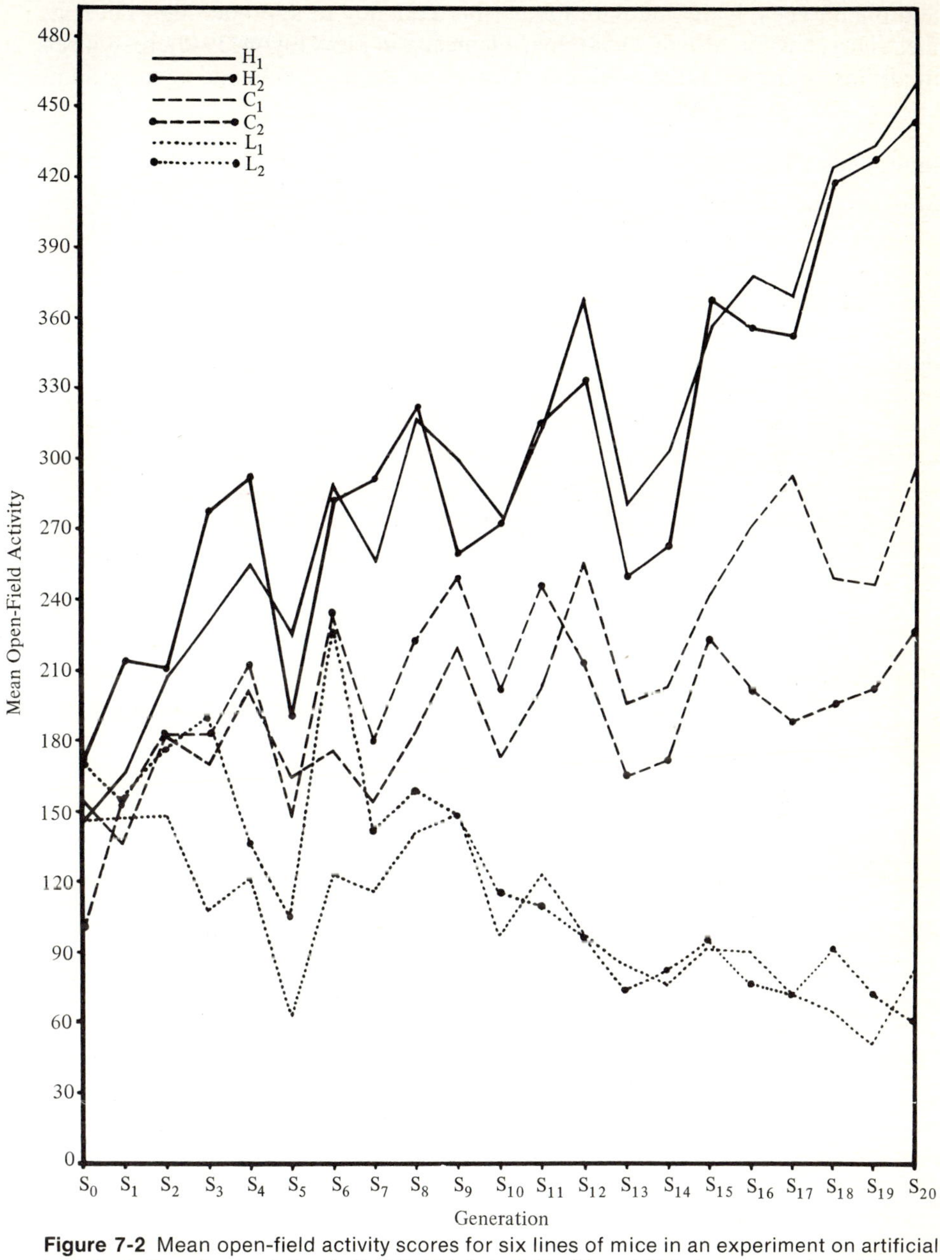

Figure 7-2 Mean open-field activity scores for six lines of mice in an experiment on artificial selection. Two lines were selected for high activity (H_1 and H_2), two were selected for low activity (L_1 and L_2), and two were randomly mated within the line (C_1 and C_2). *(From DeFries, Hegmann, & Halcomb, 1974.)*

indicate an important role of genes in the control of open-field activity. Similar results have been obtained in selection experiments in a wide array of species and phenotypes, including phototaxis (the tendency to approach light) in fruit flies (Hirsch & Boudreau, 1958), maze learning in rats (Tryon, 1940), responses in an imprinting situation in chicks (Graves & Siegel, 1969), and aggression in mice (Lagerspetz, 1969).

Interspecies Hybrids

Most behavior-genetic studies utilize animals of differing genotypes of the same species. However, animals of different but closely related species can be crossed in the laboratory. Often the behavior of the offpsring of such matings presents strong evidence of genetic influence. An excellent example of this approach was provided by Dilger (1962). Dilger studied the nesting behavior of different species of lovebirds. One species, *Agapornis personata*, cuts strips of nesting material, grasps them in its beak, and carries them to the nest. Another species, *A. roseicollis*, tucks the cut nesting material under its flank feathers as a means of carrying them to the nest. The F_1 hybrids studied by Dilger showed a rather confused pattern in which they would move to tuck nesting material, take it back into the beak again, and then repeat the pattern. Their behavior thus represented a rather nonfunctional intermediate between the two functional patterns of the two parental lovebird species.

Similar intermediate forms of behavior have been found in studies of display patterns in finches (Hinde, 1956) and of cricket song (Bentley and Hoy, 1974).

HOW MANY GENES ARE OPERATING?

Once it is established that genotype is important in controlling behavior, one can ask how many genes are acting. One can begin with the question of whether there is one functional gene or more than one.

The Mendelian Cross

To determine whether one or more than one gene is acting to control the behavior of interest, we shall use the same methods as those used by Gregor Mendel, the Augustinian monk who is credited with founding the science of genetics. Mendel worked with garden peas and noted that the plants of some lines would *breed true* for some traits. For example, all the offspring of peas with round seeds had round seeds, whereas all the offspring of peas with wrinkled seeds had wrinkled seeds. Mendel began his experiments by crossing plants from lines breeding true for different traits in order that he might observe the offspring. Typically, *all* the offspring resembled one or the other parent. When plants producing round seeds were crossed with plants producing wrinkled seeds, all the resulting offspring produced round seeds. When the offspring of such a cross resemble one of two different parental phenotypes, we refer to the phenotype which is recovered in the F_1 cross as the *dominant* trait, whereas that which is lost is the *recessive*.

Mendel next crossed different plants of the F_1 genotype to produce a second filial generation (F_2). The results were reliable; both parental phenotypes, round and wrinkled, were recovered. There were three times as many plants with the dominant round seeds as with the recessive wrinkled ones. If bred within lines, F_2 plants showing the recessive phenotype always bred true, whereas just one-third of those showing the dominant phenotype bred true.

These data are interpretable in terms of the actions of genes, which we shall treat as indivisible, particulate units of inheritance of the sort envisaged by Mendel. It will be recalled that genes occur in pairs. Different alleles are given different symbols; dominant alleles are assigned capital letters and recessive alleles get lowercase letters. We might designate the plants with round seeds as having a RR genetic complement, as the phenotype is dominant and the plants breed true. The plants with wrinkled seeds would have the rr complement. Because of segregation in the formation of gametes, each F_1 individual has an Rr genotype. In the F_2 generation, one-quarter of the plants are RR, one-half are Rr, and one-quarter are rr. This model generates predictions completely compatible with Mendel's data.

The Mendelian Cross in the Study of Audiogenic Seizures

Mendel's methods can be applied to the study of behavior, as in research by Collins and Fuller (1968). When exposed to a loud sound, such as the ringing of a doorbell in a metal washtub, some mice convulse. Because such convulsions, or seizures, are generated by auditory stimuli, they are termed "audiogenic seizures." Collins and Fuller observed the proportions of mice from two strains, C57BL/6 and DBA/2J, that convulsed and the proportion that were seizure-resistant (did not convulse). Their data are presented in Table 7-1. All the C57 mice were resistant, whereas hardly any of the DBA mice were resistant (almost all convulsed). In the F_1 generation, nearly all animals were

Table 7-1 Observed and Expected Percentages of Seizure-Resistant Mice in a Mendelian Cross

Generation	N	Percent resistant observed	Percent resistant expected
P_1 (C57BL)	45	100	—
P_2 (DBA)	58	2	—
F_1 (C57xDBA)	89	99	—
F_2 (F_1xF_1)	105	75	75
B_1 (C57xF_1)	115	96	100
B_2 (DBAxF_1)	119	49	50

Source: After Collins and Fuller, 1968. Original copyrighted 1968 by the American Association for the Advancement of Science.

seizure-resistant. With these data, we can use the model just presented to test the possibility that the differences in seizure susceptibility between the two strains are controlled by a single gene. Let us establish a set of hypotheses concerning what we would expect *if* the difference were controlled by a single gene. We shall designate the dominant (resistant) allele as A and the recessive (nonresistant) allele as a. As the two parental lines breed true, the C57s would have an AA genotype, whereas that of the DBAs would be aa (see Figure 7-3). The F_1 mice would all have the Aa genotype and all would be expected to be resistant (Figure 7-3). The model is consistent with the data, and we can now make predictions concerning additional crosses. In the F_2 generation there should be one-quarter AA mice, one-half Aa mice, and one-quarter aa mice. Three-quarters of the mice should be seizure-resistant. This is exactly what Collins and Fuller found.

The correspondence between observed and predicted frequencies in the F_2 generation suggests but does not prove that the difference is controlled by a single gene. More crosses are necessary to establish this possibility more convincingly. Collins and Fuller performed five additional crosses; all were consistent with the hypothesis of a single gene. In creating a ***backcross***, animals from the F_1 generation are mated to animals of the two parental strains. As can be seen in Figure 7-3, we would hypothesize that in the backcross to the dominant C57 line, one-half the mice should have the AA genotype and one-half the Aa genotype; all should be resistant. In the backcross to the recessive DBAs, one-half should have the Aa genotype and one-half the aa; 50

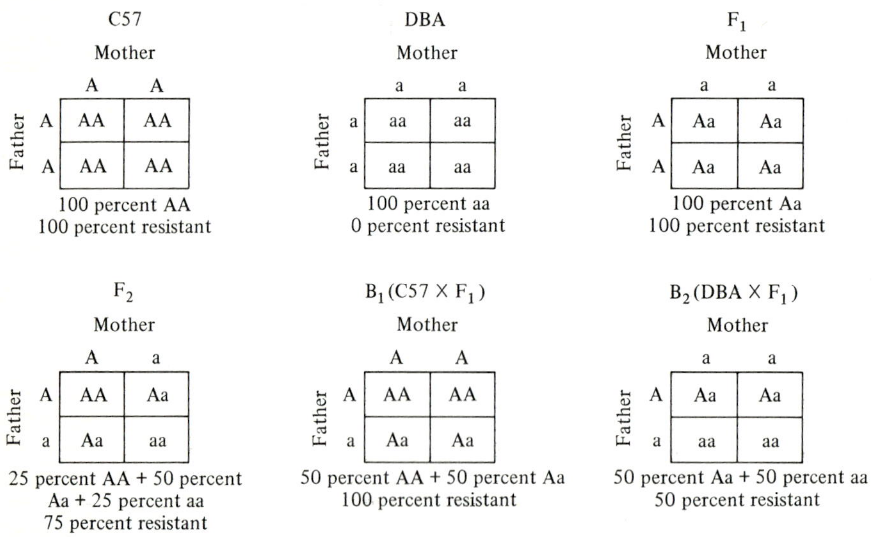

Figure 7-3 Set of diagrams to illustrate the predictions regarding seizure resistance that stem from a single-gene model of inheritance. For each of six combinations of parental genotypes, all possible offspring genotypes are shown in their expected proportions.

percent should be resistant to convulsions. The observed proportions (Table 7-1) matched the predictions quite closely—96 percent of the backcross to the dominant parent resisted and 49 percent of the backcross to the recessive parent resisted. We can conclude that the difference between these two strains in their tendency to display audiogenic seizures (under the testing conditions of Collins and Fuller) is controlled by a single gene. It should be noted that as one gives repeated tests, tests at different ages, or alters the test situation in other ways, effects of additional genes are detected.

Other Tests of Single-Gene Hypotheses

Other behavioral patterns have been shown to be affected by single genes. Whitney (1969) studied a simple phenotype, the tendency for inbred mice to vocalize when they were lifted and permitted to climb on a small platform. Few C57 mice vocalized, whereas many mice of the JK strain did. The results of the Mendelian cross were compatible with a single-gene hypothesis. One can also test single-gene hypotheses using genetically heterogeneous populations rather than inbred strains. The model is somewhat more complex than that associated with the simple Mendelian cross. The results of a study on vocalization in a heterogeneous population of mice were consistent with the results of the study of inbred strains—a single gene was implicated (Whitney, 1973).

A number of phenotypes in humans have been demonstrated to be controlled by single genes. Many forms of mental retardation—including phenylketonuria (PKU), gargoylism, and Wilson's disease—are controlled by single recessive genes. Another human phenotype affected by a single gene is the ability to taste phenylthiocarbamide (PTC). PTC tastes quite bitter to some individuals, while to others it is virtually tasteless except in very high concentrations. Your instructor may be able to obtain some PTC-impregnated paper, which permits a somewhat crude but useful assessment of ability to taste PTC. Snyder (1932) tested the single-gene hypothesis by comparing parents and their children with regard to their ability to taste PTC. The results were compatible with the single-gene hypothesis (see Fuller & Thompson, 1960). Heston (1970) proposed that there is a single gene that predisposes humans to schizophrenia.

Not all tests of single-gene hypotheses produce results indicative of the actions of a single gene. An example is provided in studies by Rothenbuhler (1967) on the hygienic behavior of honeybees. Hygienic behavior consists of the removal from the hive of larvae that have succumbed to American foulbrood disease. Some strains of bees uncap the larval cell and remove the larvae from the hive. Others do not engage in such behavior. Breeding experiments revealed that the behavior was controlled by two separate genes which assorted independently and thus appear located on different chromosomes. One gene controls the behavior of uncapping the cell, while the other controls the removal of the larvae.

For still other traits, data on observed frequencies do not fit any reasonable hypotheses regarding the actions of one or a few genes. Such traits appear to be controlled by many genes, each having a small effect on the phenotype.

Pleiotropism and Continuously Distributed Traits

Thus far, all the traits we have considered for single-gene analyses have had *discontinuous distributions*. When a trait is discontinuously distributed, individuals can be placed into a few discrete categories with respect to phenotype; intermediate values are impossible. Sex is a discontinuously distributed trait. Individuals can be classified as male or female; there are few intermediates. Seizure resistance, vocalization, PTC tasting, and hygienic behavior all can be treated as discontinuously distributed. Because we can classify individuals into discrete categories, analyses are relatively simple. Unfortunately, most traits of interest to animal behaviorists are *continuously distributed*. With continuously distributed traits, an individual can score anywhere along a continuum. Height, weight, number of square entries in an open field, and latency to retrieve a pup are all continuously distributed traits.

Can influences of single genes be detected in dealing with traits having continuous distributions? The answer is yes—but there are some problems. One problem is that of identifying which animals possess the gene of interest. Because genes cannot be directly observed, we must rely on indirect assessments of genotypes. The most common method relies on *pleiotropism*. Pleiotropism is the name given to the phenomenon whereby a single gene can affect many different characters. The presence or absence of a gene can be determined on the basis of one of its phenotypic effects. One can then search for correlated behavioral effects. For example, albinism in mice is controlled by a single recessive gene. It has been determined that the gene controlling albinism has several effects on different behavioral phenotypes. For example, the albino gene produces a reduction in open-field activity among mice which bear it (DeFries, 1969). More sophisticated methods of detecting single-gene effects are available for use with *Drosophila* (e.g., Kessler & Kraemer, 1975).

Behavior geneticists disagree concerning the value of single-gene studies of continuously distributed traits (e.g., Wilcock, 1969; Thiessen, 1971). It appears that most of the traits of interest to animal behaviorists are both continuously distributed and affected by the action of many genes, each producing relatively small effects on the phenotype.

WHAT PROPORTION OF OBSERVED VARIABILITY IS ATTRIBUTABLE, RESPECTIVELY, TO THE GENES AND TO THE ENVIRONMENT?

When dealing with continuously distributed traits that are affected by many genes, it is necessary to use the methods of biometrical or quantitative genetics (see Falconer, 1960). With these methods, it is possible to obtain estimates of a number of characteristics or parameters of the genetic structure affecting the behavior of interest. However, the computations rest upon a number of assumptions regarding the characteristics of both the genes and the measurement scale of the dependent variable. As these assumptions are sometimes difficult to justify when working with behavioral data, a healthy skepticism is appropriate when some of the more complex models are being considered.

Heritability

Much interest has been shown by behavior geneticists in calculating *heritabilities*. Essentially, heritabilities provide an estimate of the proportion of the total variance on some trait in a population that is attributable to the genotype. Perhaps we should first review the measurement of variance. Variance provides a measure of the amount of dispersion or variability in a sample. Suppose that we have two groups of three animals each; the animals in one group enter 49, 50, and 51 squares in an open field, while those in the other group enter 30, 50, and 70 squares. The two groups have equal means but differ with respect to the variability of individual scores around those means; the second is more variable than the first. *Variance* provides a way of quantifying this variability.

Variance is the average of squared deviations from the mean. To calculate variance, one finds the difference between each score and the mean, squares it, and takes the average. The variances for the two samples just described would be 0.7 and 266.7 respectively.

The essence of many inferential statistical methods and the basis for determining heritability both lie in the process of dividing or partitioning variance into its sources. For example, part of the variability or variance in a group of scores in open-field tests may be due to genetic differences and part may be due to environmental factors. If we let V_T stand for the total or phenotypic variance in a group, V_G stand for variance attributable to variations in genotype, V_E for the variance attributable to the environment, and V_I for the variance caused by their interaction, then:

$$V_T = V_G + V_E + V_I$$

It is customary to assume V_I to be negligible and thus to reduce the formula to:

$$V_T = V_G + V_E$$

"Heritability in the broad sense" or the *coefficient of genetic determination* (CGD) is defined as the percent of total variance that is attributable to genetic factors:

$$CGD = \frac{V_G}{V_T}$$

CGD is a proportion that varies between zero and 1. It will be recalled that all individuals within an inbred strain may be considered genetically identical.

Within an inbred strain, then, there is no genetic variability and both V_G and CGD must equal zero. This does not mean that there is no genetic influence on the trait but only that none of the variability *in that group* is due to variation in genotype. If we could raise a group of genetically heterogeneous individuals in identical environments, CGD would approach 1. This does not mean that environmental factors are not important in the development of the trait but only that they do not contribute to the differences among individuals in this particular group under these conditions.

If we use the symbol V_A to represent *additive genetic variance,* then *true heritability* or *"heritability in the narrow sense"* (h^2) may be represented with the following formula:

$$h^2 = \frac{V_A}{V_T}$$

The difference between V_A and V_G is simply that V_A, additive variance, is but a part of V_G, the total genetic variance. Part of the total genetic variance is attributable to dominant genes suppressing their recessive partner alleles and part is due to epistasis, the interaction of genes from different loci. Because such genetic variance is not transferred directly to the next generation, it is removed statistically to get additive genetic variance and true heritability. The value of h^2 must always be equal to or lower than CGD, because V_A is only part of the total genetic variance, V_G. "The heritability of a character is one of the most important population parameters that may be estimated by the methods of quantitative genetics" (DeFries & Hegmann, 1970, p. 27).

Calculating Heritability

There are many ways in which to calculate heritability or CGD. Estimates can be obtained in humans from twin studies, parent-offspring correlations, or sibling correlations. In studies with nonhuman species, it is common to use selection studies, Mendelian crosses, or diallel crosses (see below) to obtain estimates of h^2 or CGD.

Fuller (1960) provided an example of the calculation of CGD from Mendelian cross data from an experiment on the exploratory activity of inbred mice placed in a maze. The data are presented in Table 7-2. Means and variances for the exploratory activity scores for two inbred strains, BR and A,

Table 7-2 Exploratory Activity Scores in C57BR, A/Jax Mice, and Their Hybrids

Strain of mouse	N	Mean score	Variance (V)
C57BR	40	22.9	9.74
A	40	1.0	16.48
F_1 (C57BRxA)	38	17.0	12.23
F_2	92	16.1	29.60

Source: Fuller, 1960.

and their F_1 and F_2 crosses are presented. In order to estimate CGD, we need estimates of V_G and V_T. The F_2 generation contains all possible genotypes and thus can provide our estimate of V_T. Because all individuals within each of the two inbred strains and within the F_1 are genetically identical (all F_1 animals are heterozygous at all loci at which the inbred strains differ), there should be no genetic variance within those groups. Each score should be an estimate of V_E. Our best estimate of V_E would be the mean of the three variances:

$$V_E = \frac{V_{P_1} + V_{P_2} + V_{F_1}}{3} = \frac{9.74 + 16.48 + 12.23}{3} = 12.82$$

The genetic variance can be obtained by subtracting V_E from V_T ($V_G = V_T - V_E$; $29.60 - 12.82 = 16.78$). Then:

$$CGD = \frac{V_G}{V_T} = \frac{16.78}{29.60} = 0.57$$

We thus estimate that a little over half the variance in activity scores in these groups is attributable to genetic factors. Similar values were obtained in an analysis of C57BL and BALB/c mice by DeFries and Hegmann (1970).

While this method provides a reasonable approximation of CGD, it is often inapplicable in behavior-genetic studies. This is because, the F_2 variance often fails to exceed the variance of the F_1 or parental strains, a phenomenon referred to as the "Tryon effect" (Hirsch, 1967; DeFries & Hegmann, 1970). This is probably because behavioral traits are generally affected by a large number of genes.

Estimates of heritability or CGD have been made using a wide array of strains, methods, and behavioral phenotypes. Fuller and Wimer (1973) summarized some such measures, which ranged from an estimate of 0.22 for aggressiveness in chickens (estimated using the selection method) to 0.89 for a measure of sexual behavior in guinea pigs (obtained from a cross of inbred strains).

Interpretation of Heritability and CGD

The primary value of the concept of heritability lies in predicting the response to artificial selection. Given the heritability and the selection differential (a measure of how selectively one chooses breeding pairs from the base population), one can predict the response to selection that should be apparent in subsequent generations. Alternatively, by determining the selection differential and measuring the response to selection, one can determine the heritability. For example, in selecting for open-field activity, DeFries, Wilson, and McClearn (1970) obtained a "realized" heritability of just 0.31.

Heritabilities (and CGDs) are often used to compare the relative proportion of genetic variance in the determination of different traits. While this usage is generally reasonable, it is fraught with potential difficulties. First, heritability estimates will change from population to population. Heritabilities apply only to the population in which they were measured. If one moves to work with a different set of inbred strains, for example, the heritability may change markedly. A trait that has a high heritability within a genetically heterogeneous group should have a heritability approaching zero within a single inbred strain.

Further, heritability will change as a function of the environment. If a group of BR and A mice were tested for maze exploration after being reared in an environment quite different from that of the animals whose data are presented in Table 7-2, the resulting heritability would probably be quite different. Henderson (1970a) compared performance in a food-seeking task of inbred mice reared in standard and enriched environments. He obtained a heritability of 0.04 for mice reared in standard laboratory cages and of 0.37 for mice reared in enriched environments. An experimenter rearing mice under standard conditions might conclude that there was little genetic variance, a conclusion that would differ sharply from that of an experimenter who had reared the mice in a more complex environment. The complete picture emerges only when mice from both kinds of environments are studied.

Finally, the fact that one trait has a higher heritability than another, even when measured in a particular group reared in a particular environment, does not necessarily mean that the trait with the lower heritability is more easily modified. There are a number of reasons for this. For example, it may be that the environmental factors producing variance are more accessible for the trait with higher heritability than for the one with lower heritability.

Thus, while it is interesting to calculate and compare heritabilities, it is easy to misuse such estimates. For these reasons, contemporary behavior geneticists tend to place less emphasis on the question of the proportion of variability due to genetic factors and more emphasis on questions of developmental and physiological mechanisms, adaptive significance, and gene-environment interactions.

WHAT DEVELOPMENTAL AND PHYSIOLOGICAL PROCESSES INTERVENE BETWEEN THE GENES AND THEIR EXPRESSION IN OVERT BEHAVIOR?

Genes do not affect behavior through magic. Rather, the pathways from genes to behavior represent extremely complex routes of biochemical, anatomical, and physiological activity. The search for an understanding of these routes is just beginning, but it represents one of the current, exciting challenges of behavior-genetic research. Gould (1974) has named this research area "molecular ethology."

Genes and Biochemistry

Biochemically, the gene may be regarded as a portion of a DNA molecule that codes for the formation of a particular polypeptide. DNA molecules of the cell nucleus convey specificity to RNA (ribonucleic acid). The RNA directs biochemical activity in the cell's cytoplasm that ultimately results in the formation of structural proteins and functional enzymes (see Thiessen, 1972;

McClearn & DeFries, 1973). The enzymes are catalysts for biochemical reactions, whose rates of activity they alter. These chemical reactions affect the skeletal system; muscles; endocrine glands; the autonomic, peripheral, and central nervous systems; and indeed all systems of the body. It is clear that there must be a precise set of pathways leading from genes to enzymes to brain chemistry and behavior. Unfortunately, these pathways are not yet fully understood. The latter link in this chain will be discussed in Chapter 11.

Not all genes have the same action in the formation of RNA. Genes that function directly to produce RNA are called *structural genes*. According to the "Jacob-Monod model," there are two other important kinds of genes. *Operator genes* function by switching the structural genes on and off. *Regulator genes* act by increasing or decreasing the activity of operator genes. There thus exists a complex relationship of stimulatory and inhibitory effects of some genes on other genes. These activities are sensitive to biochemical factors in the organism. Thus, certain genes appear to be switched on or off at particular developmental stages or under particular conditions. With this conceptualization of gene action, the notion that genes and environment interact in a continuous, dynamic system becomes more than mere rhetoric or theory; it is the inevitable result of complex biochemical reactions.

Methods in Physiological Behavioral Genetics

We shall discuss four basic methods of studying the physiological-chemical routes from genes to behavior: correlation, selection, inbred strains, and genetic mosaics.

Correlation The correlational method rests in a search for a physiological correlate of some behavioral phenotype. A good example can be found in research on PKU. Phenylketonuria (PKU) is a form of mental retardation controlled by a single gene. An important event that led to the recognition of PKU as separate from other forms of retardation occurred when a Norwegian dentist with two retarded children noted that they exuded an odor that aggravated his asthmatic condition. Some chemical tests were performed and the children's urine samples were found to contain excess phenylpyruvic acid. The accumulation of this acid and other biochemical and behavioral effects are a consequence of the genetically determined absence of a single enzyme, phenylalanine hydroxylase.

It is now common to screen newborn children for PKU. If the condition is discovered early, a low-phenylalanine diet appears to minimize its effects (see McClearn & DeFries, 1973). The biochemical basis of PKU was discovered as a direct result of a search for a correlate of retardation in affected individuals.

Selection In conducting an experiment in artificial selection, one selectively breeds animals of particular phenotypes. However, one has little control over the pathway or pathways producing these phenotypes. For example, although one can select for efficiency in maze learning, intending to develop strains of "bright" and "dull" rats, one may in fact develop strains that are efficient or inefficient in maze learning because of "emotional" reactions to the test apparatus, responses to food deprivation, or some other factor (e.g., Searle, 1949). Fortunately, at the end of a selection experiment, the resulting animals are available for an analysis of the correlates of the phenotypic differences that have been created. These animals represent an important resource in the search for mechanisms mediating genetic effects on behavior.

Lagerspetz (1969) selectively bred mouse lines that were high and low in aggressive behavior. Subsequent analyses showed the two lines to differ with respect to testis weights, adrenaline content of the adrenals, and serotonin concentrations in the forebrain. Further work is required to determine which, if any, of these factors is related to aggressive behavior in a causative way. A complementary approach was taken by Collins (1970). She studied aggressive behavior in mice that had been selected for high and low brain weights and found the animals with the lower brain weights to be the more aggressive.

Inbred Strains Inbred strains of mice differ with respect to numerous biochemical, physiological, and anatomical phenotypes just as they differ on behavioral phenotypes. Discovery of such differences renders possible a whole series of research approaches designed to establish which, if any, of the correlations between strain differences in behavior and other variables bear causative relationships to one another. There are many examples of this approach. Wimer, Wimer, and Roderick (1969) found nine inbred strains to differ with respect to various indexes of the relative volumes of different portions of the brain. Hegmann (1972) found inbred lines to differ with respect to the conduction velocity of caudal nerves. Both findings have led to further work designed to discover functional pathways.

Schlesinger, Boggan, and Freedman (1970) found differences in the concentrations of various neurotransmitters in the brains of mice of the DBA and C57BL strains, the two strains differing so sharply in audiogenic seizure resistance (Table 7-1). The likelihood that this correlational relationship is also a causative one is increased by findings regarding developmental changes in both neurotransmitter concentrations and seizure resistance and by results of experiments in which drugs were administered to experimentally alter levels of these neurotransmitters.

Genetic Mosaics An important and ambitious research program is being conducted by Benzer and his associates (e.g., Benzer, 1973) in which they are attempting a "genetic dissection" of the behavioral patterns of *Drosophila*. By exposing flies to mutagens—radiation or chemicals that increase mutation rates—a considerable number of mutant phenotypes has been created. One mutation, "stuck," results in a difficulty for a male in disengaging after a normal 20-minute copulation. Another mutant, "coitus interruptus," causes males to disengage in half the normal time and to leave no offspring. Through sophisticated genetic techniques, Benzer and his associates have been able to go one step further and create animals that are genetic mosaics; they are composite individuals in which some tissues are mutant and some have a normal genotype. The scientists then use a number of mathematical techniques to determine exactly which part of the animal has to be mutant in order to produce the mutant behavioral pattern. In some cases, the relevant structure may be quite distant from the effectors producing the overt behavior. One mutation that occurred is "wings up," in which flies raise their wings straight up and thus cannot fly. The animals behave normally in other respects. It is possible that the "wings up" trait might be due to a defect in the wing itself, an alteration in the muscles, or some change in the cells in the central nervous system that control wing movements. The results of Benzer's analysis suggested that the location of the mechanism underlying "wings up" was in the fly's thoracic muscle tissue. Such methods are being applied to the study of sexual behavior and learning.

WHAT CAN BEHAVIOR-GENETIC ANALYSIS TELL US ABOUT ADAPTIVE SIGNIFICANCE?

In asking questions about the adaptive significance of traits in populations, we are attempting both to determine which traits might be important to fitness and to assess the ways in which they contribute to fitness. While the method is not infallible, some indication of probable adaptive significance can be obtained through consideration of the complementary phenomena of *inbreeding depression* and *heterosis* (Falconer, 1960; Bruell, 1964, 1967). Inbreeding is the mating together of individuals that are related through common ancestry. As inbreeding occurs—as, for example, in the repeated brother-sister matings performed in the development of inbred strains—the viability of the offspring typically decreases. Inbreeding depression may be defined as the reduction in mean phenotypic value shown by characters connected with reproductive capacity, physiological efficiency, or normal behavior (adapted from Falconer, 1960). Heterosis or "hybrid vigor" is the complement of inbreeding depression. When inbred lines are crossed, the offspring show an increase in those traits that previously had suffered a reduction from inbreeding depression.

It appears that the only traits to show substantial effects of inbreeding depression and heterosis are those that contribute substantially to fitness. While inferences about fitness can be made from studies of inbreeding, such a procedure is rather tedious. Most studies aimed at making inferences about adaptive significance using behavior-genetic data rely on crosses among inbred strains. The essence of the procedure is to cross two inbred strains and compare the scores of the F_1 generation with the scores of the two parental strains. For some traits and crosses, the F_1 values will fall near the *midparent value*, that value midway between the mean scores of the two parental strains. We refer to this as *intermediate inheritance*. In other cases, the F_1 will lie significantly closer to one parental value than the other, or show *directional dominance* (e.g., Table 7-2). In some cases, the F_1 will show true hybrid vigor or heterosis and outscore both parents. Where there is intermediate inheritance, it is likely that either selection has not acted on the trait or there has been stabilizing selection (selection favoring intermediate values). Where there is directional dominance, it is likely that natural selection has acted to produce individuals that score in the same direction as to the F_1s relative to the midparent score. Thus, if one crosses inbred strains and finds the F_1s to depart from intermediate inheritance, it is likely that that trait is related to fitness and that scores in the direction of the F_1 were adaptive in the natural habitat.

In fact, it is risky to make inferences from a single cross of two inbred strains, as such a cross may not be a representative one. For that reason, the *diallel-cross* design is typically used. A diallel cross is simply a design in which one produces all possible F_1 crosses among three or more inbred strains. An example appears in Table 7-3. The genotype of the mother appears along the side of the table and that of the father along the top. The scores of the inbred animals thus appear along a center diagonal. The scores off the diagonal represent F_1 crosses. When all or most of the F_1s depart significantly from intermediate inheritance, one can infer that the trait is relevant to fitness. In Table 7-3, it will be noted that in all 12 crosses, the score of the F_1 is below that of both parents. It is therefore likely that this trait contributes to fitness in rats and that low scores are adaptive.

It should be noted that the diallel cross is an extremely powerful design from which one can estimate a wide range of genetic parameters including heritability, magnitude of maternal effects, directional dominance, and epistasis (Broadhurst, 1967).

Diallel-cross studies have been performed on a large number of traits. Traits such as alcohol preference and open-field defecation tend to show intermediate inheritance (see Bruell, 1967). From such studies we would infer that the following traits, which show directional dominance, are adaptive: high levels of wheel running and maze exploratory behavior in mice (see Bruell, 1967), fast mating speeds in *Drosophila* (Fulker, 1966), high brain weights in mice (Henderson, 1970b), low mount frequencies in rats (Dewsbury, 1975a), and efficient performance in most learning tasks (see Wahlsten, 1972).

Table 7-3 An Example of a Diallel-Cross Design—Data on the Mount Frequencies in the Second Series of the Second Test for Rats of Four Inbred and Twelve F_1 Genotypes

	Strain of father			
Strain of mother	**ACI**	**F344**	**LEW**	**WF**
ACI	5.2	1.6	2.3	2.8
F344	1.3	3.1	2.4	2.0
LEW	2.9	2.4	5.3	4.5
WF	2.9	1.7	5.0	6.4

Source: After Dewsbury, 1975.

We shall return to a consideration of the diallel cross and of the probable evolutionary explanation for the kinds of phenomena just discussed in Chapter 14.

HOW DO GENES AND THE ENVIRONMENT INTERACT TO PRODUCE BEHAVIOR?

Interactions between genes and environment have been mentioned several times already and will be mentioned again; they are important. The crux of this matter is that the same genes may produce very different effects in different environments and, conversely, the same experimental treatment may produce very different effects on different genotypes.

An excellent example of gene-environment interaction appears in a study of avoidance learning in mice by Bovet, Bovet-Nitti, and Oliverio (1969). Inbred mice were required to learn a two-way avoidance task (see Chapter 16). The mouse is placed in a chamber that is divided into two compartments by a wall with a door in it. The mouse must learn to go through the door from one compartment to the other within 5 seconds of the onset of a light lest it receive shock. Various inbred mouse strains learn such tasks, although—as we should expect—they differ in rate of acquisition.

The most interesting part of the study by Bovet and coworkers (1969) came when they manipulated the interval between trials. The problem of spaced versus massed practice is a classical one in experimental psychology. In general, it is believed that the learning of most tasks is more efficient when practice is spaced out rather than massed into one or a few sessions. Bovet and coworkers compared C3H and DBA mice in an avoidance learning task in which the overall intertrial intervals were varied. Each mouse received 250 trials. The intertrial interval was generally 30 seconds. For one group of mice from each strain, all trials were massed in a single 250-trial session (called the "0-minute interval condition"). For other groups of each strain, trials were given in blocks of 50, with intervals between blocks set at 5, 15, 30, 60, or 120 minutes or 24 hours for different groups. The remarkable results are presented in Figure 7-4. Mice of both strains learned the task. For the C3H mice, there was an orderly relationship between intertrial interval and rate of learning; mice learned better when trials were closely spaced (0 minute) than when they were widely spaced (24 hours). The data were just as orderly for the DBA mice. However, the functional relationship for DBA mice was in exactly the opposite direction from that for the C3Hs. Mice learned *more poorly* when trials were massed than when they were spaced. Thus, the same treatment produced diametrically opposite effects when given to mice of different genotypes.

The magnitude of strain differences in open-field activity is greatly influenced by the level of illumination of the open field during the tests (e.g., McClearn, 1960).

There is an important message for the animal behaviorist in data such as these. Much research on animal behavior consists of a search for *general laws* of behavior. Such laws would relate various factors that can affect behavior to

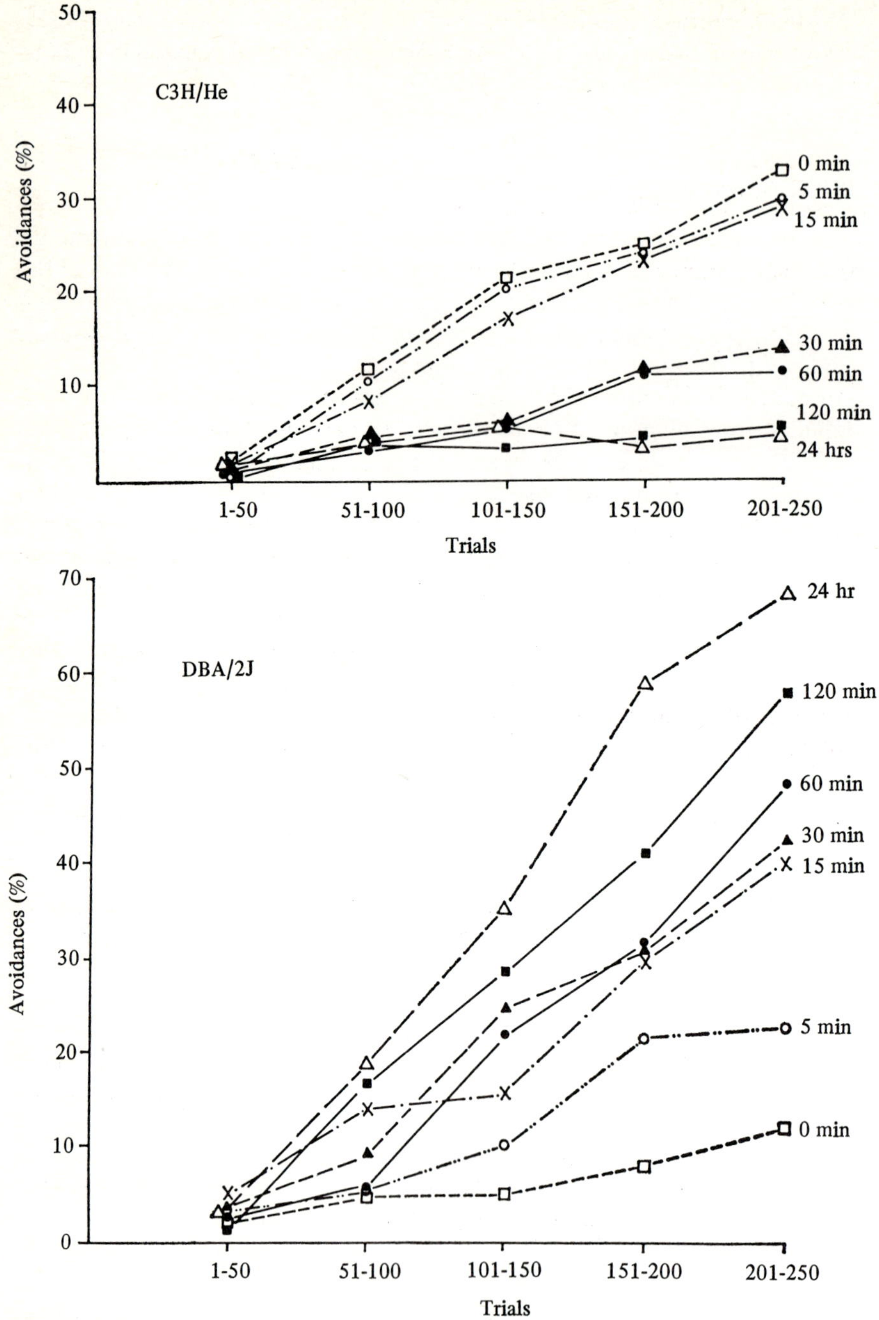

Figure 7-4 Comparison of the performance of C3H and DBA mice over 250 trials in an avoidance learning task when the length of the interval between blocks of trials is systematically varied. *(After Bovet, Bovet-Nitti, & Oliverio, 1969. Copyright 1969 by the American Association for the Advancement of Science.)*

their behavioral consequences in lawful ways. Such laws generally are written so as not to include the nature of the organism (e.g., "Spaced practice is more efficient than massed practice."). The important point is that for many factors influencing behavior, the nature of the organism must be an integral part of any behavioral "law." The same factor may produce very different effects on animals of different genotypes. The nature of the organism must be considered in the formulation of laws of behavior (Hirsch, 1963; Vale and Vale, 1969; Fuller and Wimer, 1973). The interactive effects of genotype and environment are particularly important in developmental studies, a point to which we shall return (Vale and Vale, 1969; Henderson, 1968a).

ROLE OF BEHAVIOR GENETICS IN ANIMAL BEHAVIOR

Behavior genetics can play an important role in the future development of the science of animal behavior. There are at least three ways in which this can occur. The first is through the impact of behavior-genetic research itself. There are many exciting phenomena whose genetic bases are under study and whose impact will be felt by the field of animal behavior as a whole. The second is in providing valuable experimental preparations. The behavior-geneticist can provide tools—such as genetic mosaics, strains with different brain weights, and techniques for analyzing developmental effects—that can be of great value to animal behaviorists. Finally, behavior-genetic research provides important cautions regarding the kinds of orderliness we can expect in formulating the laws of behavior.

SUMMARY

Evidence that genes affect behavior comes from a wide range of sources, including studies of strain differences, artificial selection, and interspecific hybrids.

By using Mendelian crosses for traits with discontinuous distributions or by capitalizing on pleiotropic effects of genes of known major function, effects of single genes can sometimes be detected.

Techniques are available that allow estimation of the proportion of total variance for a behavior in a particular population that is due to genetic or additive genetic factors.

The pathways from genes to behavior lead through a complex but potentially understandable route including biochemical, anatomical, and physiological steps.

Heterotic effects derived from studies using the diallel cross provide an indication of the probable adaptive significance of behavior.

Environmental treatments interact with genotype to produce gene-environment interactions, the understanding of which is critical to the formulation of laws of behavior.

Chapter 8

Experience and Behavioral Ontogeny

The development of behavior represents the product of the interaction of genotype and environment. Genetic factors were discussed in Chapter 7; we shall discuss experiential factors in this chapter.

A large number of contemporary animal behaviorists have adopted an epigenetic approach to behavioral development.

> An epigenetic approach holds that all response systems are synthesized during ontogeny and that this synthesis involves the integrative influence of both intraorganic processes and extrinsic stimulative conditions. It considers gene effects to be contingent on environmental conditions and regards the genotype as capable of entering into different classes of relationships depending on the prevailing environmental context. In the epigeneticist's view, the environment is not benignly supportive, but actively implicated in determining the very structure and organization of each response system (Moltz, 1965, p. 44).

Behavioral ontogeny, then, is conceived as resulting from a complex and ever-changing interaction between the organism and its environment. One difficult aspect of such a position is that it is sometimes difficult to determine where to begin. If one begins with a consideration of environmental influences, one must recognize that this ignores the ever-changing nature of the organism.

If one begins with the changing organization of the organism, one may be criticized for ignoring the environment and the fact that developmental patterns will necessarily differ in different environments. The fact is that if we are ever to understand these complex processes, we must simplify and abstract. However, in doing so it is important that we remember the simplifying assumptions we have made and that we later attempt to synthesize the various dynamic relationships into a complete developmental pattern.

Because it represents an ever-changing baseline upon which environmental factors act, we shall first consider behavioral ontogeny as it unfolds in several species within environments that are fairly typical for the species. We shall then proceed to the effects of specific environmental events.

Two additional qualifications are in order before we proceed. First, behavioral development represents a *continuous process* in which the changes over time may be both subtle and gradual. However, for purposes of convenience, most developmental behaviorists tend to divide behavioral development into discrete stages. It should be remembered that while these stages represent useful conceptualizations, they are abstractions from continuously changing patterns with transitions between stages that are often quite gradual. Second, it should be noted that we often think of developmental processes as preparing an organism for adult life. It must be remembered that in order to become a functional adult, the organism and its behavior must be viable at all stages; behavior in the neonate is just as important to overall fitness as that in the adult.

ONTOGENY OF BEHAVIOR

Included in the study of behavioral ontogeny are analyses of the changing nature of the organism before hatching or birth, at the time of hatching or birth, and throughout its entire developmental history.

Embryonic Behavioral Ontogeny

Domestic birds have been used as subjects in much of the research on the behavior of embryos. The reason is easy to see. One can easily cut a window in the shell of a developing egg and, if careful, observe directly the course of the development of behavior. Such a manipulation is much more difficult in a live-bearing mammal.

The course of the development of motor patterns in chick embryos has been well described (Hamburger, 1963, 1971; Oppenheim, 1974). Hamburger distinguishes three kinds of motility (movement):

Category A (spontaneous motility): Spontaneous motility consists of movements that result from the autonomous (self-governed) discharges of nervous tissue and which are capable of occurring in a steady-state, stable environment (e.g., in the absence of sensory input).

Category B (endogenously stimulated motility): Endogenously stimulated motility consists of movements that are triggered or caused by humoral agents (body fluids) or other changes within the internal environment.

Category C (evoked motility): Evoked motility consists of movement patterns which are initiated as a result of stimulation of the sensory systems from the external environment.

Spontaneous motility in the chick embryo begins at about 3 1/2 to 4 days after the beginning of incubation, with a slight bending of the head. Episodes of such movement occur at irregular intervals. The duration of bouts of spontaneous movements increases gradually over days as the interval between such bouts decreases. A peak of activity is reached at about day 13. Movements during this period involve the head, trunk, limbs, toes, eyeballs, eyelids, and beak and generally appear to be aimless, jerky, and convulsive.

Beginning at about day 17, a very different pattern appears, characterized by highly integrated movements. While these are somewhat less frequent than the earlier nonintegrated movements, they eventually orient the body into the hatching position and lead to pipping and the act of hatching at about day 20. Vocalizations are emitted by the embryo during the last few days before hatching.

More recently, a generally similar developmental pattern has been described for laboratory rats (Narayanan, Fox, & Hamburger, 1971).

Hamburger (1971) believes that "spontaneous activity takes precedence over stimulated activity in embryonic behavior development" (p. 63). He offers five lines of evidence to support this conclusion:

1 The motor system differentiates in advance of the sensory system.

2 Many embryos thus perform movements *before* they become sensitive to stimulation.

3 No influence of sensory input on embryonic spontaneous motility could be detected in the chick embryo up to day 17.

4 All embryos perform spontaneous movements from the very onset of motility, whereas only mammalian embryos are sensitive to stimulation from the outset.

5 The capacity for the autonomous electrical discharges that are presumed to underlie spontaneous motility seems to be a primitive property of all nerve tissue.

Although sensory input appears not to be necessary for the development of spontaneous motility, it remains possible that both sensory input and early spontaneous motility may be of importance to the development of later behavior (Gottlieb, 1971).

Little is known about the ontogeny of endogenously stimulated motility (category B). Evoked movements (category C) begin at about day 7 of incubation in chicks. The chick embryo responds to light tactile stimulation, delivered with a small brush, and to more vigorous stimulation as well. Such movements are first elicited only in the region about the mouth, but within 1 to

2 days the sensitive region spreads to encompass the entire body surface. Sensitivity to proprioceptive stimulation begins at about day 10.

The development of sensitivity in the sensory modalities of chicks proceeds in the order of tactile, vestibular, auditory, visual (Gottlieb, 1968, 1971). Remarkably, this same sequence of ontogeny in sensory systems appears to occur in a wide range of mammalian species as well. Gottlieb (1971) proposed that this developmental sequence may be common to all birds and mammals and may have remained unchanged from the earliest ancestral reptile groups or even earlier. Gottlieb (1968) concluded that, while deprivation or enrichment of sensory input to one modality would be unlikely to alter the order in which senses develop in chicks, such manipulations might affect the relative rate of development in different systems.

Peking duck embryos are able to respond selectively to the maternal call of their own species at about 5 days before hatching. This ability is not dependent on prior exposure to either their own vocalizations in the egg or those of their siblings. However, variations in the amount and timing of normally occurring auditory stimulation predictably accelerate or decelerate changes in the responses of the embryo to maternal calls (Gottlieb, 1971).

Behavior in Newly Hatched and Newborn Animals

Animals of different vertebrate species differ greatly with respect to the stage of development of their behavioral capacities at birth or hatching. The young of some species are *precocial* and able to engage in a wide range of behavioral patterns, including locomotion, soon after hatching or birth. Such a state is typical of many flock-living birds, such as ducks and chickens; flock-living ungulates, such as sheep, goats, and moose; and guinea pigs. The young of other species are *altricial*, or relatively helpless and requiring time for the development of locomotion and other behavior. The young of many songbirds, such as wrens and cardinals; colony-nesting marine birds, such as gulls; and many mammalian species including rats, mice, cats, and most primates are altricial.

There is a dramatic change in the behavior of the chick around the time of hatching (Balaban & Hill, 1969; Clemens, 1973). The newly hatched chick shows greatly increased vigilance and a dramatic increase in muscle tone. During the few minutes of hatching, the chick changes from a flaccid, spongelike embryo into an alert, active young organism.

Although most studies of behavioral development have concentrated upon vertebrates, it is well known that the newly emerged young of many invertebrate species are capable of relatively complex, integrated movements. For example, butterflies of the genus *Pieris* are capable of flying within less than 2 hours of emergence; "neither maturation nor practice is of any great importance for flight capacity" (Petersen, Lundgren, and Wilson, 1956, p. 338).

The two most characteristic behavioral patterns of newborn mammals are sleep and sucking. The newborns of many species spend two-thirds or more of their time asleep. Rates at which the young of different species suck for milk vary from species to species but tend to be consistent within species (Wolff, 1968).

Postnatal Behavioral Ontogeny

The early part of the life of many vertebrates is a period in which dramatic changes occur; typically, this period is characterized by the most dramatic changes in the lifetime of the organism. Three kinds of changes can be distinguished (Clemens, 1973). Some behavioral patterns are present at birth and disappear as the animal matures. Reflexes present in the newborn human include the grasp reflex, tight closing of the hand when stimulated by contact or pressure; the Babinski reflex, extending of the big toe and fanning out of the other toes when the sole is stimulated; and the rooting reflex, a turning of the head and forming of the mouth when the cheek is stimulated. These reflexes disappear during development, although they may reappear in old age, with organic brain damage.

A second kind of change is the appearance of new behavioral patterns. Many patterns—such as climbing, walking, and rearing on the hindquarters—may be absent at birth but appear at a later stage of development.

Finally, the relative frequency of performance of behavioral patterns present at birth changes with development. For example, while sleep and sucking are present at birth, their frequencies and durations of occurrence change markedly as the animal develops.

All ontogenetic changes in behavior can be viewed in relation to the developing nervous system. Although the greatest part of neural development occurs prior to hatching or parturition, many changes take place postnatally. These include changes in the total number of neurons in particular parts of the nervous system, in the number of dendrites, or branches of individual nerve cells, in the thickness of cerebral cortex, in the degree of myelinization (some nerves are covered with fatty myelin), and in electrophysiological and chemical activity (Fox, 1966; Bekoff & Fox, 1972).

The direction of neural development is generally caudal-rostral. By that we mean that the caudal, or lower brain structures, develop before the rostral, or higher regions. The disappearance of some of the reflexes present in the neonate appears to be a function of such development. Excitatory centers for such reflexes appear to be located caudally, with inhibitory centers located rostrally. Thus, a behavioral pattern may be displayed in the neonatal stage, when the higher inhibitory centers have yet to mature, but the behavior may be lost as these inhibitory influences mature. The newborn guinea pig shows a pattern of lordosis similar to that of the sexually mature, receptive female. The newborn displays this posture in response to tactile stimulation. This response appears to function as the mother licks the young, stimulating urination and defecation. Lordosis makes the relevant body parts more accessible. The response is lost after several days. Beach (1966) attributed the disappearance of this response to the delayed maturation of inhibitory centers in the higher portions of the brain.

The literature on developmental changes in behavior is considerable. The development of a few response systems will be summarized briefly.

Sleep Sleep is a dominant activity of the newborn animals of many species, but the amount and characteristics of sleep change dramatically with development. In newborn laboratory rats, for example, as can be seen in Figure 8-1, about two-thirds of the day is spent in sleep. However, virtually all the sleep is paradoxical (REM) sleep. Beginning at about the tenth or twelfth day after birth, the percent of time spent in paradoxical sleep decreases, while the time spent in slow-wave sleep increases until adult levels are attained (Jouvet-Mounier, Astic, & Lacote, 1969). Somewhat similar changes were observed in cats and guinea pigs, although developmental changes in the more precocious guinea pig were less substantial.

Reflexes and Locomotion The postnatal development of certain reflexes and various aspects of locomotor ability were studied by Altman and Sudar-

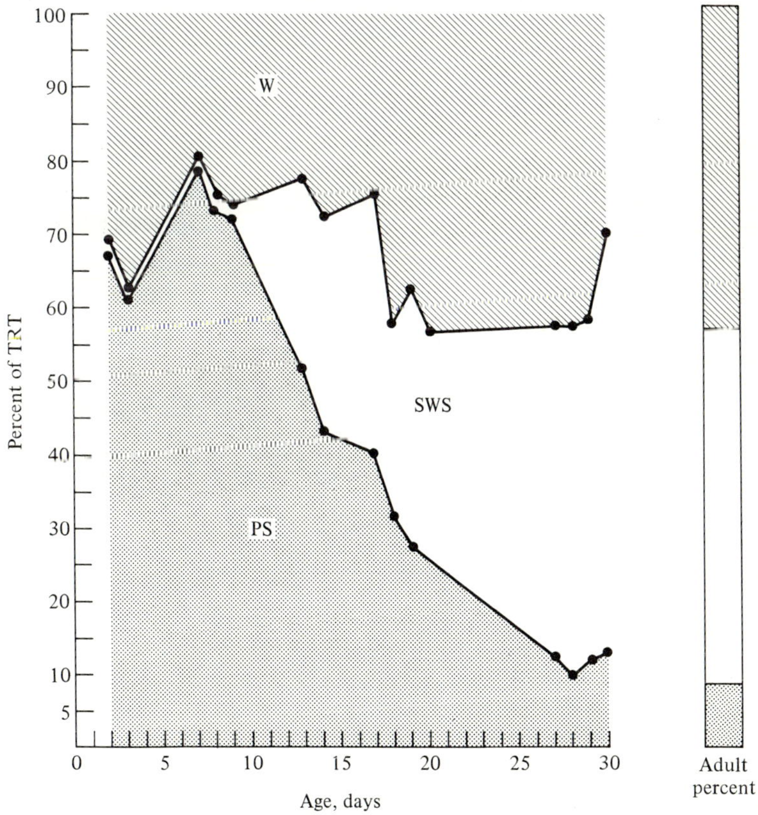

Figure 8-1 Changes in 8-hour recordings of mean percentages of wakefulness (W), slow-wave sleep (SWS), and paradoxical sleep (PS) as a function of age for young rats. *(After Jouvet-Mounier, Astic, & Lacote, 1969.)*

shan (1975) and are portrayed in Figure 8-2. It is apparent that different response systems mature at different rates. Placed on its back, the newborn rat displays a "righting" response at birth. However, other behavioral patterns only appear later (e.g., elevation of the head, righting in midair, and ascending a rope). The behavioral patterns of descending a rope and jumping across a cliff are not fully developed even at weaning.

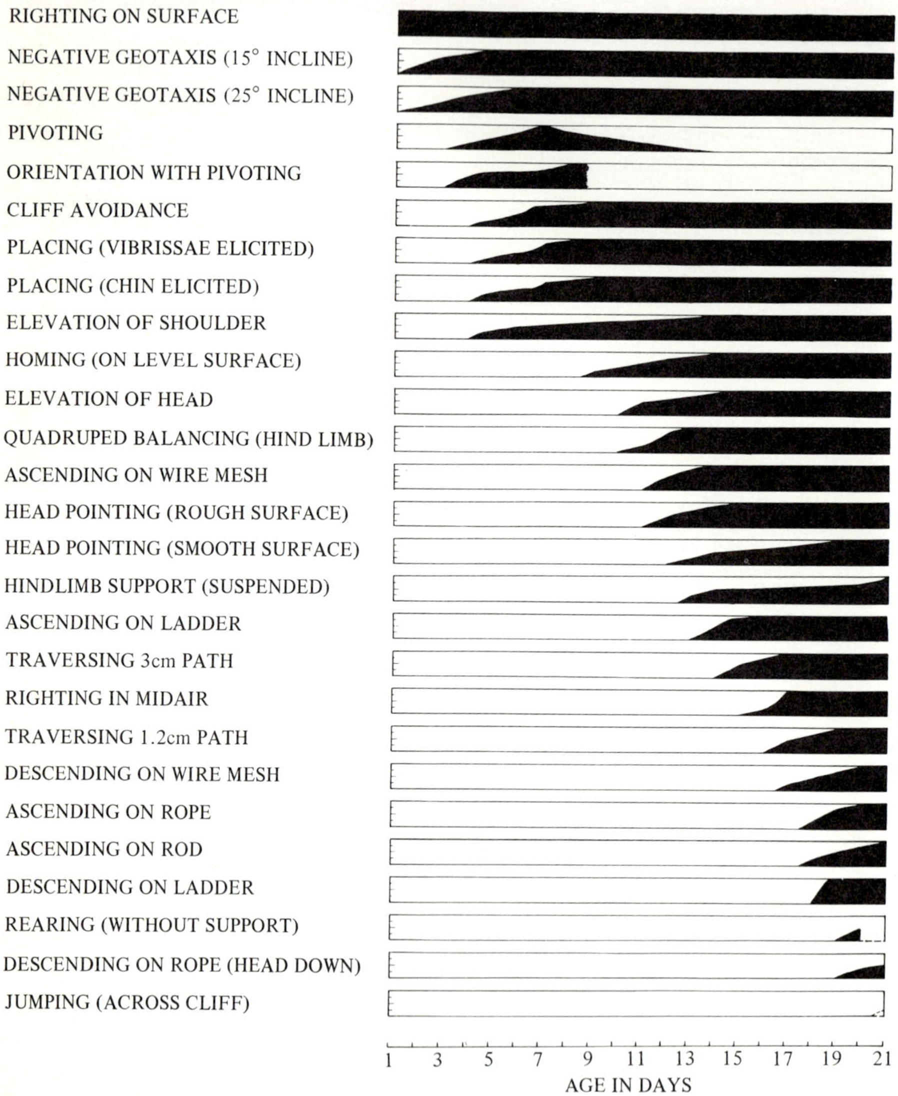

Figure 8-2 Ontogeny of locomotion and reflexes in young rats. Each diagram shows a point for 0, 25, 50, 75, and 100 percent. In the majority of instances, these numbers refer to the percentage of animals successful in the full display of the response. In some instances, reference is to the level of performance relative to asymptotic frequency. *(After Altman & Sudarshan, 1975.)*

On the basis of a similar study, Fox (1965) divided the development of reflexes and behavior in house mice into five stages:

1 *Perinatal (birth to 3 days):* Weak rooting, righting, geotaxis, and limb placement.

2 *Neonatal (3 to 9 days):* Strong and stereotyped reflexes, pivoting, and circular locomotion.

3 *Transition (9 to 15 days):* Primitive reflexes such as rooting disappear; adult locomotor activities begin; sensory systems develop rapidly.

4 *Prejuvenile (15 to 26 days):* Refinement of locomotor abilities and sensory capacities.

5 *Juvenile (26 days onward):* Further development of sensory and motor responses.

The transitional period (9 to 15 days) is the period of greatest change in house mice.

Activity Measures Campbell and his associates have found an interesting developmental pattern for different measures related to activity and arousal. Campbell and Mabry (1972) studied activity levels in stabilimeter cages (cages under which microswitches are mounted so as to be activated every time the animal moves from one end of the cage to the other). Developmental changes are portrayed in Figure 8-3. Guinea pigs, a precocial species, showed little change over time. By contrast, hamsters showed a sharp peak of activity between 10 and 20 days, followed by a decline to adult levels. Similar results were obtained with rats (Moorcroft, Lytle, and Campbell, 1971). The authors

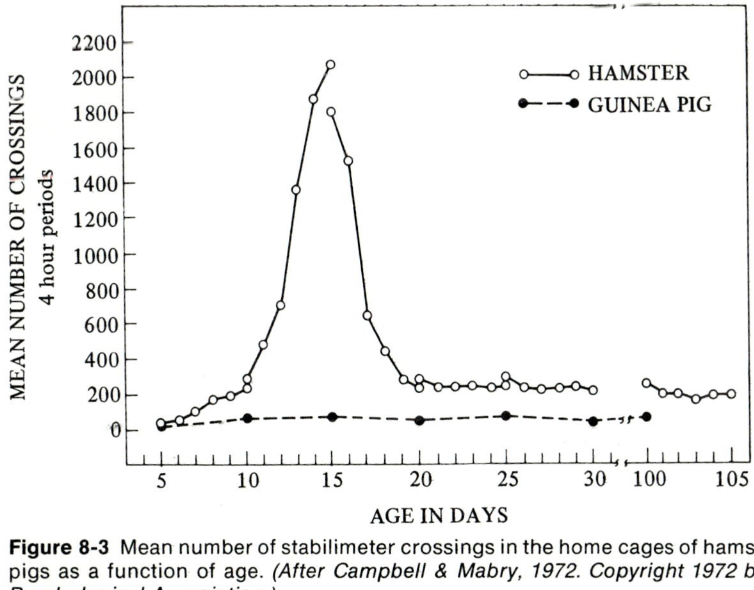

Figure 8-3 Mean number of stabilimeter crossings in the home cages of hamsters and guinea pigs as a function of age. *(After Campbell & Mabry, 1972. Copyright 1972 by the American Psychological Association.)*

originally attributed these peaks to the pattern of brain development in which the excitatory system of the brainstem matures before inhibitory centers located in the forebrain (Campbell and Mabry, 1972). As the peak in activity occurs only when rats are isolated from their mothers and siblings, it is possible that "isolation distress" is an important factor in producing this result (Randall & Campbell, 1976).

Candland and Campbell (1962) studied activity and defecation in the open field in rats from age 20 days to adulthood. Both the number of squares entered and the amount of defecation showed sharp increases between days 30 and 50 and remained at these levels through adulthood. Home-cage defecation and activity did not change. Results were interpreted as representing the development of emotionality or fearfulness.

Social Behavior With all this ontogenetic change in individual behavioral patterns, it is obvious that there must be dramatic changes in social interactions. The most widely studied species in this regard are probably the canids. Scott and Fuller (1965), for example, conducted a long-term project on the development of social behavior in dogs. Bekoff (1974a, b) analyzed the development of social play in a variety of canid species. Scott (e.g., 1973) divides the development of behavior in dogs into four periods:

1 *Neonatal (birth to 2 weeks):* Establishment of neonatal nutrition
2 *Transition (2 to 3 weeks):* Transition from neonatal to adult capacities for sensory perception, locomotion, learning, and social behavior
3 *Socialization (3 to 12 weeks):* Formation of primary social relationships and site attachment
4 *Juvenile (12 weeks to 6 months or older):* Development of motor skills

The socialization period (3 to 12 weeks) is a particularly important time for the puppy, as it is during this period that it typically forms associations with both other organisms and new places. If it is to be introduced into a new home or a conspecific group, this is the ideal time.

EFFECTS OF ENVIRONMENTAL EVENTS ON BEHAVIORAL ONTOGENY

The effects of environmental events acting at particular times in development can be of profound significance for later behavior. Many kinds of environmental events are of importance at different times and in different species. The phenomenon of imprinting may represent the archetype of an experiential effect. Much research also has been devoted to the role of early experience in the development of mammalian behavior. We shall discuss each in turn.

Imprinting

The behavioral patterns of birds in nature are not directed randomly at other organisms but rather are generally directed at biologically appropriate targets.

On excursions from the nest and throughout much of their early lives, the young of precocial bird species follow their mother. At sexual maturity, the birds of most species choose to mate only with members of their own species. Much interest has been directed at understanding the nature of the processes through which this selectivity in the direction of behavior comes about.

History Scholars have uncovered many early references to phenomena similar to imprinting (e.g., Kevan, 1976). The most substantial early report was that of Spalding (1873), who noted, among other things, that chicks would follow any moving object as soon as they were able to walk. Heinroth extended observations on related phenomena in a variety of avian species. Interest in imprinting received its greatest impetus from the work of Konrad Lorenz (1937a). Lorenz cited numerous examples of cases in which birds would not mate with animals of their own species as a result of having been deprived of contact with them during their early lives. For example, Lorenz kept a greylag goose isolated from other geese for the first week of its life and noted that the goose would follow human beings in preference to other birds. The bird would wait at the front door of Lorenz's house and attempt to follow humans as they emerged. Lorenz named the phenomenon *Prägung*, which means "stamping in" in German, but which has been translated to "imprinting."

Lorenz proposed that imprinting differed from true associative learning in four ways:

1 Susceptibility to imprinting was thought to be confined to a very limited period in the bird's life—the *critical period.*

2 Once accomplished, imprinting was thought to be *irreversible.* If a young bird was imprinted to a bird of another species, later contact with its own species would not eliminate the effects of the earlier experience.

3 Imprinting was thought to be unique in that the object toward which behavior was to be directed was determined well in advance of the first occurrence of the behavior itself. Thus, early imprinting, occurring well before sexual maturity, was shown to affect the selection of mates that would take place much later in life.

4 Lorenz viewed imprinting as a form of *"superindividual conditioning,"* in that behavioral patterns become directed not just toward the specific individual animal to which a young bird is exposed but toward the whole class of organisms of which the stimulus object is a member.

It should be noted at this point that many later researchers question whether some or all of these distinctions are in fact valid ones.

Since 1937, imprinting has proved a fascinating topic to both psychologists and zoologists, who have devoted many hours and printed pages to attempts to understand the phenomena. Because the methods and results differ somewhat, we shall discuss filial and sexual imprinting separately.

Filial Imprinting In referring to filial imprinting, we shall address the problems of the phenomena and underlying processes that determine the orientation of the following response of young, precocial birds. We are interested, to put it somewhat loosely, in how a young bird comes to know whom it is to follow around.

Methodology Most research on filial imprinting has been conducted in laboratories such as that of Eckhard H. Hess (e.g., Hess, 1959). Hess's basic apparatus is pictured in Figure 8-4. A young bird is hatched in the dark and kept isolated in a dark box inside an incubator until ready for testing. The bird is introduced through an automatic door into the apparatus, a circular runway 5 feet in diameter. In most experiments, a model of some sort—a male mallard as in Figure 8-4, a green box, or a red ball—is moving at a fixed speed on the runway. A speaker is embedded in some models, so that they appear to emit sounds. The bird is left in the apparatus for a fixed period of time—the imprinting period. If conditions are appropriate, the bird will follow the object about. The bird is removed from the apparatus by way of a trap door. Later, the bird is reintroduced to the apparatus for a period of testing. Typically, the animal is given a choice between a pair of stimulus objects, the one to which it was previously exposed and some other object. Usually, it shows a preference for the object to which it was exposed. Many kinds of manipulations can be made with this basic apparatus. We shall review the results of a few experiments.

Stimulus objects The young bird is not a *tabula rasa,* or "blank slate" in the sense used by John Locke. While a broad range of stimuli can be used successfully in imprinting experiments, some stimuli are much more effective than others (Bateson, 1971; Graves, 1973). Stimuli that are generally conspicu-

Figure 8-4 Apparatus for laboratory experiments on imprinting. A young bird is placed in a circular runway 5 feet in diameter and exposed to a model moving at a constant speed. *(After Hess, 1959.)*

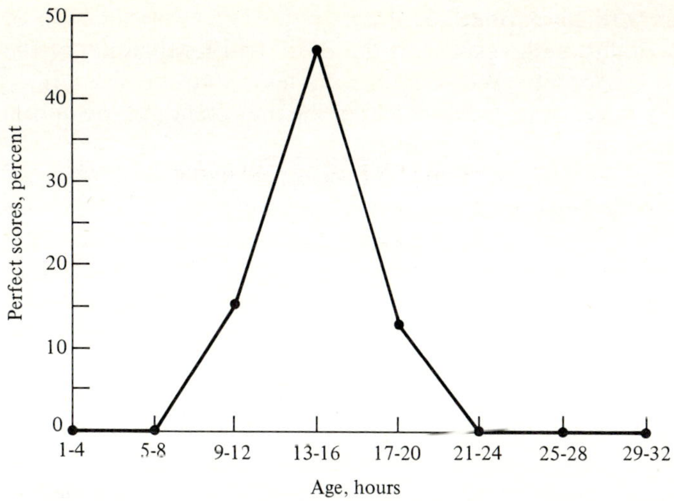

Figure 8-5 The sensitive period for imprinting. Percent of birds obtaining perfect scores in imprinting tests as a function of the age of the bird at the time of initial exposure. *(After Hess, 1959.)*

ous to humans appear most effective in imprinting. Klopfer (1967), for example, found much better imprinting when the training model was striped with a complex pattern than when it was painted white. Indeed, birds that had followed the white model during imprinting generally preferred the striped model during testing. Thus, although birds generally prefer the object to which they were imprinted, powerful visual stimuli can override the effects of early exposure.

Sensitive period The notion of a "critical period" for imprinting was fundamental to Lorenz's conceptualization. In general, experimental data have been consistent with the view that imprinting occurs most effectively during a restricted period in the development of the organism. Data from Hess (1959) are presented in Figure 8-5. Imprintability reaches its peak in ducklings at 13 to 16 hours of age. Imprinting exposure given too early or too late is ineffective.

By "critical period," scientists generally have meant that there is a particular time in development at which particular systems are most susceptible to modification. Presumably, relevant underlying mechanisms are maturing most rapidly during such periods. The concept of the critical period has generated much controversy. Virtually all workers agree that imprinting, for example, does not occur equally well at all stages of the organism's life. However, some workers contend that the changes in susceptibility are somewhat more gradual than one would expect from the concept of the critical period or the data in Figure 8-5. Further, it has been possible to modify the timing of the "critical period" in a number of behavioral systems. For example, rearing conditions can affect the timing of the susceptible period for imprinting (Bateson, 1971). For these reasons, many workers have come to prefer the term "sensitive period" to "critical period," as it carries less of an implication of suddenness and inflexibility with respect to the onset and termination of the period of maximal susceptibility. We shall use the less controversial term—"sensitive period." However, this does not lessen the importance of the notion of a restricted period of maximal imprintability.

Imprinted objects as reinforcers Imprinted objects have been shown to be effective reinforcers for young birds. Birds will learn arbitrary responses under conditions in which presentation of the imprinted object is contingent upon the response. For example, Hoffman, Searle, Toffey, and Kozma (1966) found that young chicks would peck a key when the response led to the presentation of the imprinted object, a plastic milk bottle.

Underlying processes Much attention has been paid to and much debate generated by the problem of the nature of the mechanisms underlying the imprinting phenomenon. Space limits prohibit a detailed discussion. Rajecki (1973) has reviewed six different viewpoints regarding the nature of the processes underlying imprinting: the contextual, ethological, conditioning, epigenetic, information-processing, and neuronal-model interpretations.

Reversibility The property of irreversibility, together with that of the sensitive period, was basic to Lorenz's conceptualization of the imprinting phenomenon. Because it is expensive to maintain large numbers of animals over long periods of time, few long-term experiments have been conducted on the reversibility of filial imprinting. The ease with which filial imprinting is reversed appears to vary with the nature of the stimulus object and the conditions of imprinting. Much of the filial imprinting studied in the laboratory may be quite reversible. For example, Hess (1972) exposed young ducklings to humans for a period of 20 continuous hours. The ducklings followed the humans everywhere. Yet the ducklings followed a live mallard mother after just 1 1/2 hours of exposure to her at a later time. "Clearly laboratory imprinting is reversible" (Hess, 1972, p. 25). By contrast, Hess was unable to reverse the imprinting of ducklings that had been exposed to their natural mother for 16 hours after hatching.

Species differences It is worth pointing out that different species may well respond differently in similar situations. While it may seem obvious that a domestic chick and mallard duck may show different lawful phenomena of imprinting, failure to consider such factors has contributed to some confusion in the imprinting literature.

Auditory factors It has long been known that combined auditory-visual stimuli are more effective in imprinting than visual stimuli alone. Gottlieb (1973) reported a series of experiments directed at the problem of the relative importance of auditory and visual cues in imprinting. He studied birds that were hatched and reared under conditions in which they had heard but not seen their siblings and had neither seen nor heard adults. Young ducklings of three taxa readily followed models that were characterized by the species-characteristic maternal appearance and maternal call. However, substitution of the visual component of one species for another produced little change in the following response. By contrast, altering the maternal call created substantial deficits in the following response. Birds that had never heard the maternal call responded selectively to the calls of their own species. When Peking ducklings had been following one of two similar maternal visual replicas and the calls were switched between replicas during a test, the ducklings either reversed their preference or simply stopped following the first replica. These observations suggest that during the early life of ducklings, auditory selectivity may be the primary determinant of the orientation of the following response. Young ducklings "come into the imprinting situation already 'tuned' to the distinctive audiometric characteristics of the maternal assembly call of their own species" (Gottlieb, 1973, p. 371).

Hess has shown that a virtual dialogue takes place between the mother mallard and the embryo as the time of hatching approaches. Hess further suggests that young birds may learn the particular auditory characteristics of *their* mother and thus be able to discriminate her call from that of others of the species.

Sexual Imprinting In dealing with "sexual imprinting," we refer to the process whereby the individual comes to direct its sexual behavior selectively at some individuals but not others. Loosely, the bird must determine with whom it should mate. Often it is assumed that sexual imprinting and filial imprinting result from the same process occurring at the same time. It appears that this may not always be the case. Filial imprinting and sexual imprinting are sometimes separable.

Methodology Methods in the study of sexual imprinting differ from those of studies of filial imprinting. Generally, exposure to the stimulus object is of much longer duration in studies of sexual imprinting. In studies of *intra* specific sexual imprinting, a bird is *reared* by a member of its own species but of different color from itself (e.g., white). Such studies have been conducted using species such as ducks, pigeons, and finches. In *inter*specific studies, birds are reared by members of another species. Such studies have been performed using various species of ducks, fowl, and sparrows. It will be noted that altricial as well as precocial birds can be used in studies of sexual imprinting. A special case of interspecific sexual imprinting includes studies in which humans have served as the imprinting object. Although some of the evidence is quite anecdotal, reports of sexual imprinting to humans resulting from hand-rearing have been reported in 25 species of birds (Immelmann, 1972).

In studies of sexual imprinting, mating preferences are tested in adulthood. While some studies have used hormones to accelerate sexual maturation, such procedures may distort normal developmental processes.

Stimuli Success of sexual imprinting varies greatly with the stimulus used. In some studies, for example, imprinting to humans appears somewhat weaker than that to heterospecifics somewhat closer to the imprinted bird in both appearance and behavior. There have been few, if any, tests of sexual imprinting directed at inanimate objects.

Species differences "The distribution of function between the innate perceptory patterns and the acquisition of imprinting is very different in different species" (Lorenz, 1937a, p. 268). Imprintability can also vary as a function of sex within species.

Sensitive periods Sensitive periods for sexual imprinting appear much less rigidly limited in time and often occur somewhat later in the animal's lifetime than the sensitive periods for filial imprinting. In most studies where strong sexual imprinting has been found, exposure has been relatively prolonged. Sexual imprinting may occur during a completely different sensitive period, later in life than that for filial imprinting. Immelmann (1972) estimates the sensitive period for sexual imprinting in greylag geese to begin at about 50 days of age and to end about 140 days of age; that for mallards appears to begin between 5 and 19 days of age and to last for several weeks. Klinghammer (1967) notes that sexual imprinting occurs relatively early during the nesting period in jackdaws, zebra finches, and hawfinches, but that the final choice of sexual partner occurs near the time of sexual maturity in bullfinches, crows, and ravens. In doves and pigeons, choice of sexual partner can be influenced long after the young are independent of their parents.

Reversibility It is a common finding that birds that have been imprinted to an alien species will mate with *either* their own or the alien species if tested with just one species at a time. Given a choice, however, they often display preferences that appear quite persistent. For example, Schein (1963) studied three male turkeys that had been reared in the absence of other turkeys for the first 32 days of their lives and were tested for mating preferences at 5 years of age. The animals would court humans when in the presence of humans and turkeys when in the presence of turkeys. Given a choice, the males would court humans. Although results vary with the species studied, the duration of contact, and the nature of the stimulus, sexual imprinting appears quite persistent.

Imprinting in Birds: A Viewpoint There is a reason for going into the imprinting literature in some detail. Imprinting is a very complicated process. There is a tendency for some authors to draw analogies between imprinting in birds and early learning phenomena in other species, particularly young humans. Often in such analogies, avian imprinting is conceptualized as occurring during a rigid critical period early in life, occurring instantaneously as a "stamping in," and as being completely irreversible. In fact, one sees dramatically different functional relationships regarding imprinting depending on the species, the stimuli, the kind of imprinting under study, and a host of other factors. One can demonstrate a complete gamut of phenomena, from a lack of preference for the imprinted object to a lifelong atypical mating preference. As with virtually any well-analyzed behavioral phenomenon, the determinants are complex. Facile generalizations to other species and other situations are inappropriate.

Imprinting in Mammals Phenomena very much like filial imprinting in birds have been demonstrated in certain precocial mammals. For example, Sluckin (1968) was able to demonstrate differential preferences for a black-and-white cube versus a tennis ball in young guinea pigs as a function of differential early exposure.

A different phenomenon, "maternal imprinting," has been studied by Klopfer (1971). Mother goats respond selectively to their own young in preference to the young of other females. Of 15 mothers that were deprived of their young immediately after birth, all rejected their young when tested 1 to 3 hours later. By contrast, 14 out of 15 mothers that had been permitted just 5 minutes of contact with their young accepted them in similar tests. Maternal imprinting appears to be quite a rapid process.

Studies of Early Experience in Mammals

The past quarter of a century has seen a veritable explosion of research on the consequences for later behavior of manipulations of the early experience of mammals. Spurred on by exciting data from studies of imprinting, by application of concepts from Freudian theory, and from the obvious potential importance of information on effects of early rearing for later behavior in humans, a regiment of experimental, comparative, and developmental psychologists has produced literally hundreds of research and review papers on the effects of early experience in mammals. These studies have used a great variety of different subjects and procedures. While it is impossible to provide a comprehensive survey of this research in a few pages, we shall attempt to convey the general flavor of the area. We shall discuss research on prenatal manipulations, early enrichment, and early deprivation.

Prenatal Manipulations Typically, in studies with prenatal manipulations, a pregnant female is exposed to some sort of stress. Persistent effects of prenatal maternal stress on the behavior of the offspring have been repeatedly observed. The most common hypothesis regarding the mechanism of such an effect is that the stressors create alterations in the hormonal environment of the female, and these alter the fetus in ways that are detectable later in its life.

Thompson (1957) exposed a group of females to paired presentations of a buzzer and shock. They learned to avoid the shock by running to the safe side of a shuttle box when the buzzer sounded. The females were then mated. During their pregnancies, the females were exposed to the buzzer, with no shock or opportunity for avoidance, three times per day. Offspring were cross-fostered to different females to control for postnatal effects. Prenatal maternal stress was found to produce decreases in activity and increases in the latency to activity in an open field, as well as longer latencies to leave the home cage and to reach food in a runway after 24 hours of food deprivation. Results were interpreted in terms of increased "emotionality" in the offspring of the stressed mothers.

As is true with many experimental manipulations, genes and environment interact in producing experimental effects. DeFries, Weir, and Hegmann (1967) found effects of strain of the mother, strain of the father, and strain by treatment interactions in a study of the effects of physical and chemical stress applied to pregnant mice on the open-field behavior of their offspring.

Ward (1972) found prenatal maternal stress to result in lowered levels of male copulatory behavior and higher levels of female-like lordotic responding in male rats.

Early Enrichment The vast majority of studies of early experience in mammals have been concentrated on the effects of early enrichment. The rodents in most animal laboratories are reared in relatively sterile metal cages. By "enrichment," we mean the introduction of some kind of environmental manipulation or event above and beyond the normal early environment of the laboratory. This literature is very complex. For almost any experiment in which a given effect is obtained, one can find another experiment in which a similar treatment appears to produce either no effect or the opposite effect. There are a number of reasons for this.

The major problem lies in the large number of variables that can affect later behavior and the considerable range of different parameters that have characterized different experiments in this area. King (1958) reviewed the early research on early enrichment and noted that age at the time of the experience, age at the time of testing, duration or quantity of the early experience, type or quality of the early experience, and type of performance task all affected the kind of result obtained. Different effects have often been found in different species, usually laboratory rats and house mice. Within species, we should not be surprised that there are many interactions between genotype and environment (Henderson, 1968a). What this means is that there is no simple way to make broad generalizations across all situations. We must consider all relevant factors. As was seen in the discussion of genetics and psychological laws in Chapter 7, the nature of the organism is an important part of all laws. These complexities might best be regarded as a fascinating research problem rather than as a nuisance (Henderson, 1968a). We shall consider a few of the major factors and results of research in this area.

Manipulations In many experiments, the early manipulation seems quite minimal. The experimenter may simply remove the pup from its mother and place it in a tin can for a few minutes a day for the first few days of life. In other experiments, the pup may be shocked, systematically handled or "gentled," or given some other treatment while away from its mother (e.g., x-irradiation or convulsions). A variety of stimuli can be delivered without removing the pup from its mother. Temperature stress has been the most often studied manipulation, although auditory and other stressors have been used. Postweaning, animals may be reared in a variety of environments of varying complexity.

Effects of early enrichment A wide range of effects of early enrichment have been reported. We shall concentrate on five of the most commonly reported phenomena: (1) accelerated growth and development, (2) reduced emotional reactivity, (3) improved performance in learning tasks, (4) more appropriate physiological responses to stress, and (5) altered neural development (see Daly, 1973).

Early handling, shock, and other treatments often have been reported to produce increases in the weights of young rodents at weaning (e.g., Denenberg and Karas, 1961; Levine and Otis, 1958). Many studies have failed to obtain such effects, however (see Daly, 1973).

Early handling or shock has often been reported to affect "emotionality." Changes in emotionality are often inferred on the basis of behavior in the open field. While interpretations involving the emotionality concept are very controversial, it is true that the bulk of the literature indicates that early environmental enrichment results in alterations of defecation in the open field, activity in the open field, or both (e.g., Denenberg and Smith, 1963; Denenberg, Morton, Kline, and Grota, 1962). Early shock and handling have been reported to produce similar effects. Both generally produce increased activity, decreased defecation, or both.

Early enrichment has repeatedly been shown to produce improved learning in a variety of tasks but especially in avoidance learning (e.g., Levine, 1956). The reason for such effects is not always clear. It is possible that effects of preweaning enrichment on learning are secondary to other alterations, as in emotion-related responses to the test apparatus.

Several studies have demonstrated that rats given early handling appear to show more appropriate responses to stress than unhandled controls. Thus, they respond to novel environments with a more moderate output of steroid hormones from the adrenal glands, but physically threatening situations evoke a greater adrenal response in them than in unhandled controls (Levine and Mullins, 1966).

Manipulations of early experience can alter brain chemistry and anatomy. Tapp and Markowitz (1963) compared rats that were stimulated to others that were not stimulated during the first 10 days of life. Handled animals had higher subcortical brain weights and lower subcortical acetylcholinesterase levels. Rats reared in a complex postnatal environment showed a number of changes in the brain, including increased weight of the cerebral cortex, increased acetylcholinesterase activity, increased cortical depth, and increased branching of the dendrites of nerve cells (Rosenzweig, Krech, Bennett, & Diamond, 1968; Greenough, 1975).

Effects of the time of manipulation Virtually all workers in the area of early experience agree that the age of the animal at the time of manipulation is of great importance to the result obtained. Many of the effects of handling during the first few days of life are not obtained when older animals are handled in an identical manner. Prenatal and postnatal enrichment can produce different effects on brain chemistry and anatomy as well as on behavior. While it is clear that age is an important variable and the first 5 days of life constitute a period of particular sensitivity to many manipulations in rats and mice, some effects are obtained with stimulation given at later times. Thus, Denenberg (1968) concludes that there is no truly critical period for early enrichment. The period of maximal sensitivity can only be determined in relation to the quality and quantity of stimulation, dependent variables, and related factors.

Mechanisms underlying enrichment effects Numerous proposals have been made regarding the mechanisms underlying observed effects of early enrichment (Russell, 1971; Daly, 1973). Three hypotheses will be considered: (1) temperature changes, (2) maternal behavior, and (3) early stress.

According to the hypothermia (low-temperature) hypothesis as espoused by Schaefer (1968), the effects of early experience arise as a result of temperature changes occurring during such experience. When the animal is removed from the nest for handling, its body temperature changes. This temperature change, rather than more direct effects on the nervous system, may be critical to later effects. Effects similar to those of handling have been produced by cold stress without handling (see Schaefer, 1968). Schaefer argued that the neonatal nervous system is not sufficiently developed to mediate a more direct effect of early stimulation and that therefore a temperature-mediated mechanism appears more plausible.

A second explanation is that effects of early enrichment are mediated indirectly through alterations in the behavior of the mother. Handling of pups results in altered patterns of maternal behavior, including increases in licking of the pups (Priestnall, 1973; Lee and Williams, 1974). Animals that receive a variety of treatments emit ultrasonic cries at increased rates, and these function further to increase maternal attention. Thus, alterations in behavior may be more a function of changes in attention by the mother than of direct effects on the pups (Richards, 1966; Barnett & Walker, 1974; Thoman & Levine, 1970).

A third view of the effects of early enrichment stems from observations related to stress. It was noted earlier that handled animals appear to respond more "appropriately" to both mild and severe stress than unhandled control animals. The regulation of hormone levels appears to be under the control of a negative feedback system in the hypothalamus and pituitary gland (see Chapter 12). As hormone levels increase, the activity of the control centers for hormonal secretion decreases. This "negative feedback system," which resembles the thermostat that probably regulates temperature in your house, has been referred to as a "hormonostat" (Levine & Mullins, 1966). It is postulated that, as a result of increased adrenal activity resulting from early stimulation, certain characteristics of the hormonostat are altered so that it functions more efficiently in adulthood.

These three and other explanations of the effects of early enrichment are not mutually exclusive. All could be acting in some situations, with varying degrees of importance according to the situation.

Mother-mediated effects We have already seen that manipulation of the pups affects maternal behavior and that alterations in maternal behavior may be important mediators of many effects of early enrichment. A further question is that of possible alterations in the behavior of pups determined by the experiences their mothers had when they (the mothers) were pups. Denenberg and Whimbey (1963) showed that the offspring of mothers that had been handled in infancy differed from offspring of control mothers with respect to both body weight at weaning and open-field performance. This phenomenon was carried one step further by Denenberg and Rosenberg (1967), who showed that handling of female rats in infancy could affect the activity and weaning weight of their grandchildren. A number of handling and housing variables interacted to produce the total effect.

Interpretation There is general agreement that early enrichment has profound effects on adult behavior. Further, the nature of the variables affecting such modifications are not subject to major controversy. However, these variables appear to interact in different ways under different conditions to produce a bewildering array of specific results and to generate much controversy regarding interpretation.

A further caution is in order. It is common to regard the effects of early experience as "beneficial" and the laboratory environment as "impoverished." As Daly (1973) has pointed out, one must be careful when making such generalizations. There is no sure way to generalize from most laboratory tests to the factors that are critical to survival and reproduction in the natural environment. Thus, we should be cautious in making inferences about what is likely to be beneficial to the animal in nature. Similar caution is in order in evaluating the laboratory environment. Many rodents live in burrows. Because of the construction of the burrow, temperature fluctuations in the typical laboratory are likely to be much greater than in a burrow. In fact, it is likely that laboratory cages provide a *more* stimulating sensory environment for preweaned rodents than do burrows. Daly (1973) suggests that "laboratory conditions are *over*-stimulative for the very young" (p. 435).

Environmental Deprivations The obvious complement to studies of environmental enrichment are studies of environmental deprivation. Typically, young animals are reared under conditions in which they are isolated either from conspecifics or from particular stimuli. Many of these studies have been directed specifically at the ontogeny of perceptual systems and will be discussed in Chapter 10. Here, we shall emphasize effects on general emotional, social, and reproductive behavior.

Dogs Thompson and Heron (1954) found Scottish terriers reared in various restricted environments inferior in problem solving to dogs reared as pets. Melzack (1968) summarized results of several experiments in which dogs were reared in environments that drastically restricted sensory input. On removal from restriction, deprived dogs showed (1) extremely high levels of activity; (2) abnormal perception of visual, auditory, olfactory, and noxious stimuli; and (3) difficulty in inhibiting inappropriate responses. For example, dogs reared in restricted environments had difficulty in learning to inhibit the behavior of approaching a lighted match.

Fuller (1967) found similar effects of early isolation on adult behavior. He suggested three alternative explanations for these effects. The first depends on the idea that environmental stimulation is necessary for the development of normal organization of the nervous system. Abnormal behavior in this view is a result of the lack of sensory-induced maturation of the nervous system. The second view is that the nervous system is organized at birth but undergoes some kind of degeneration due to the lack of sensory input. As a third alternative, Fuller suggested that the sudden influx of sensory input that bombards the isolated animal when it first encounters a normal environment may be responsible for abnormal behavior. On this view, effects of rearing under restricted conditions are due not so much to the rearing conditions themselves as to the conditions under which the sudden transition from the impoverished to the normal environment occurs. Fuller found that just 10 minutes of contact with the normal environment every 2 weeks was sufficient to reverse many of the effects of deprivation. Further, special handling techniques and tranquilizing drugs administered at the time of emergence from deprivation counteracted the effects of deprivation.

Rhesus monkeys After the studies of imprinting, perhaps the best-publicized early experience studies have been those of Harry F. Harlow and his associates on rhesus monkeys (e.g., Harlow, 1962, 1971; Harlow, Harlow, and Suomi, 1971; Mason, 1960). Harlow and his associates have reared many rhesus monkeys under varying conditions of deprivation. Many were reared in complete isolation from conspecifics. Model surrogate "mothers," covered with terry cloth, were provided in some experiments. The surrogate mothers were effective for young monkeys as foci for many of the clinging and other "affectional" responses that would normally be directed at the mother. However, when the deprived monkeys matured, they were found to be grossly deficient in their reproductive behavior. Some monkeys not only refused to mate but showed exaggerated oral activities, self-clutching, and rocking movements early in life and "apathy and indifference to external stimulation" (Harlow, Harlow, & Suomi, 1971, p. 543). After much effort, a few females were successfully bred. However, these "motherless mothers" were totally ineffective, ignoring their infants, chewing on infants' feet and fingers, and crushing the infants' faces on the floor. If isolate monkeys were placed with socially normal age-mates, they were the recipients of severe aggression. However, contact with monkeys younger than themselves produced "therapeutic" effects on the isolates.

Rodents Studies of early environmental deprivation in rodents have been concentrated on reproductive behavior as a dependent variable. Although earlier studies suggested otherwise, it is now clear that rats and guinea pigs reared in isolation show deficits in copulatory behavior as adults (Valenstein, Riss, & Young, 1955; Gerall, Ward, & Gerall, 1967; Gruendel & Arnold, 1969). Typically, animals reared under deprived conditions display abnormal motor patterns, a lack of appropriately oriented mounting, and a general ineffectiveness in copulating. Although isolation-reared females showed no deficits in observable maternal behavioral patterns, their offspring showed higher mortality and delayed eye opening compared to the offspring of normally reared females (Thoman & Arnold, 1968).

SUMMARY

The ontogeny of behavior can be followed through the entire life-span of the organism. In embryonic development, the onset of spontaneous motility precedes the onset of evoked movements. Postnatal maturation is associated with continued and sweeping changes in a variety of behavioral patterns including sleep, reflexes, activity, and social behavior.

Filial imprinting can occur to a wide range of stimuli, and is most effective in a fairly restricted sensitive period. However, effects of filial imprinting appear to be both reversible and subject to the overriding influences of auditory preferences, which appear not to be learned. Sexual imprinting has a much broader and later sensitive period but is often less easily reversed.

Early environmental enrichment and deprivation both have profound effects on later behavior. Specific effects obtained in experiments with early environmental manipulation vary as a function of variables such as the age at the time of the experience, the age at the time of testing, the type of early experience, the quality of the early experience, and the nature of the dependent variable.

Chapter 9

Nature and Nurture

Having discussed both genetic and environmental factors in the development of behavior, we are now in a position to discuss one of the more persistent problems in the study of animal behavior—the "nature-nurture issue." In broad perspective, the nature-nurture issue is concerned with the relative role of genes and the environment in the development of behavior. More particularly, debate has focused about the development and evolution of those behavioral patterns that are highly stereotyped and species-characteristic. Can we ask the question, "Is this behavior learned or innate"? Can we say that behavioral patterns are innate?

One often reads that the nature-nurture issue is a "pseudoquestion," whose resolution is now obvious: All behavior is the product of both genes and environment. Surely most animal behaviorists would agree that both genes and environment interact in behavioral development. However, when one goes beyond this facile statement, disagreement still exists. Further, while many students "know" the resolution to the nature-nurture issue, many do not understand the issues underlying that "resolution." One need only observe popular reaction to books suggesting that some human behavior may have some genetic determination to become aware of the importance of the issue in the eyes of the general public. The fact is that the nature-nurture issue is very much with us and is likely to be so for some time.

The nature-nurture issue is a difficult one with which to deal for several reasons. First, the issues are often complex and esoteric. The mere substitution of one word for another in an otherwise innocuous sentence is enough to incur the emotional wrath of a large number of animal behaviorists. Second, the positions of different scientists have changed over time. Thus, if we speak of "Lorenz's view" of the issue, it is not clear whether we mean Lorenz's view as expressed in 1937 or in 1965. They differ in important ways. Fortunately, scientists' views of the nature-nurture issue appear not to be completely innate; they are modified as the result of experience. A third reason for the difficulty in dealing with this issue is that discussions of nature and nurture can often become deflected into matters of personality and "school" rather than remaining based on conceptual issues. We like to view the scientist as a dispassionate and objective observer. The fact is that scientists are living human beings whose feelings get hurt. An exchange of papers between Lehrman (1953, 1970) and Lorenz (1965) provides a case study. Finally, and most importantly, different scientists are trying to answer different questions. We noted in Chapter 1 that much confusion can be caused when animal behaviorists interested in one class of question (e.g., immediate causation) criticize others for not providing answers to their question, when indeed the other scientist is trying to answer a very different class of question (e.g., evolution). Nowhere is this problem more apparent than in dealing with nature and nurture (Lehrman, 1970). Individuals concerned primarily with the amazing adaptedness of behavior and with the intricate developmental patterns of behavior have often appeared to disagree on the nature-nurture issue, when the real difference between them was that they were trying to do different things in order to answer very different questions.

We shall first discuss the nature of the issues involved in the nature-nurture question in a historical perspective. Such a perspective is essential to understanding the roots of the disagreements and the nature of the partial resolution that has taken place. We shall then discuss the development of some behavioral patterns that are particularly illustrative of the actions and interactions of nature and nurture in behavioral development.

EARLY HISTORY

> My task is to give an account of a Four Hundred Years' War which is still not concluded. . . . Its central issue has been man's status as a demigod, enjoying closer kinship to spiritual than to corporeal creatures. . . . But much of the action has taken place on two flanks: on the left, the contested position was that instincts do not exist at all, not even in animals; on the right, that many specifically human characteristics have an instinctive basis (Diamond, 1974, p. 237–238).

This is the way in which Diamond opened a discussion of the history of the controversy over instinct.

As was pointed out in Chapter 1, the origins of the concept of instinct can be traced back well beyond 400 years. These origins lay in an attempt to contrast the behavior of humans with that of the "irrational brutes." The prescientific controversy over instinct was centered on the issue of the human constitution in relation to other species and the rest of the universe. Prescientific usage has been reviewed by Warden (1927), Beach (1955), and Diamond (1974). According to the prescientific concept, instincts were complex, unlearned, adaptive behavioral patterns that functioned quite teleologically (see Ghiselin, 1973). The hand of God often was invoked.

Charles Darwin took the concept of instinct from philosophy and tried to strip it of its teleological character in order to deal with the evolution of behavior. Darwin was followed by a group of behaviorists—including Herbert Spencer, C. Lloyd Morgan, and William McDougall—whose conceptions of instinct represented a hybrid between the older philosophical view and the more natural-historical view of Darwin. McDougall included characteristics such as parental feeling and self-abasement in his list of instincts. These views, in turn, led to an anti-instinct revolt within psychology, with prominent behaviorists contending that there were no instincts in any animal. The war was truly active on both flanks.

CLASSICAL ETHOLOGICAL VIEWS

The classical ethologists sought to accomplish the same goal as Darwin, that of stripping the instinct concept of much of its "excess baggage" and developing it as a viable construct in the biological study of the evolution of behavior. In order to accomplish this, it was necessary to divorce the natural-science concept of instinct from that associated with Spencer, Morgan, and McDougall. Indeed, Lorenz (1937b) devoted many pages to protecting this flank. Lorenz sought to develop a concept of instinct that would be biologically meaningful and of explanatory value in the study of behavioral evolution.

The Classical Ethological View of Instinctive Behavior

For Lorenz, the central feature of species-characteristic behavior was its adaptedness. Yet the individual organism developed complex behavioral patterns in the apparent absence of learning. How could one account for the development of adapted behavior without invoking teleological notions of the sort that would give the animal credit for understanding and anticipating features of the environment that clearly were biologically incomprehensible? The solution proposed by Lorenz was his concept of instinctive behavior. A listing of some of the characteristics of the classical ethological view of instinctive behavior follows:

1 Instinctive behavioral patterns were genetically based as the result of natural selection. The reason that instinctive behavioral patterns could occur was that natural selection had operated so that the bases for the behavior have been encoded in the organism's genes.

2 There was an absolute dichotomy between instinctive behavioral patterns and those that are individually acquired. There was no gradation between instinctive and learned patterns; such patterns belonged to two completely separable categories.

3 There was an "instinct-conditioning intercalation." Lorenz believed that bits or units of behavior were either learned or innate. These units were intermixed and interlaced in much behavior, so that an instinctive element could follow a learned element and vice versa. Gross behavioral patterns represented the action of both the learned and the innate. However, if the gross patterns were analyzed into their constituent elements, both instinctive and conditioned elements would be found to be interlaced.

4 Instinctive elements were believed to be unmodifiable. Experience was important in the expression of instinctive behavior. However, experience acted only to determine *which* response would occur and what its intensity would be. "The result is never a *novel* behaviour pattern which has not been hereditarily fixed and predetermined in exactly the combination of movements found" (Lorenz, 1937b, p. 268).

5 While instincts were finely adapted to the environment, they occurred without any conscious purpose on the part of the behaving organism. Instincts were marvelously adapted responses but were performed "blind." One could see this when animals were tested in unusual environments and the instinctive behavior which was appropriate for the normal environment, yet inappropriate to the situation at hand, nevertheless was run off in typical fashion.

Most of the other characteristics of fixed action patterns, as discussed in Chapter 2, were incorporated into the notion of instinctive behavior. Such behavior patterns were independent of orienting stimuli, stereotyped, characteristic of all appropriate members of the species, completely functional on their first occurrence, centrally coordinated, and endogenous (see Konishi, 1966).

Evidence for the Classical View of Instinctive Behavior

The classical "view" of instinctive behavior really represented two views. Two different sets of evidence were used to support the views of two different sets of characteristics of instinctive behavior (Lehrman, 1970; Beer, 1973). The difference between the two views centers on the meaning of the word "innate." On the one hand, "innate" meant that behavior had a genetic basis. Natural selection has acted to create species of organisms in which *a distinct heritable component* enters into the development of behavior. The geneticist can detect this heritable component and analyze it, using the tools of behavioral genetics that were discussed in Chapter 7. On this view, nothing is said about the lack of environmental input, only that there is some heritable component. On the other hand, "innate" was taken to mean that behavior was *developmentally fixed* and independent of the actions of certain environmental factors. This is a much stronger view, and it implies a rigid pattern of development in the individual rather than simply a detectable heritability. While it is clear that Lorenz sometimes recognized this distinction between the treatment of instinctive behavior patterns "both in the phylogenetic and in the ontogenetic sense" (Lorenz, 1937b, p. 260), the two senses have often become confused in subsequent debate in what Beer (1973) has termed "the mischief of ambiguity." It is this confusion of questions about ontogenetic and phylogenetic characteristics that has resulted in some of the problems with the instinct concept.

Evidence for Instinctive Behavior (Phylogenetic Sense) A number of lines of evidence led classical ethologists to the view that phylogenetic processes had produced a genetic component to behavior:

Behavioral homologies Homologous behavioral patterns are behavioral patterns of several different species all of which were derived from a common ancestral species. An example comes from Lorenz's studies of the display patterns of ducks and geese. Lorenz (1941) found the motor patterns of different species of ducks and geese to be highly similar; differences increased as one examined species progressively more distantly related to one another according to standard taxonomical criteria. This would imply that the behavioral patterns were genetically based as the result of phylogenetic patterns of inheritance.

Adaptedness The very adaptedness of many instinctive patterns implied that they must be organized as the result of phylogenetically originated genetic foundations. A male salticid spider, for example, must emit the appropriate display pattern toward a female to inhibit her prey-catching behavior or else be eaten. It is critical that the behavior be complete and functional on its first occurrence. Clearly this behavior has to originate in phylogenetic adaptation.

Occurrence in inappropriate situations The occurrence of instinctive behavior in situations in which it was inappropriate implied that natural selection must have acted elsewhere to produce the behavior. For example, ground-nesting doves of the Galapagos Islands show a "distraction display" when a human approaches the nest. Yet there are no natural predators on these islands against which such a display would be effective (Eibl-Eibesfeldt, 1961). This implied that the behavior was functional in another geographic location and preserved on the Galapagos by heredity.

Selective breeding and cross-breeding The results of studies of selective breeding, considered by Darwin, together with those from studies of cross-breeding of different species provided abundant evidence of a genetic basis for behavior. Lorenz (1937b) cited work by Heinroth, who bred a shelduck with an Egyptian goose. The offspring showed a courtship pattern unlike that of either of its parents but similar to the "ancestral form" typical of other species of Anatidae. Other examples of cross-breeding were discussed in Chapter 7.

Evidence for Instinctive Behavior (Ontogenetic Sense) Several lines of evidence implied a lack of learning in the ontogeny of instinctive behavior patterns.

Appearance of behavior where learning is unlikely Some behavioral patterns were observed to develop in ways such that learning explanations seemed unlikely. For example, when provoked, young ducklings were found to display the same motor coordination of the wing and shoulder as that seen in the fighting responses of adult birds. However, these patterns were shown at a time when the young ducklings still had disproportionately small and entirely nonfunctional wings. If the wings were not yet functional, how could the duckling have learned functional motor patterns?

Many insects display functional behavioral patterns shortly after emergence. Again, learning is unlikely.

The isolation experiment In arguing for the independence of instinctive behavior from learning, ethologists have relied very heavily on an experimental design that is variously called the "isolation experiment," "deprivation experiment," or "Kasper Hauser experiment." Essentially, this is an environmental deprivation study of the sort discussed in Chapter 8. If the animal is raised in the absence of conspecifics and many environmental stimuli, how could it have learned the behavioral pattern? Grohmann raised domestic pigeons in very narrow, tube-shaped cases that prevented the birds from spreading their wings. When released, the development of normal flight was so fast as to render a learning explanation unlikely. Similar experiments have been conducted with a wide array of species and behavioral patterns.

CRITICISMS OF CLASSICAL ETHOLOGICAL VIEWS

It was some 15 years after Lorenz's formulations before the first really notable criticism of the classical ethological view of instinctive behavior appeared in print. The first salvo was released by a group of North American animal behaviorists who had been trained in psychology departments. These people had learned to be very skeptical of the concept of instinct. Their interests lay in the study of the control and development of behavior. Little attention was paid to the proposal that instinctive behavior had a genetic component that was rooted in phylogeny. Questions of evolution were not foremost in the minds of American psychologists, but questions of ontogeny were. However, the classical ethological view of the ontogeny of instinctive behavior was sharply criticized by American psychologists (e.g., Lehrman, 1953; Hebb, 1953; Beach, 1955; Schneirla, 1956; Jensen, 1961), and later by some "English-speaking" ethologists as well (e.g., Tinbergen, 1963a; Hinde, 1970). In general, these criticisms centered on two problems: the ethological conceptualization of environment and learning and the general implications of adopting the instinct-learning dichotomy.

Criticisms of Ethological Conceptualizations of Environment and Learning

1 Environmental influences on behavior can be much more diverse than implied by the learning-instinct dichotomy. Among the environmental factors that can influence behavior are maintenance procedures (e.g., feeding schedules), circadian rhythms, electrical brain stimulation, brain lesions, hormones, drugs, diet, and maturation (Jensen, 1961). To contrast just learning with instinct is to neglect the role of these and other factors.

2 The environment is necessary for all behavior. Critics contended that both the environment and heredity are necessary for all behavior. To ask how much of a behavior is dependent upon heredity and how much on the environment is akin to asking how much of the area of a field is due to length and how much to width. The area is fully dependent on both (Hebb, 1953). In its most obvious (and trivial) form, there would be no organisms without both genes and an environment.

3 Only genes and cytoplasm are inherited. On this view, behavior, and any other phenotype for that matter, cannot be said to be "inherited," because all that the organism inherits is a set of genes and cytoplasm.

4 All genetic effects must be mediated through their interaction with environmental factors. Thus, all genetic influence must be indirect.

5 The innate is defined in negative terms. In the classical ethological view, the innate was often defined in negative terms as that which is unlearned. Defining a major construct solely in negative terms, in a two-class system, and in relation to a concept as poorly understood as learning causes major problems in identifying behavior as innate. One must "accept the null hypothesis" that learning has not occurred, and this is a logically difficult process, especially when learning is poorly understood.

6 Deprivation experiments do not exclude the possibility that there are important environmental factors in the ontogeny of behavior. American psychologists were very much concerned about the interpretation given deprivation experiments. They pointed out that such experiments could be used to exclude the possibility of certain factors influencing behavioral ontogeny. However, many other factors remain uncontrolled. In interpreting a deprivation experiment, one must consider just what it is that the animal is deprived of; one does not by any means exclude the entire environment.

7 One can only say that *differences* between organisms are due to heredity or environment. Because all behavior is dependent upon both heredity and environment, it is impossible to say that a behavioral pattern is due to one or the other. It is meaningful, however, to compare two organisms and determine whether the *difference* between them is attributable to genetic or environmental factors. If two organisms of different heredity are reared in identical environments, *differences* in behavior between them must be regarded as of genetic origin. If two organisms of identical genotype are reared in different environments, behavioral *differences* are attributable to environmental factors.

Evaluation Several of these criticisms are rather minor and others can be either countered or sidestepped. The most pervasive of these arguments was that both genes and environment are necessary for the occurrence of all behavior. This argument has led to a reformulation of the concept of instinct on the part of most behaviorists. Few today would seriously propose that behavior is not jointly dependent on both heredity and environment.

Still at issue is the role played by factors other than genes and learning in behavioral development. Ethologists such as Lorenz still contend that most of the environmental factors discussed by Jensen (e.g., drugs, brain lesions, and diet) cannot impart the specificity of action to truly adaptive behavioral patterns. While brain lesions, drugs, and other related factors can interfere with behavioral development and can alter the rate and probability of occurrence of behavioral patterns, there are few examples of the creation of specifically adapted behavioral patterns resulting from such manipulations.

Criticisms of Ethological Conceptualizations Related to the General Implications of Adopting an Instinct-Learning Dichotomy

Critics of classical ethology pointed to several unfortunate implications of the adoption of a dichotomous view of nature and nurture.

1 Adoption of such a view can lead to premature categorization of behavior and a consequent shutting off of research activity. If we adopt a learned-instinctive dichotomy, it will be easy to classify a behavioral pattern as instinctive after a few deprivation experiments. Such premature classification may very well shut off research directed at understanding the steps involved in the ontogeny of the behavior. It was pointed out that the degree of assurance with which a behavior is considered as instinctive bears an inverse relationship to the amount of research attention paid the behavior. The more a behavior is studied, the less likely we are to classify it as instinctive. The implication again is that classifying the behavior as instinctive shuts off research. Classifying or naming a behavior does not explain it.

2 The development of "instinctive behavior" may be very different at different "phyletic levels." This argument, made by Schneirla (1956), stemmed from a general concern with the problem generalizing from one "phyletic level" to another. Schneirla believed that we can divide the animal kingdom into various "levels," and that the processes underlying behavior will differ at different levels. His fear was that if we use the same word to apply to behavior at these different levels, we may easily slip into believing that the same processes always underlie the behavior.

Evaluation Both these points provide valuable cautions when dealing with the problem of instinctive behavior. However, both might better be conceived as cautions to be remembered when using a concept such as that of instinctive behavior rather than as arguments against such usage.

RECENT FORMULATIONS

Since the time of the original classical ethological formulation and the first salvo of criticisms directed at it, much has been written regarding the controversy. The fine points of each argument have been dissected out for careful examination and positions have occasionally been modified on the basis of such feedback (e.g., Tinbergen, 1963a; Lorenz, 1965). Several more recent formulations or restatements will be considered.

The Heritability View

One approach to the heredity-environment problem is the application of the concept of heritability (e.g., Whalen, 1971). If both genes and environment influence behavior, it might be valuable to know the extent to which each influences a given behavioral pattern in a given population. Whalen points to several important advantages that accrue from using heritability as a solution to the nature-nurture problem. Unfortunately, there is a serious difficulty with this approach. Heritability estimates tell us only the percent of variance among individuals in a population that is attributable to genetic factors, not the relative proportion in which genes and environment act in the ontogeny of behavior (see Hebb, 1953, and Chapter 7). Thus, for example, if we study an inbred strain, heritability ought to be near zero, even though genetic factors may be just as important in the development of behavior in that strain as in an outbred strain.

An Information Approach

For Lorenz the central problem of instinct theory lay in understanding the adaptedness of specific behavioral patterns that are displayed in particular situations. As noted earlier, Lorenz believes that the adaptive specificity characterizing behavior can arise only from the genes or through learning. In his book *Evolution and Modification of Behavior*, Lorenz (1965) adopts an information approach to the nature-nurture issue. He contends that the only sources of "information" for the guidance of behavior are phylogeny and learning. This usage of "information" has generally been met with disfavor. At one level, it represents an analogy that does more to obscure the problem than to clarify it. At another level, the meaning of "information" can be interpreted in the context of a whole mathematical treatment of so-called "information theory." The problems of measuring information in this context and of relating it to the mathematical specifications of information theory appear insurmountable (see Hinde, 1970).

Completeness of Form on First Occurrence

One of the criteria that ethologists have long used for characterizing fixed action patterns (which were taken to be innate) was that they are essentially complete on their first occurrence. Often, this characteristic went hand in hand with a belief that the reason behavior appeared so complete was that it was controlled by genes with no modification from learning. However, the latter characteristic is by no means an essential feature of such a formulation. Recently, several ethologists have taken the criterion of completeness of form on first occurrence and elevated it into the sole criterion for instinctive behavior. Of course, the environment enters into all instinctive patterns; specific learning may even be critical to some instinctive behavioral patterns. Regardless of these factors, however, behavioral patterns would be considered innate if they are essentially complete on first occurrence. On this view, behavior that is instinctive need not depend on particular ontogenetic processes in behavioral development but includes all behavioral patterns that are complete in form on first occurrence. Thus, Ewer (1971) explains that when using the term "innate," she means "simply behaviour that appears in its typical adapted form the first time an animal of suitable age is faced with the correct situation" (p. 803). Alcock (1975) writes "Instincts can be considered species-specific behavior emerging in reasonably complete form the first time an animal of a certain age and motivational state encounters a particular stimulus" (p. 66).

This position appears to have merit and to bypass many of the problems encountered by other views. It is important that the behavior under consideration be in "adapted form" or "species-specific." The first time a human places a coin into a slot machine, the behavior appears in reasonably complete form. However, most behaviorists would prefer not adding such behavior to the corpus of the innate. The formulation is most difficult to apply in situations where the stimuli that elicit the behavior are not clearly defined, as in the ontogeny of birdsong.

A POINT OF VIEW

Throughout this tortuous discussion various viewpoints have been presented. The reader should make up his or her own mind as to the proper resolution of the issues presented, perhaps after consulting some of the primary sources cited above.

If there is a majority view on the nature-nurture question, it appears to go as follows. There is no doubt that many behavioral patterns have genetic components and that these are the product of natural selection acting to produce an organism whose behavior is adapted to the environment in which it finds itself. However, it is not proper to propose a dichotomy between the learned and the innate. Both genes and the environment are necessary for all behavior. Because both genes and the environment are important to all behavior, the nature-nurture issue is a pseudoquestion. We have no legitimate means with which to deal with the precise definition of instinctive behavior in the ontogenetic, as opposed to the phylogenetic, sense.

epigenetic view

The view of the present author diverges somewhat from what would seem to be the dominant position. The nature-nurture issue is not a pseudoquestion. The dominant view appears to beg the question. The fact is that there are many species-characteristic behavioral patterns that appear in essentially complete form on their first occurrence and seem to be very difficult to modify in ontogeny. While it may be easy to alter the probability or frequency with which such patterns occur, when they do occur they occur in a species-characteristic form. There would seem to be no compelling reason not to apply the term "instinctive" to such behavioral patterns. Instinctive behavior, on such a formulation, would be viewed not as one side of a learned-innate dichotomy but as one end of a continuum between behavioral patterns that are clearly learned and those which appear to achieve specificity without specific learning.

innate - occurs fullblown on first occurance + released by stimulus

Learning may be critical for the ontogeny of an instinct. However, the specificity of eliciting stimulus and completeness of response of first occurrence would still lead one to classify the behavior as instinctive. Consider the stickleback reared in a deprived environment and which selectively and completely responds with aggressive displays to models with red undersides. Environmental factors including specific learning may be very important in the development of motor coordination, swimming ability, and the functioning of the visual system. However, such factors do not explain the specific selective responsivity of the fish to stimuli with *red* undersides and not to other models. It is this specificity in an adaptive context that seems critical to instinctive behavior.

While it thus may be that the term "instinctive" *can* be used legitimately if used carefully, there is a very real danger in such usage. The animal behaviorist is concerned with four classes of questions: mechanism, development, evolution, and adaptation. All are important. The fine points of behavioral ontogeny may not be critical in the study of adaptation. However, when we are asking developmental questions, it is far more important that we study the complexities of behavioral ontogeny than that we classify a behavioral pattern as instinctive or not. The important point is that we must not let classification take the place of analysis. Instincts can have very interesting ontogenies.

STUDIES IN BEHAVIORAL DEVELOPMENT

We shall now proceed to consider the development of several behavioral patterns, the analyses of which are representative of the kind of information that can be gained from a comprehensive study of the factors influencing behavioral ontogeny.

Pecking Preferences in Young Herring Gulls

An excellent example of comprehensive developmental research on instinctive behavioral patterns is provided by work on the pecking preferences in young herring gulls. Hailman (1967, 1969) published papers on this research with two illustrative titles, "The Ontogeny of an Instinct . . .," and "How an Instinct Is Learned."

Young herring gulls are altricial and are fed by their parents, who regurgitate partially digested food for them. When returning to the nest, the parent lowers its head in front of the gull chick. If it has not recently been fed, the chick will aim a peck at the parent's bill, grasping and stroking it. The parent then regurgitates the food. If the young do not eat the food immediately, a complex interplay between parent and chick usually stimulates the chick to eat.

Adult herring gulls have a distinctive red spot near the tips of their bills. Tinbergen and Perdeck (1950) constructed hand-held models of gull heads and presented them to young chicks in order to study the nature of the sign stimuli that release pecking in herring gull chicks. In some models the red spot was placed on the forehead rather than the bill. Tinbergen and Perdeck found that models with the red spot on the head received fewer pecks than those with the red spot on the bill. This result was interpreted as reflecting the operation of a highly specific *innate* releasing mechanism (IRM) that was activated only by highly specific, configurational stimuli.

There were a number of problems with the Tinbergen-Perdeck experiment. First, as models were hand-held, little control could be exerted over the presentation of models to chicks. More importantly, Schneirla (1956) noted an important uncontrolled difference between the spots on the two models. As the spot on the bill was lower than that on the forehead, it was nearer to the animal and moved at a faster rate than that on the forehead. As the model was swung in an arc about a fixed point near the end of the handle, the spot near the tip moved farther and faster than that higher up on the forehead (see Figure 9-1). Thus, it was possible that the birds were responding to the more conspicuous stimulus rather than to a configurational stimulus.

Hailman's solution to the problem was both simple and important; he got a longer stick. Using an apparatus in which the models could be presented in a

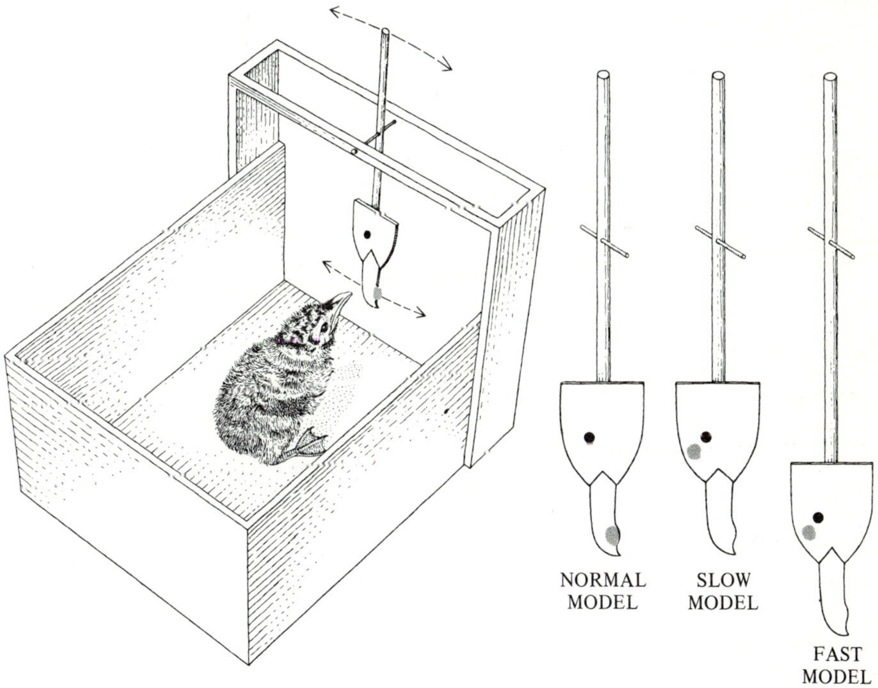

Figure 9-1 Essentials for the study of the ontogeny of pecking preferences in herring gull chicks. The experimental apparatus is shown at left. Three models of the head of an adult herring gull are shown at the right. The normal model is patterned after the adult herring gull, with a spot on the bill. The slow and fast models are altered, as the shaded spot is placed on the forehead; the two models differ in that the fast model is lower and thus the spot is nearer the chick and moves at a faster rate through a greater distance than the slow model. *(From J.P. Hailman, "How an Instinct is Learned." Copyright © 1969 by Scientific American Inc. All rights reserved.)*

hand obscured face

controlled fashion, Hailman used three models instead of two (see Figure 9-1). Two models were similar to those of Tinbergen and Perdeck. The new model had a spot on the forehead and a longer handle, so that the movement of the spot on the head was equivalent to that on the bill of the other model. The results were clear. Newly hatched chicks responded to the new model with just as many pecks as they directed at the model with the spot on the bill. The pecking of newly hatched chicks thus appears controlled by relatively simple stimuli. When Hailman tested older chicks with experience in nests, however, a specific preference was shown for the model with the spot on the bill over both models with the forehead spot.

The pecking response and its control by a rapidly moving spot appear instinctive by the criteria discussed earlier. However, the selective responsivity to highly configurational stimuli (i.e., to models with bill spots) appears to require experience. Environmental influences on other aspects of the behavior, including pecking accuracy and parental identification, were also detected. Thus, the pecking behavior of gulls is an "instinct" with an interesting ontogeny.

Squirrels as Nutcrackers

A mature, experienced European red squirrel is truly adept at opening nuts, which make up a major portion of its diet. It gnaws a furrow from base to tip on the broad side of the nut. Sometimes a second furrow is gnawed on the opposite side. The squirrel then wedges its lower incisors into the crack and breaks the nut open. Eibl-Eibesfeldt (1963) reared red squirrels under conditions in which they had opportunities to gnaw and thus develop their jaw muscles, but they were never permitted to open nuts. Such squirrels typically gnawed furrows, but with random orientation, until the nut eventually broke open. Attempts to wedge the nut open using the teeth succeeded only if the furrow had been gnawed in the proper orientation. Gradually, the squirrels learned to gnaw furrows parallel to the grain of the nut, and they then concentrated on the broad side. The squirrels continued attempts to pry the nuts open and eventually perfected this aspect as well. Some squirrels never did perfect the typical pattern of opening nuts but developed their own idiosyncratic methods (e.g., gnawing a hole or series of furrows near the end of nut). Eibl-Eibesfeldt interpreted the movement patterns of gnawing as innate, whereas their orientation and integration into a complete, functional pattern were interpreted as learned.

Habitat Preferences in Deer Mice

Deer mice, *Peromyscus maniculatus*, live in an extremely broad range of different habitats throughout much of North America. Of the 66 subspecies of deer mice, two have received the most study. *Peromyscus maniculatus bairdi* lives in fields, while *P. m. gracilis* is a forest dweller. Wecker (1964) reported the results of a series of studies on the role of genotype and environment in the development of the habitat preferences of *P. m. bairdi*, the field dweller. He constructed a large (100 by 16 feet) outdoor test enclosure that was located so that it was half in a field and half in an oak-hickory woodlot. With this test enclosure, Wecker was able to assess the preferences of different groups of *P. m. bairdi* for field and woods habitats. Wild-trapped mice showed a distinct preference for the field, as would be expected. When offspring of wild-trapped mice were reared in a laboratory environment, they still preferred the field. This and other information makes it seem likely that habitat preferences in deer mice can develop in the absence of experience in the habitat. Next Wecker constructed a pen in the woods in which he confined a group of deer mice that were permitted to rear young animals. These animals, reared in a woods environment, still preferred the field habitat. Early rearing experience was not sufficient to reverse habitat preferences.

Quite different results were obtained with mice that were the offspring of a line of *P. m. bairdi* that had been in the laboratory for 12 to 20 generations. No strong habitat preference was found in laboratory-reared animals. However, mice of laboratory stock that were reared in a field pen displayed a marked preference for the field. Mice of laboratory stock that were reared in a pen in the woods displayed no strong preference. It appears that, as the result of 12 to 20 generations in the laboratory, the developmental patterns underlying habitat preferences have been changed. Mice with experience in neither habitat show no preference. However, mice of the laboratory stock benefit differentially from experience in the woods and field habitat. Reared in the field, they come to prefer the field habitat; reared in the woods, they show no strong preference.

The ontogeny of habitat preference thus represents an interaction of both genotype and early experience, with the nature of the interaction between the two factors being variable.

Development of Grooming Patterns in House Mice

Rodents of many species show distinctive stereotyped movement patterns in face grooming. In house mice, for example, complex movements of the forelimbs, head, and tongue are coordinated into regular grooming patterns. These grooming patterns are not present at birth in this altricial species but begin to appear when young mice are about 10 days of age. Because face grooming involves such a complex pattern of contacts between the forepaws and the face and tongue, it seems reasonable that the development of normal grooming patterns might be contingent upon such contact.

In order to assess the role of forepaws in the development of grooming patterns, Fentress (1972, 1973) studied three groups of inbred mice: (1) normal control animals, (2) mice from which a single forelimb had been painlessly amputated at birth, and (3) mice from which both forelimbs had been amputated. Fentress was able to identify grooming movements on the basis of the animals' distinctive sitting posture, tongue movements, and movements of the shoulder and upper arm. The results were striking. Grooming patterns developed in the mice with amputated forelimbs in ways very similar to the development in normal mice. A striking example of the "endogenous coupling" of limb and tongue movements was seen in the analysis of movements which, in normal mice, are used in licking the forepaws. Mice with amputated limbs showed characteristic synchronization between shoulder movements and tongue movements, even though the limb stub that was moved by the shoulder motion was nowhere near the tongue at the time it was extended. The mice "licked" air. Similarly, in normal mice, the near eye is closed just before the forelimb contacts the head in gross grooming movements. The eye was similarly closed in mice with amputated limbs, even though there was no limb from which the eye had ever needed protection. Mice with amputated limbs did show subtle alterations in their behavior that appeared to represent adjustments to their unusual condition. The preeminent role of an endogenous coordination in the development of this grooming pattern is nevertheless clear.

Birdsong

We have saved what may be the best example for last. Nowhere is the complexity of the interaction between genetic and experiential factors more clear than in the study of the ontogeny of the songs of birds. Is birdsong learned or innate? As we go through a few examples from the massive literature on the development of birdsong, consider once again the many problems involved in asking such a question.

Bird vocalizations are generally divided into two types: calls and songs. Calls are short, rather simple signals that generally are uttered by either sex at any time of the year and with little individual variation. By contrast, songs have a more complex acoustical pattern than calls; are often emitted primarily by males in courtship, territorial defense, or both; are often seasonal; and can have considerable individual variation.

Minimal Environmental Involvement It appears that in most orders of birds, vocal ontogeny generally does not require specific environmental influence; thus genetic factors are predominant (Nottebohm, 1972). The development of the vocal repertoire of both domestic fowl and ring doves is of this sort (Konishi, 1963; Nottebohm and Nottebohm, 1971). In both cases, the vocal repertoire of the adults develops normally in birds that are reared in total isolation from others of their species. Indeed, even deafening of the individual has little effect on the ontogeny of vocal behavior. The offspring of European cuckoos and cowbirds, species which lay eggs in the nests of other species (brood parasitism), do not need to hear the normal song. This makes good evolutionary sense.

Imitation in Song Learning There are many species in which individuals are capable of learning a broad range of songs through imitation. Species such as mockingbirds and Australian lyrebirds are noted for their imitation of the calls of other species in the natural habitat. Other species, including several species of parrots, imitate a broad range of song patterns in captivity but appear rarely to do so in nature.

Another variation on this theme is found in a group of species which learn their song from the male that raises them, be he father or foster father. "The social bond determines the correct model" (Nottebohm, 1972, p. 119). Bullfinches and zebra finches are of this sort. Bullfinches that are hand-reared by humans can be taught to whistle; those reared by canaries sing canary songs.

Males of the Indian Hill mynah bird imitate the calls of other males of the area, whereas females imitate the calls of other females of the area (Bertram, 1970). Obviously the young male and female are able to recognize the call appropriate for their sex, even though the species lacks sexual dimorphism.

Self-Tutoring Song sparrows represent an interesting intermediate case between the ontogeny of vocalization in species with minimal environmental involvement, such as ring doves, and those with considerable learning, such as mynahs. Song sparrows that are reared in complete isolation from other birds, even in sound-isolated chambers, develop a song that is virtually indistinguishable from the normal song of the species. Even song sparrows reared by canaries sing a normal song-sparrow song (Mulligan, 1966). However, if a bird is deafened and reared in isolation, it displays only a rudimentary version of the song. Mulligan suggests that song sparrows inherit a kind of *template*. A template is a presumed "recognition system" that permits the bird to recognize the song that is appropriate. It is proposed that when vocalizations are first produced, they are "evaluated" against the template and gradually refined by reference to it until the adult song is perfected. The bird is able to perfect its species-characteristic song in isolation, but only if it can hear the songs it produces.

Complex Ontogenetic Patterns While several of these examples of the ontogeny of birdsong may seem quite complex, others seem even more so. Consider the white-crowned sparrow (Marler, 1970). White crowned sparrows have relatively complex songs that include distinct dialects, wherein the song patterns of birds from different geographical regions are quite distinguishable. Marler noted three quite different dialects among birds living in different parts of the San Francisco Bay area.

Birds reared in isolation from other white-crowned sparrows develop a song that is recognizable as a white-crown song but which lacks the detail and dialectical variation typical of normal birds. Deafened birds fail to produce a song that at all resembles that of the typical normal bird. However, if a white-crowned sparrow is permitted to hear the full song of the males of its species between 10 and 50 days after hatching, it will develop the full song, with the dialect of its tutor, when it first sings from 5 to 6 months later. Exposure to the normal song before 10 days or after 100 days has no effect; there appears to be a sensitive period for exposure. The song heard during the sensitive period must be that of a white-crowned sparrow; songs of other species have no effect.

There is one more important step in the ontogeny of the full male song. At about 150 days after hatching, the young sparrow displays a variable "subsong." At that time it perfects the song very much in the manner described above for song sparrows. Eventually, the white-crown is able to produce the song that it heard as a nestling. It must be able to hear itself during this second important interval. The process of self-tutoring can take place in isolation as long as the animal can hear its own song. Once perfected, the song is not subject to further modification. Deafening the bird at this stage has little effect on the continuation or form of song production.

In order to explain the ontogeny of song in white-crowned sparrows, Marler proposed a more complex template model than that proposed for song sparrows. At the start of the sensitive period, the template is "only a crude specification of normal song, but sufficient to exclude songs of other species" (Marler, 1970, p. 1). During the sensitive period, input from white-crowned songs renders the template more precise. When the bird starts to sing on its own, vocalizations are matched to the perfected template by auditory feedback. No "extrinsic reinforcement" is necessary.

Chaffinches are another species whose vocal repertoires have been much studied. The ontogeny of song in chaffinches resembles that in white-crowned sparrows in many respects. Whereas the sensitive period of white-crowns ends at about 50 days, that of chaffinches lasts until their first breeding season at about 10 months of age (Nottebohm, 1970).

SUMMARY

The nature-nurture issue is presented not as a pseudoquestion but as an important issue in the study of animal behavior whose full resolution is yet to be attained. The early history of the instinct concept, the classical ethological view of instinct, criticisms of the classical ethological view, and some recent formulations are presented. It is concluded that there exist many behavioral patterns that appear essentially complete on first occurrence and that appear not to require specific learning for the development of their specific stimulus orientation or motor patterning. While environmental influences affect the ontogeny of all behavior, the careful student of behavior would appear justified in referring to such behavioral patterns as instinctive.

Several studies in behavioral development are presented in order to illustrate the diversity and complexity with which genetic and environmental factors can interact in the ontogeny of adaptive behavioral patterns. The ontogeny of pecking preferences in young herring gulls, nutcracking in squirrels, habitat preferences in deer mice, grooming patterns in house mice, and bird-song all illustrate these complexities. The *important* task for the student of behavioral ontogeny is the complete analysis of all factors influencing behavioral development, not the classification of behavioral patterns as learned or innate.

Part Four

Mechanisms Correlated with Behavior

In this part we shall consider the second of the four sets of questions to be asked about behavior—questions relating to mechanisms. Three kinds of mechanisms will be considered: sensory-perceptual, neural, and hormonal. Each of these systems bears an important relation to animal behavior—its development, evolution, and adaptive significance.

This part is entitled "Mechanisms *Correlated* with Behavior" for an important reason. The relationship between these systems and behavior is a reciprocal one. Sensory and perceptual mechanisms not only affect the behavior of an organism but change as a result of the interaction of the organism and its environment. Neural systems not only play a role in the control of behavior but are altered as a result of behavior. Endocrine systems not only enter into the control of behavior but are often modified as a result of behavior. The study of mechanisms, then, must be viewed as one in which behavior and its correlated mechanisms interact in a dynamic and ever-changing reciprocal relationship rather than as a static one-way street.

The omission of two topics, motivation and mathematical models, requires explanation. By tradition, most textbooks in animal behavior include a chapter dealing with the motivation of behavior. Typically, such chapters include material relating to short-term, nonstimulus variables influencing behavior.

Increasingly, however, animal behaviorists are recognizing that motivational constructs are unnecessary. The concept of motivation tends to be used as a garbage pail for a variety of factors whose nature is not well understood. In writing a definition of "motivation," Verplanck (1957) noted, "The term usually indicates that the controlling variables of a set of behaviors are unknown. Its wide use as an explanatory concept suggests that, for some, ignorance is a virtue—it is admission of ignorance that is a virtue" (p. 22). Bunnell (1973) put it differently: "If we had no genes, no anatomy, no physiology, no ecology, and no past experience, we might need to invent motivation. Since we have all these things, together with a set of evolutionary principles, to guide us, we have all we need in terms of a framework within which to work at the problem of the *how* of behavior" (p. 117). The reader will find that most of the phenomena that might have been covered under the rubric of motivation have been discussed elsewhere—under patterns of behavior, development, mechanisms, or other topics.

A particular set of problems is associated with the major construct used in motivational analyses—the concept of "drive" or "arousal." Such a construct rarely leads to the kind of economy of description in the relationships between independent and dependent variables for which it is intended. For an excellent discussion of the problems inherent in such a construct, see Chapter 8 of Hinde (1970).

Another topic that might have been included is that of mathematical modeling of behavioral patterns (e.g., Heiligenberg, 1974; McFarland, 1974). The use of mathematical models and control-systems theory to represent complex interactions among several variables affecting behavior has been growing in recent years—especially among individuals trained as ethologists. If some of the pitfalls that have troubled their usage in psychology can be avoided, such models hold promise for increased precision in relating behavior to its antecedents and consequences and in the suggestion of important experiments to be performed. As most models tend to be rather complex, they are presently regarded as beyond the scope of a text such as this. If their development continues at its present rate, some means of incorporating mathematical and systems-theory approaches into future books will have to be found.

Chapter 10

Sensory-Perceptual Systems and Function

We live in a complex and ever-changing world. All organisms must possess the ability to detect changes in that world and to respond to them in ways that ultimately will contribute to fitness—survival and reproduction. Sensory-perceptual systems function in detecting changes in the environment (stimuli) and initiating changes within the organism that can lead to adaptive responding where appropriate. Once activated, sense organs produce barrages of electrical activity (action potentials) that are transmitted within the central nervous system and ultimately lead to adaptive responding.

The number of potential stimuli in the environment is enormous. Were the organism to respond to all potential stimuli in its world, it would not only waste energy but be bombarded with an indiscriminable mixture of the relevant and the irrelevant. Therefore, organisms have become adapted to be responsive to some kinds of stimuli and nonresponsive to others. The fact that the organism is insensitive to some potential stimuli may be almost as important as the fact that it is sensitive to others. The selective nature of the organism's sensory machinery and the "blind spots" it necessitates are the result of the operation of natural selection acting on the organism's progenitors. As a result, the organism meets "a surfeit of stimuli" with "a paucity of receptors" (Dethier, 1971, p. 706).

As a result of the operation of different sets of selective forces acting in different ways in different organisms, the sensory systems of different species have come to differ substantially. If we are to understand the behavior of the individuals of a species, we must understand the nature of the stimuli to which they are and are not responsive—their unique perceptual worlds. Each organism has a unique perceptual world, a world which Jakob von Uexküll, in a monograph of 1934 ("A Stroll through the Worlds of Animals and Men: A Picture Book of Invisible Worlds"), termed the *Umwelt*. "The first task of *Umwelt* research is to identify each animal's perceptual cues among all the stimuli in its environment and to build up the animal's specific world with them" (von Uexküll, 1934, p. 13).

Humans, like other species, have developed their own set of sensory systems. This creates a problem for the researcher, because the stimuli which are critical to the *Umwelt* and behavior of a species under study may be very different from those of the investigator. Bats and many species of rodents utilize sound frequencies too high for humans to detect unless they are aided by special instruments. Honeybees are sensitive both to light of wavelenths humans label "ultraviolet" and to the plane of polarization of light waves. If we are to be effective in studying animal behavior, we must be cognizant of the sensory capacities of different species as well as our own sensory limitations. Special instrumentation is generally available to help investigators increase their sensitivity to stimuli that are important to the organism.

Species differ not only with respect to the stimuli to which they are sensitive but also in relation to the uses to which they are able to put sensory information. Often these capacities differ markedly from those of humans. Consider the octopus (Wells, 1961). We know that octopuses are sensitive to differences in the weights of different objects because they compensate for such differences when handling objects of differing weights. However, octopuses appear unable to learn a discrimination task that requires them to respond to differences in weight. Somehow, the information that is in their sensory systems cannot be utilized to learn a discrimination. Octopuses are also poor at discriminating the shapes of objects. Their ability to discriminate surface textures is odd. Wells studied the ability of octopuses to discriminate among objects which were of similar weight but which differed with respect to the pattern of grooves that had been wedged into the surface. The animals were perfectly able to learn a discrimination based on the percent of the surface that was grooved but appeared unable to learn to discriminate the direction of the grooves—a characteristic that humans discriminate readily. In working with different organisms, we must be careful about the assumptions we make regarding not only the sensitivity of their sensory systems but also the uses to which sensory input can be put.

In this chapter, we shall consider sensory-perceptual capacities and function in a variety of animal species. Although some would separate material on sensory and perceptual research, we shall consider them together. Sensory-perceptual systems function as a whole. Often there is considerable control of sensory receptors by structures higher in the brain. The line between sensation and perception is a fuzzy one. The entire afferent sensory system may function in "filtering" sensory input and permitting only portions of it to reach higher centers. Thus, Marler (1963) suggested that a significant portion of the selectivity that so characterizes much species-characteristic behavior may be the result of mechanisms operating at the periphery rather than in the central nervous system. The "innate releasing mechanism" (IRM) may be scattered throughout the entire sensory system rather than being confined to the central nervous system.

Because most universities offer whole courses in sensation, perception, and physiological psychology, we shall concentrate in this chapter on a comparative approach to the functioning of sensory-perceptual systems rather than on the mechanisms of sensation and perception.

METHODS IN THE STUDY OF SENSATION AND PERCEPTION IN ANIMALS

A first task in research on sensation and perception, as pointed out by von Uexküll, is that of identifying the stimuli to which animals of different species are responsive. We begin with a discussion of the various methods that are available to aid in attempts at understanding *Umwelten*. As animals cannot simply tell us what they perceive, indirect methods must be used.

A second task is to determine just which of the stimuli to which the animal is responsive are controlling particular behavioral patterns. Slightly different methods are required in searching for answers to such questions.

Methods in the Study of Sensory Capacity

Anatomical Methods A preliminary clue to an animal's sensory capacities lies in the anatomy of its sensory systems. While it is sometimes difficult to infer function from structure, anatomy does provide information regarding the limitations and adaptations of particular systems. We might expect different visual functioning from nocturnal species with very large eyes and fossorial species with degenerate visual systems. The eyes of predators and arboreal species tend to be located at the front of the head; the overlap in visual fields of the two eyes is important to depth perception. In prey species, the eyes are generally located near the side of the head, thus providing a maximal field of view. The eyes of nocturnal lizards, birds, and mammals have large lenses which concentrate light; those of diurnal lizards, birds, and mammals are smaller and tuned to make finer discriminations (Marler and Hamilton, 1966).

Electrophysiological Methods Because stimuli initiate changes in electrical activity in the nervous system, it is possible to record electrical activity and thus determine the nature of the stimuli to which the organism is potentially sensitive. Ralls (1967) recorded auditory evoked potentials in the brains (posterior colliculi) of mice of several different species. She found that they are responsive to tones of much higher frequency (perceived as pitch) than are humans. Responsivity up to the range of 70 kHz (kiloHertz), compared with the 20 kHz or so of humans, was not uncommon. Care must be exercised in interpreting electrophysiological data. While they show that the nervous system responds to certain stimuli, behavioral data are needed to show that the organism can actually utilize this information. In the case of mice, the behavioral significance of ultrasound has been established (e.g., Brooks and Banks, 1973).

Measurement of Species-Characteristic Responses Information about sensory capacities can be gained by assessing the effects of various stimuli on naturally occurring, species-characteristic behavioral patterns. The advantage of these techniques is that they usually do not require extensive periods of training or habituation.

King and Vestal (1974) utilized the *optokinetic response* to study visual acuity in mice of the genus *Peromyscus*. The mice were placed inside the center of a cylindrical metal drum (Figure 10-1), the inside of which was painted with vertical black-and-white stripes. As the drum is rotated and the stripes move across the visual field, animals of a wide range of species will slowly move their heads, eyes, and bodies to follow the direction of movement. With wide, easily discriminated stripes, the mice showed a clear optokinetic response. When very narrow stripes are used, they cannot be detected and the drum is perceived as a homogeneous gray field; no following occurs. The minimal visual angle perceived by these mice was about 5 minutes of arc (60 seconds of arc = 1 minute; 60 minutes = 1 degree; 360 degrees = 1 circle).

Measurement of Learned Responses The measurement of various learned responses is quite commonly used in determining the sensory capacities of animals. In many experiments, the presence or absence of conditioned responses to stimuli of varying physical characteristics is recorded. Alternatively, one may train an animal to respond differently as a function of the presence or absence of a stimulus. One can then probe the limits of the animal's sensory capacities. Lissmann and Machin (1958) used such a method to demonstrate that electric fish could use their electrical senses to detect the presence or absence of metallic objects in the absence of cues to other sensory modalities. Schusterman and Balliet (1970) trained two sea lions to emit a burst of click vocalizations (short-duration sound pulses) when they saw a striped target and to remain silent when they saw a gray target. Thus, the investigators were able to study the sea lions' ability to resolve narrow stripes. Visual acuity was quite close to that reported for King and Vestal's mice—5.5 minutes of arc.

More complex operant learning techniques have also been applied to the study of sensory-perceptual systems. Henton, Smith, and Tucker (1966) used the *conditioned suppression* technique to study odor discrimination in pigeons. The first step in using the conditioned suppression technique is to train the

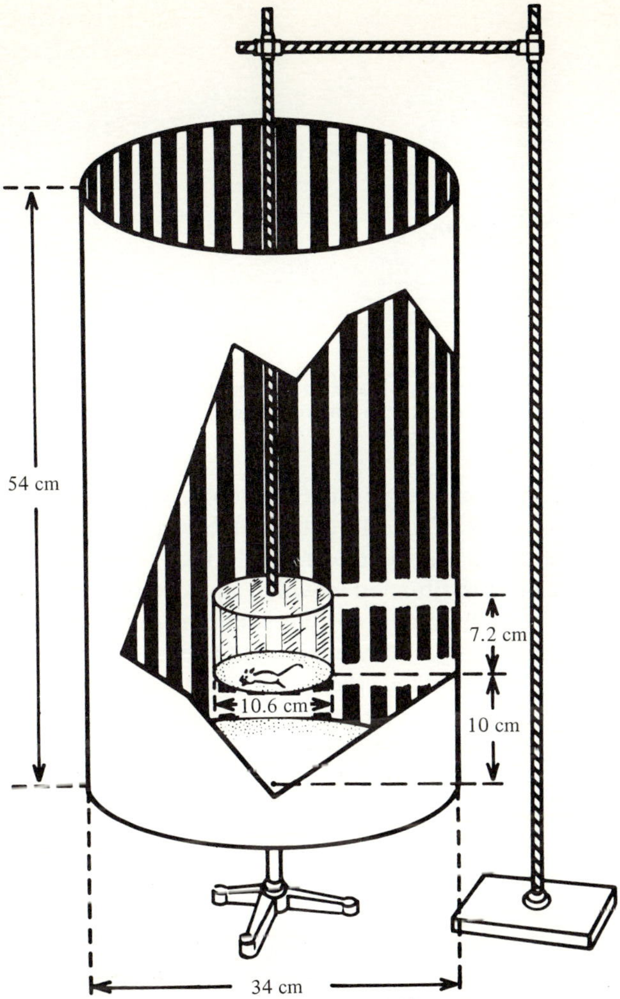

Figure 10-1 Apparatus for measuring the optokinetic response in mice. A cylinder containing a mouse is suspended independently of a rotating drum, so that a striped pattern rotates about the stationary mouse cage. *(From Vestal & King, 1968.)*

animal to perform a learned operant task. Pigeons can easily be trained to peck at a key (small disk) if their pecks are rewarded with food. When the schedules of delivering reinforcement are adjusted appropriately, stable rates of key pecking are established. When the animal reaches a stable performance level, suppression training begins. A stimulus is presented, and shock is delivered a short time later. Even though the shock is unavoidable, the animal ceases key pecking when the stimulus appears. By varying the parameters of the stimulus and determining whether or not the animal's responding is suppressed, one can determine the sensory capacities of the organism. Why use such a complicated technique? The conditioned suppression technique can be quite sensitive.

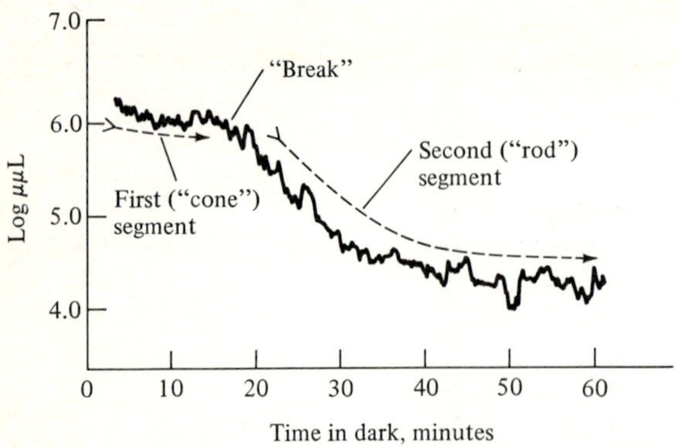

Figure 10-2 Dark-adaptation curve secured from a pigeon during the first hour it was in the dark. The luminance of the stimulus patch, in log micromicrolamberts, is on the ordinate. *(After Blough, 1955.)*

Henton et al. found that pigeons were responsive to odors, whereas most earlier studies using simpler techniques failed to detect such responsivity.

An equally fascinating technique was developed by Blough (1961) and is termed "tracking." Let us return to the situation in which a pigeon is pecking a key for food reinforcement. We shall complicate the situation somewhat by giving the bird two keys and adding some relay equipment behind the scenes. The bird is trained to peck one of the two keys (key A) when it sees a light and at the other (key B) whenever there is no light. The relay equipment is so programmed that every time it pecks key A, the intensity of the light decreases slightly; every time it pecks key B, the intensity increases. What will happen? If we start with a bright light, the pigeon will steadily peck at key A and the light will grow slowly but steadily dimmer. Eventually, the bird will no longer perceive the light. It will then switch to key B and the light will get progressively brighter with each peck—but only until the pigeon again perceives the light and shifts to key A again. With this technique and appropriate recording devices, the pigeon plots a graph as it "tracks" its own threshold. The method is particularly useful in tracking changes in threshold over time, as in dark adaptation (Figure 10-2) or in response to administration of a drug. Changes in sensory thresholds are continuously plotted.

Methods of Determining Effective Stimuli for Particular Behavioral Patterns

As an important function of animal sensory systems lies in communication, it should come as no surprise that the methods used in determining effective stimuli for particular behavioral patterns overlap substantially with those used in the study of animal communication, as discussed in Chapter 6.

Correlational With correlational methods, one observes the reliability of association in time of a particular signal and a particular behavioral pattern. For example, Sales (1972) noted an association between the occurrence of ultrasonic calls during courtship and mating in rodents and specific behavioral patterns occurring at the time of the call.

Experimental Manipulation of Stimuli or Signals A common method in the study of effective stimuli is the experimental production or manipulation of a signal. The ethological use of models to determine effective stimuli, as in the example of the red belly of the stickleback, provides a classical example. The elements of a complex birdsong can be rearranged in order to analyze the effectiveness of various aspects of the song in eliciting behavior. For example, in the study of indigo buntings, Emlen (1972) played both tape-recorded normal songs and others that had been doctored in various ways on a tape recorder and speaker located in a male's territory. An interesting way to produce altered signals is by breeding hybrids between two species, as in studies of various species of frogs (Bogert, 1960; Gerhardt, 1974).

Interruption of Sensory Systems One method of studying effective stimuli is to interrupt a sensory system by causing lesions or by otherwise removing neural tissue and observing the consequences. In order to investigate a possible role of olfactory stimuli in controlling mating behavior in hamsters, Murphy and Schneider (1970) removed the olfactory bulbs (see Figure 11-2) from a group of males. The operation eliminated both olfactory sensitivity and mating behavior. Such results are difficult to interpret because the operation removes not only sensory input but eliminates brain tissue as well. Olfactory sensitivity can be eliminated peripherally by injecting toxic chemicals into the nasal passages. Results of the application of such methods to the problem of the control mating behavior in hamsters have been controversial (Powers & Winans, 1973, 1975; Doty & Anisko, 1973).

A BRIEF SURVEY OF SOME SENSORY SYSTEMS

We shall now briefly survey some of the sensory systems and their functions in a variety of species. We shall present a brief introduction to the different sensory modalities together with a few particularly interesting examples of their function. For a more complete survey of animal sensory systems, see Hess (1973) and the references therein.

Vision

Many protozoans and coelenterates possess only a diffuse light sensitivity in which an eyespot is capable of detecting only overall illumination levels. More complex organisms have evolved a considerable variety of light-sensitive organs. The compound eyes of some insects are constructed of bunches of units, *ommatidia*, that are oriented parallel to one another, with a light-sensitive portion at one end and afferent connection to the rest of the nervous system at

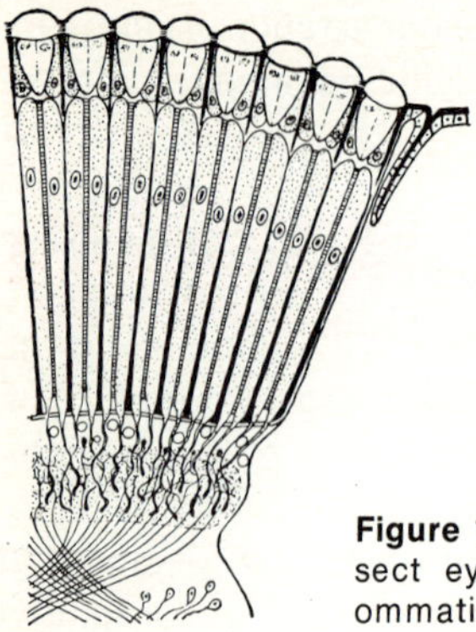

Figure 10-3 Structure of a compound insect eye, showing the organization of ommatidia. *(From Snodgrass, 1935.)*

the other end (Figure 10-3). A remarkable example of convergent evolution can be found in the eyes of some cephalopod molluscs, such as octopuses, and those of vertebrates. Both are organized in the manner of a camera, with lenses, irises, and retinas.

Species differ greatly in visual acuity, or the ability to detect small stimuli. Whereas the *Peromyscus* mice and sea lions discussed earlier can detect angles of about 5 minutes of arc, humans detect about 1 minute. The visual acuity of certain birds, such as falcons, appears to be several times greater than that of humans (e.g., Fox, Lehmkuhle, & Westendorf, 1976). Albino rats appear not to detect visual stimuli until they approach a full degree of arc. It is odd that a mammal with such poor visual acuity should be the subject of so many psychological studies of visual discrimination learning.

Color vision involves the ability to discriminate among visual stimuli of different wavelengths. The range of effective wavelengths for different species is variable, with some species sensitive to ultraviolet waves and others insensitive to wavelengths in the red region. The ability to discriminate among different wavelengths (color vision) is equally variable. By using a "checkerboard" technique in which honeybees had to fly to food dishes resting on patches of differentially colored cardboard, von Frisch (e.g., 1971) demonstrated that honeybees discriminate four color groups. Color vision has been demonstrated in some species of cephalopod molluscs, fishes, amphibians, reptiles, birds, and mammals. Most rodents and lagomorphs (rabbits, etc.), with the exception of squirrels, appear deficient in color vision. Color vision generally is better developed among diurnal mammals than nocturnal species.

A classical example of the analysis of visual systems is the study of Lettvin, Maturana, McCulloch, and Pitts (1959) entitled, "What the Frog's Eye Tells the Frog's Brain." Lettvin el al. implanted into the heads of frogs small metallic electrodes through which electrical activity originating in the frogs' retina could be detected. Stimuli of various sorts were then presented in the

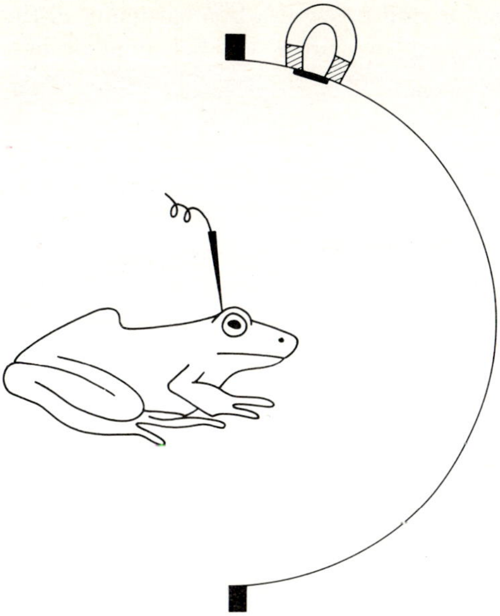

Figure 10-4 Schematic drawing of apparatus for studying the visual system of frogs. Frogs with electrodes implanted face a semicircular cylinder. Small objects are moved in the visual field through use of a magnet on the side of the cylinder not visible to the frog. *(From Maturana, Lettvin, McCulloch, & Pitts, 1960.)*

visual field of the frog (Figure 10-4). The researchers found that the frog's visual system contains five classes of cells:

Class 1—Sustained-edge detectors: These neurons respond maximally to small edges that move into the receptive field and stop there.

Class 2—Convex-edge detectors: These neurons respond maximally to small dark spots with convex edges moving toward the center of the visual field.

Class 3—Moving-edge detectors: These neurons fire maximally when an edge moves in and out of the visual field.

Class 4—Dimming detectors: These neurons respond maximally to decreases in light intensity.

Class 5—Dark detectors: The activity of these neurons is inversely proportional to light intensity; the brighter the light, the less they respond.

There are several interesting features of the frog visual system as portrayed by Lettvin et al. One often assumes that the sense organ, the retina in this case, functions to receive sensory input and pass on a relatively faithful image to the brain, where information is processed. This clearly is not the case. The frog's retina plays a very important role in information processing and the information passed on to the brain is in a highly processed form. Interest has focused about the convex-edge detectors, affectionately known as "bug perceivers." With a group of retinal neurons selectively sensitive to such stimuli, the frog can presumably make very rapid responses, as are required in catching flying insects. Such a sensory system would not only maximize speed but also filter much irrelevant information and prevent a bombardment of the brain with extraneous information. The cost, of course, is a loss in flexibility; once information is lost in the retina, it can never be regained in the brain. Thus, some of the flexibility in utilization of visual input, such as that in mammalian species, is missing.

Visual systems function in a wide variety of contexts, including food acquisition, predator avoidance, visual exploration, and regulation of circadian rhythms. In the social context, visual signals are an important aspect of the communication system of many species, particularly diurnal species living in open habitats.

Audition

Auditory systems are those that are selectively responsive to the relatively high-frequency vibrations that occur in different media, including air and water. Insects detect sounds by way of relatively simple hair sensilla, complex tympanic organs, the antennae, and other organs. Tympanic organs are located in the thorax, leg, or base of the wing in different species of insects. The evolution of complex auditory systems in vertebrates began with fishes, and considerable hearing abilities have been demonstrated in many species of fishes, amphibians, reptiles, birds, and mammals. Birds respond to sounds of higher frequencies and are better at localizing sounds than fishes, amphibians, and reptiles. The mammalian ear is characterized by the existence of a pinna (often called "the ear"), three bones in the middle ear, and a coiled cochlea.

In a study of the evolution of hearing in mammals, Heffner, Ravizza, and Masterton (e.g., 1969) studied the auditory sensitivities of oppossums, hedgehogs, tree shrews, and bush babies using the conditioned suppression technique. They concluded that high-frequency sensitivity, at least up to 32 kHz, is a pervasive characteristic of most nonhominoid auditory systems; chimpanzees and humans were alone of 19 species surveyed in lacking high-frequency sensitivity. Humans are more sensitive than many species to tones of low frequency. Heffner et al. (1969) concluded that "man's ancestral lineage must have been exposed to strong and persistent selective pressures for low-frequency sensitivity" (p. 21).

Moths are specially adapted for the detection and avoidance of approaching predatory bats (Roeder, 1967; Roeder & Treat, 1961). Roeder and Treat hauled some 300 pounds of equipment to the hills of Massachusetts to enable them to record electrical activity from individual cells in the nerve leading from the moth's ear. By amplifying the output of the moth's auditory nerve and feeding it to a speaker, Roeder and Treat were able to monitor its activity. The moths detected bats at a distance of about 100 feet, thus demonstrating a level of sensitivity better than that of bats in detecting moths or, for that matter, better than that of the best microphones Roeder and Treat could find. On hearing a distant bat, the moths fly in the direction opposite to it. If a bat approaches quite closely in an attack, the moth displays an erratic diving maneuver that functions in the avoidance of predation.

A major function of auditory systems lies in intraspecific communication.

The example of birdsong has already been discussed. Humpback whales produce songs that travel over considerable distances, with elements lasting 7 to 30 minutes (Payne & McVay, 1971). Crickets display various calls that function in a variety of contexts, including courtship and the maintenance of territories (Alexander, 1968). By playing such sounds through loudspeakers, Ulagaraj and Walker (1973) were able to attract mole crickets to the speakers.

With the recent advent of sophisticated auditory equipment, a whole domain of *"ultrasonic communication,"* that outside the normal human hearing range, has been unveiled. Ultrasonic communication is common in rodents and functions in a variety of contexts. Brooks and Banks (1973) found that collared lemmings emit ultrasounds as neonates, during mating, when predators are detected, and in the context of agonistic behavior. Six distinct types of ultrasonic sounds were recognized. Barfield and Geyer (1972) described a postejaculatory ultrasonic "song" of 22 kHz in laboratory rats. Particular attention has been focused on the ultrasounds of neonates. Rodent neonates appear to emit two different classes of ultrasounds (Noirot, 1972). One class is given in response to cold and attracts the parent to search for and retrieve the pup. The other is given in response to unusual tactile stimulation and appears to cause adults to cease rough handling or aggressive behavior.

An interesting communication system has been described in the tree frog, *Eleutherodactylus coqui* (Narins & Capranica, 1976). Males emit a two-note call ("co-qui") each evening from about sunset to midnight. There is an apparent dichotomy in the functioning of the two notes. The "co" note functions among males in the regulation of territoriality. The "qui" note is the part of the call that attracts females. The differential function of the two notes reflects a difference in the best frequencies of the auditory systems of the two sexes. Such sexual differentiation in the auditory organ represents another example of the extent to which sensory information can be processed in the periphery and is adapted for specific functions.

Chemical Senses

General chemical sensitivity, mediated by relatively undifferentiated sense organs, is found in even the simplest of animal species. Taste, or gustation, is more sensitive than a general chemical sense and generally functions in contact reception. Olfaction, or smell, is the most sensitive chemical sense and is responsive to chemicals diffusing, often in very low concentrations, from a source distant from the animal. The senses of taste and smell are differentiated in insects and are present in most species of vertebrates. Research on chemical senses is complicated by the difficulty of preparing and controlling stimuli and by the general insensitivity, relative to other species, that characterizes human chemical senses.

The gustatory system of blowflies has been the subject of much study by V. G. Dethier and his colleagues (e.g., Dethier, 1971). Blowflies possess 245 to 257 gustatory hairs on their mouthparts, 3,120 on their six legs, and 65 to 67 on the oral surface of the mouthparts. The taste capacities of all these hairs are nearly identical; each taste hair is innervated by five sensory neurons. One of these five neurons responds to mechanical stimuli. The other four are gustatory receptors—one for water, one for sugar, and two for salt. Considerable peripheral interaction occurs among receptors of different types as complex substances are encountered. The reader who would like a highly readable and fascinating account of the early days of Dethier's research program and of the "behind-the-scenes" operation of science is advised to enjoy an evening with Dethier's book *To Know a Fly* (1962).

Different species of newborn, previously unfed snakes respond with tongue flicking and prey-attack behavior to water extracts of skin substances taken from various species of small animals. Species differences in responsivity to extracts from different animals correspond with the normal feeding preferences characteristic of the species (Burghardt, 1967).

It is well known that in order to breed, salmon return to the stream in which they were hatched. Often these migrations are made with considerable difficulty against currents, past dams, and through areas inhabited by predators. How does a salmon determine to which stream to return? Evidence in support of a chemical basis for choice appears convincing. In one experiment, Scholz, Horrall, Cooper, and Hasler (1976) were able to experimentally "imprint" young salmon to chemicals that later would characterize different streams flowing into Lake Michigan. A lasting effect of early chemical "imprinting" on the selection of a stream was demonstrated.

Pheromones Pheromones are chemical signals that convey information between different members of a species (intraspecific communication). They may be differentiated from *allomones*, which are signals used in *inter*specific communication; *hormones*, which are chemicals that communicate among different organs within a single organism; and from other chemical stimuli not functioning in communication, such as those implicated in food and habitat selection. Two broad categories of pheromones are usually distinguished (e.g., Bronson, 1971). *Signaling pheromones* result in a more or less immediate effect on the behavior of the recipient animal. By contrast, *priming pheromones* trigger hormonal activities that may become manifest in overt behavior only at a later time.

Early research on insect pheromones was reviewed by Wilson (1965). Honeybees have 11 different pheromone-secreting glands on their bodies. Perhaps the most spectacular example of insect pheromones is that of the sex attractant of silk moths, *Bombyx mori*. The male's antennae are so sensitive that just one molecule of the sex attractant ("bombykol") secreted by the female will trigger a nerve impulse. When about 200 nerve impulses are generated in a span of 1 second, the male moves upwind in search of a mate (Schneider, 1974). Numerous insect pheromones have recently been chemically identified and isolated.

There are many good reviews of information on pheromones in mammals (e.g., Gleason & Reynierse, 1969; Eisenberg & Kleiman, 1972; Thiessen &

Rice, 1976). The sources of pheromones in different species include feces and urine as well as the secretions of a bewildering array of glands located in many regions of the bodies of different species. Pheromones may be deposited by means of object marking, marking of a social partner, self-marking, substrate marking, or release of the pheromone into the air. Pheromones convey a variety of different kinds of information in different species, including species and racial identity, sexual identification and reproductive state, individual identification, age, and mood. Pheromones are functional in affecting reproductive behavior (sexual and maternal), and other forms of social behavior (withdrawal and submission, aggression and dominance, and scent marking) (Eisenberg & Kleiman, 1972).

Three classical pheromonal effects on reproduction in mice were named for their discoverers:

Lee-Boot effect: The normal estrous cycle of house mice lasts about 4 to 5 days. If females are housed in groups, they will cease regular cycling and display a spontaneous "pseudopregnancy"—the Lee Boot effect (van der Lee and Boot, 1955). Pheromones are implicated.

Whitten effect: Introduction of a male or his excretions to a cage of female mice results in the initiation of synchronized estrous cycles in the females, with an estrous peak occurring on the third night after introduction—the Whitten effect (Whitten, 1956).

Bruce effect: If females that have recently mated with one male are exposed to a strange male or the odor of a strange male, many will display a "pregnancy block," a failure of pregnancy due to a blocking of the implantation of fertilized ova into the uterine wall—the Bruce effect (Bruce, 1960). There is some evidence that blocking can occur after implantation as well.

A variety of other pheromones have been demonstrated to function in mammalian reproduction. Hamster vaginal secretions have an excitatory effect on the mating behavior of male hamsters (Johnston, 1975). Pheromones also affect rates of sexual maturation. The sexual maturation of male mice is accelerated by the presence of a male; that of female mice is accelerated by the presence of males and inhibited by the presence of females. Pheromones mediate these effects (Vandenbergh, 1969, 1971a; Drickamer, 1974). The possibility of pheromonal effects in the copulatory behavior of rhesus monkeys is quite controversial (Michael & Keverne, 1968; Goldfoot, Kravetz, Goy & Freeman, 1976). There is some evidence for menstrual synchrony and suppression in humans that may be pheromonally mediated (McClintock, 1971).

Maternal pheromone in rats is secreted in the caecotrophe that is defecated by lactating females. It functions in orienting the young pups to the mother and in synchronizing the mother-young relationship (Leon, 1974).

The urinary marking patterns of dominant and subordinate male house mice separated by a wire partition can be differentiated by allowing the males to urinate on filter paper and examining the filter paper under ultraviolet light.

Dominant males vigorously mark the entire cage floor, whereas subordinate males typically void urine in only a few small pools (Desjardins, Maruniak, and Bronson, 1973).

Active Sensory Systems

Active sensory systems differ from those discussed thus far in that with these systems the organism actively emits energy in one form or another and perceives objects in the environment on the basis of alterations in the energy patterns that return.

The best-known active sensory systems are the echolocation systems of bats (e.g., Griffin, 1958; Griffin, Webster, & Michael, 1960; Simmons, Howell, & Suga, 1975). Bats are able to determine the size, shape, distance, direction, and motion of objects using their sonar systems. The sounds emitted by bats vary according to the species and situation.

In their laboratory tests, conducted in rooms of varying sizes, Griffin et al. (1960) found that bats of the genus *Myotis* could catch mosquitoes and fruit flies in the dark at rates of 10 mosquitoes or 14 fruit flies per minute. The characteristics of the sound output of the bats change as an insect is hunted. During a search phase, before an insect is detected, pulses are emitted about once every 50 to 100 milliseconds (thousandths of a second). During the approach phase, after an insect is detected, there is a progressive shortening of the between-pulse interval. During the terminal phase, when the bat is within a few centimeters of an insect, the interval shortens further, so that there is just 0.5 millisecond between pulses.

Bats have also been shown to be quite efficient at flying rapidly in the dark through complex patterns of wires strung across an experimental room.

Echolocating systems have also been found in South American oilbirds and in porpoises (e.g., Kellogg, 1958; 1961).

A variety of species of electric fishes are able to locate objects by using an active electrical sensing system. The electric organs set up a field about the fish. The presence of objects that conduct electricity at a rate either higher or lower than that of the water is detected on the basis of the distortions created in the electrical field (see Figure 10-5). Electric organs have evolved independently at least six different times; in both elasmobranchia and teleosts and in freshwater as well as salt. The organs of different species are located in different parts of the body, from the region around the eyes in the stargazer to the tail in some African species. Some species emit strong electrical pulses; the 500-volt jolt of the electric eel can stun a horse. Other species emit currents so weak that they are imperceptible to humans without instrumentation. The latter function primarily as sensory systems (see Bennett, 1970). Lissmann and Machin (1958) demonstrated that electric fish were able to discriminate among objects solely on the basis of their electrical conductivity. While the discharge frequencies of some species are quite invariant, changing primarily as a function of temperature or jamming, the discharge rates of other species

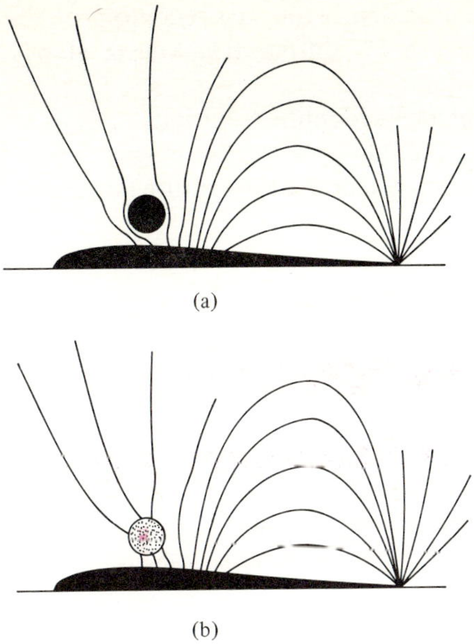

Figure 10-5 The electrical field of an electric fish in the presence of *(a)* an object of low conductivity and *(b)* an object of high conductivity. The presence of objects of conductivity differing from water can be detected from the pattern of electrical current arriving at the electrical receptors. *(From Lissmann & Machin, 1958.)*

change in response to a variety of stimuli, including light-dark cycles, the presence of objects in the water, and the availability of food (Dewsbury, 1966). Electrical signals also function in communication among individuals of various species of electric fish, probably in both reproductive and agonistic contexts (Black-Cleworth, 1970; Hopkins, 1974).

Other Sensory Systems

A variety of other sensory systems are also important in animal behavior. Sensations of *pain* are common in humans but difficult to study in nonverbal species. It is a reasonable inference that pain is felt by many species, as they often display movements or vocalizations apparently indicative of pain.

The *tactile sense*, or sense of touch, is another widespread sensory modality. Tactile sensitivity varies in different species and on different parts of the body within a species. Interference with tactile sensations from the genital region, either by nerve section or administration of an anesthetic, severely disrupts copulatory behavior in rats and cats (Carlsson and & Larsson, 1964; Adler & Bermant, 1966; Aronson & Cooper, 1968).

Proprioception is concerned with relative position or movement of body parts. Equilibration receptors in vestibular systems, such as the semicircular canals of mammals, provide information regarding orientation relative to gravity. Arthropods have a variety of statocysts subserving analogous functions.

Infrared heat systems are used by a variety of species of snakes in locating warm-blooded prey. The pit vipers, including rattlesnakes, have an infrared receptor located between the eye and the nostril, while the receptors of the Boidae, including the boa constrictor, are more diffuse (Gamow & Harris, 1973).

Although the receptors have not yet been identified, various species of animals are responsive to *magnetic waves*, including the earth's magnetic field, a fact that holds promise for some fascinating research of the future (see Russo & Caldwell, 1971; Keeton, 1974a).

DEVELOPMENT OF SENSORY-PERCEPTUAL SYSTEMS

At birth or at the first opportunity for testing (e.g., the age of eye opening), the young of many species show well-developed perceptual function. However, although this perceptual ability is quite advanced, it is typically not equal to that of the adult in all its richness; perceptual experience is often necessary for the development of full perceptual capacity. This prompted Fantz (1965) to quip, "Perception is innate in the neonate and learned in the adult." While this statement may be slightly exaggerated for effect, the advanced capacities of the neonatal system and role of experience for full adult perceptual development are thus dramatically stated.

That some perceptual capacity develops in the absence of specific learning should not be surprising. Viewed in the context of questions of adaptive significance, it may be important that the young animal be able to make discriminations, such as those of depth perception, at the time it begins to locomote. In the context of the present questions of mechanism, it appears that mechanisms known to exist are sufficient to explain the development of the "wiring" of the brain for perceptual function without experience (Sperry, 1971). It appears that the connections among neurons in the brain develop as a result of specific chemical affinities, so that one growing neuron "seeks" a particular other neuron with which to connect (synapse) on the basis of chemical cues. J. J. Bernstein (personal communication) put it colorfully. Growing axons in the nervous system might be regarded as being strawberry-seeking, vanilla-seeking, chocolate-seeking, etc. A vanilla-seeking axon would continue to grow, bypassing chocolate and strawberry neurons, until it found a delicious vanilla neuron on which to synapse.

Most research on perceptual development has focused on the visual modality. An example is provided by research on depth perception. The most common apparatus for such research is the *visual cliff* (see Figure 10-6). The essential features of a visual cliff are an elevated centerboard and two sides, a "deep side" and a "shallow side." The shallow side consists of a piece of glass with a pattern immediately underneath. On the deep side, the pattern is well beneath the glass. Thus, when an experienced animal is placed on the centerboard of a visual cliff, there will seem to be a choice between moving to a safe, shallow area or falling off a "cliff." Adults of a large number of vertebrate species have been tested on the visual cliff and most show good depth perception, descending to the shallow side far more often than to the deep side

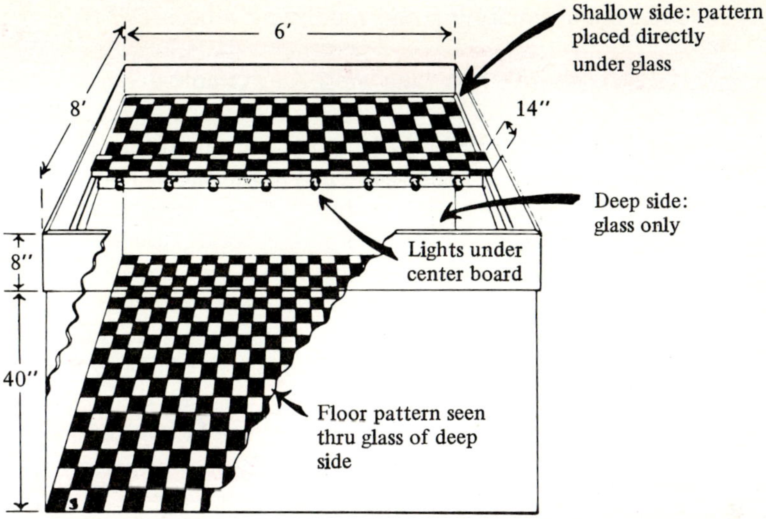

Figure 10-6 A schematic drawing of a visual cliff used for testing large animals, such as kittens, rabbits, and goats. The pattern on the shallow side is just under the glass; that on the deep side is well beneath the glass. *(From Walk, 1965.)*

(Walk, 1965). When young animals are tested, it appears that depth discrimination generally is functional at the age of first locomotion. Young chickens, goats, and lambs display good depth discrimination in the visual cliff at the age of 1 day. There are some exceptions to the general rule of innate bases for the ontogeny of depth discrimination in vertebrates, however. The performance of gerbils reared under standard laboratory conditions is quite poor and contrasts dramatically with that of species such as house mice (Thiessen, Lindzey, Blum, Tucker, & Friend, 1968). Rabbits and kittens fail to show depth discrimination in the visual cliff at the age of eye opening. If rabbits or cats are reared in darkness, the development of depth discrimination is further delayed. This may be due either to a failure of development of the mechanisms underlying depth perception or to interfering emotional or competing responses. Some sensory-perceptual systems appear functional at birth but may degenerate in the absence of normal experience (e.g., Hubel & Wiesel, 1963).

Numerous other perceptual abilities have been found to be functional in the very young animal. Oppenheim (1968) found color discrimination and color preferences in newly hatched ducklings. Vestal and King (1968) found mice of the genus *Peromyscus* to display the optomotor response (Figure 10-1) on the day of eye opening. Interestingly, one strain that had been selectively bred for early eye opening showed the onset of the optomotor response at the normal age of eye opening for the species, despite the 2.4 additional days of visual experience. Fantz (1967) found pattern discrimination and pattern preferences in young humans tested as early as under 48 hours of age.

"In the development of perceptual behavior, the contributions of gene-determined neural systems cannot be overemphasized, *except* by denial of their continuing dependence upon confirming and modifying processes" (Riesen, 1973, p. 419). Indeed, experience has been shown to be important in the full development of many visual perceptual capacities. An example of the role of experience can be found in a study of the development of visual perception in rhesus monkeys (Wilson & Riesen, 1966). Wilson and Riesen raised 12 rhesus monkeys in the absence of patterned light from birth to the age of either 20 or 60 days. Because Riesen's earlier research had shown that a complete absence of light could result in degeneration of the retina, the young monkeys were permitted 2 1/2 hours per day of input from light that had been rendered unpatterned by covering the eyes with white plastic ocular contacts. When tested, the visually deprived monkeys were similar to newborns in both their rate of learning visual discriminations and in untrained visual behaviors. Visual acuity improved as the monkeys gained experience. Several characteristics of visual development required considerable experience despite the advanced ontogenetic status of the monkeys in other respects. Visual fixation and pursuit of objects appeared after about 8 days; avoidance of the deep side of the visual cliff required about 20 days.

Some interesting insight into the role of experience in perceptual development has been gained from studies in which animals have been reared in normal light but deprived of certain kinds of experience. Held and Bauer (1967) reared infant rhesus monkeys in an apparatus that precluded their seeing their body parts. When one hand was exposed to view at the age of 35 days, visual fixation on the hand was "insistent and prolonged"; visually guided reaching was poor, but it improved with 10 hours of experience. Hein and Held (1967) reared kittens that experienced patterned light but only while wearing large, cone-shaped collars that prevented them from seeing their limbs and torsos. If a normal cat is moved toward a surface such as a table, it extends its forelimbs as if to prevent collision (called "visual placing"). This response was normal in kittens reared without sight of limbs and torso. If normal cats are moved toward an "interrupted surface" (a surface containing both solid and empty portions), they will orient their placing responses toward the solid portions of the surface. Deprived kittens failed to orient their visual placing responses toward the solid areas. Thus, while the fundamental visually guided placing response does not require experience with the sight of body parts below the neck, such experience *is* important to the full development of the response.

The role of experience in the development of visual perception is further clarified by additional research by Held and Hein (1963). They demonstrated that kittens must not only have visual experience if normal perception is to develop but that this experience must be correlated with movements that they themselves initiate. Ten pairs of cats were reared in darkness for most of their lives. Once their motor capacities were sufficiently developed, the cats began a procedure in which they received 3 hours per day of normal visual experience in a special apparatus (see Figure 10-7). One kitten of each pair was designated as the active (A) cat and the other as the passive (P) cat. As the active animal moved about in the apparatus, the passive animal was moved along 180 degrees

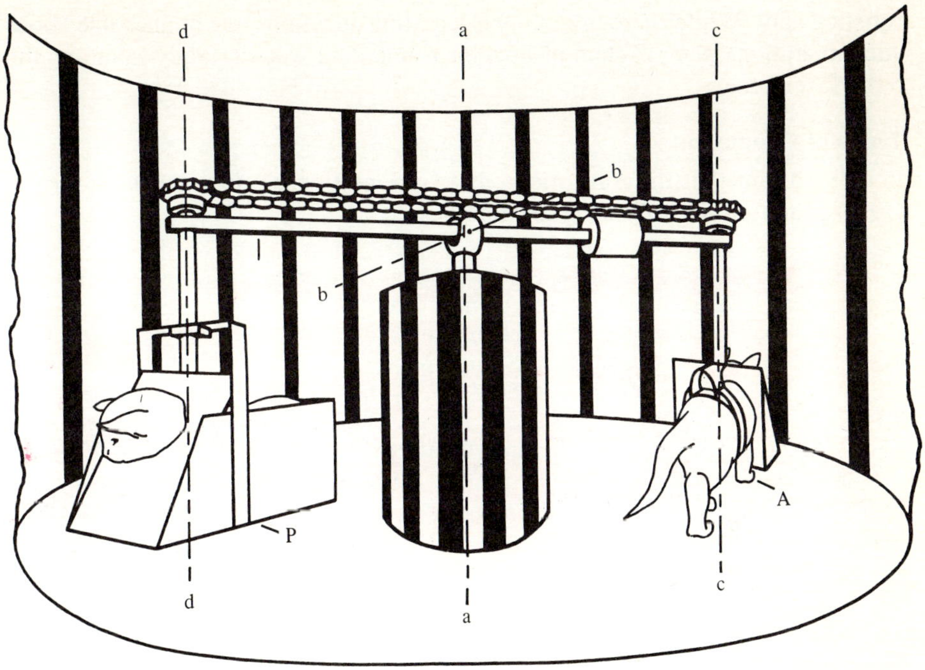

Figure 10-7 Apparatus for equating motion and consequent visual feedback for an actively moving (A) and a passively moved (P) cat. *(From Held & Hein, 1963. Copyright 1963 by the American Psychological Association. Reprinted by permission.)*

behind. Thus, both cats had nearly *identical* visual experience. However, the visual input of the active cat was correlated with movements it initiated, whereas the visual experience of the passive cats had no such correlation. The performance of A cats was found normal and that of P cats abnormal (failing) on tests of visually guided paw placement and in depth discrimination in a visual cliff. Thus, it is once more clear that, although the newborn may show evidence of well-organized visual perception, there is a subtle interplay between genetic and experiential factors in the ontogeny of full adult perceptual capacity.

NAVIGATION IN BIRDS

One of the most fascinating problems in animal behavior is that of how animals find their way in long-distance migrations. While abilities to navigate have been demonstrated in a variety of vertebrate species, it is navigation during the long-distance migrations of birds that has provided some of the greatest achievements and greatest mysteries in all animal behavior. The distances of such migrations can be extraordinary; Arctic terns breed in the Arctic and winter in the Antarctic. The accuracy of bird orientation can be equally impressive; birds may migrate to another continent yet return to the same roosting spot. While there are many interesting questions one might raise about such migrations, the problem of how birds find their way certainly is among the most fascinating.

Types of Orientation

It is important that different methods of orientation be distinguished. Griffin (1955) proposed the following terminology:

Piloting: Steering a course by using familiar landmarks. Many species can use familiar landmarks in an area to find direction.

Compass orientation: The ability to head in a given compass direction without reference to landmarks. Some species use cues to find a direction in which to fly. If they use only compass orientation and suffer a displacement of longitude, they should arrive at a site distant from the usual goal, as they would be unable to compensate for the displacement.

True navigation: The ability to orient in the direction of a particular site (goal) without reference to landmarks. An animal with the ability to orient toward a particular site would be able to compensate for a displacement of longitude and arrive at the appropriate destination.

Possible Cues for Orientation

It is true navigation that provides the most spectacular examples of bird orientation. Various possible cues for navigation have been proposed.

Stars Because human beings are able to use the stars to navigate, it was natural that they might suspect that birds too would use the stars in navigation. The hypothesis was systematically formulated by E. G. F. Sauer (e.g., Sauer, 1957). Emlen (1975) summarized the results of a long research program on stellar orientation in indigo buntings, a species of songbird that breeds in the eastern United States and winters in Central America and the Caribbean. Emlen devised a clever yet simple means of studying the preferred direction in the orientation of his birds. Each was housed in a specially designed cage, the base of which was an ink pad. The sides of the cage were made of a piece of white blotting paper shaped like a funnel, narrow at the bottom. A bird in such a cage can see only what is directly above it. Further, when it becomes active and hops up, ink from the bird's feet is deposited on the blotter paper, leaving a permanent record of its movements. When indigo buntings were placed outdoors in such cages, their orientation was appropriate—south in the fall and northeast in the spring.

In order to study orientation in migratory birds more completely, Emlen followed Sauer's lead and conducted research in a planetarium. Emlen found that the preferred direction of indigo buntings varied according to their physiological condition. By altering length of day, he was able to study simultaneously two groups of birds: one in fall condition and one in spring condition. The results were as expected; although all birds were tested in May, those in fall physiological condition (having just experienced shortening lengths of day) oriented southward, while those in spring condition headed north.

By manipulating the stellar patterns in his planetarium, Emlen found that, in determining direction, the birds used a group of stars clustered within about 35 degrees of the North Star. They do not compensate for time. How do indigo buntings come to find their way by reference to the stars about the North Star? One might suppose that such behavior was instinctive. To test this, Emlen studied orientation in three groups of hand-reared birds. A group reared in diffuse light showed intense activity levels in the fall but no consistent orientation; the ability to orient appeared learned. A group of hand-reared birds exposed to the planetarium sky on alternate nights for 2 months, and with normal celestial rotation simulated, showed normal, southward orientation. A third group was given equivalent experience except that rather than simulating normal star movements over the 2 months, the stars rotated about the star Betelgeuse. Birds in the third group oriented away from Betelgeuse. Emlen proposed that young buntings attend to the movement of star patterns and learn which star is "north" on the basis of the pattern of movements of stellar rotation. What might be the adaptive significance of the evolution of a system wherein the means of orientation is learned? Emlen pointed out that as the direction of rotation of the earth's axis is changing, the polestar changes. Thus, over the long span of evolutionary time, buntings are able to orient toward the most appropriate star, even if the identity of that star changes as a result of changes in the rotation of the earth.

The Sun Research on the cues that are used by birds in orientation by day has focused on the study of homing pigeons (see Schmidt-Koenig, 1965; Keeton, 1974a). Gustav Kramer and his associates demonstrated that birds are able to use the sun to find compass direction. Further, birds are able to compensate for the change of position of the sun during the day, change that occurs at an average of 15 degrees per hour. Birds that were kept under artificial photoperiods, shifted by 6 hours from that in nature, showed when tested outdoors errors of 90 degrees from what would be expected if the sun were used as a cue.

Although time-compensated, sun-based compass orientation may be important for navigation, it is not sufficient to explain true navigation. Various more complex hypotheses have been proposed. G. V. T. Matthews (1955) proposed a *sun-arc hypothesis*, according to which birds compare the sun's altitude extrapolated to the noontime position with that at the same time (determined by an internal clock) at the home site. Such a mechanism might provide sufficient information for correct orientation. Studies of the perceptual capacities of pigeons indicate that they are capable of detecting movement similar to the movements of the sun (Meyer, 1964). Unfortunately, most recent evidence has been incompatible with the sun-arc hypothesis. The opinions of most researchers on bird orientation appear to lean toward Kramer's *map-*

and-compass hypothesis. To navigate, a bird would need not only a compass, as it has been demonstrated they have, but a map as well. While the nature of the map is not understood, results of a number of studies, primarily those of time-shifting using altered photoperiods, suggest that somehow the animal must use its compass information in conjunction with some kind of "map" to engage in true navigation.

Magnetic Cues As some birds are able to home under conditions of substantial overcast, use of the sun cannot provide the full explanation of all homing orientation in pigeons. Evidence that birds use information from the earth's magnetic field is now quite convincing (Keeton, 1974a, 1974b; Walcott, 1974). Keeton attached either magnets or pieces of brass to pigeons and studied their orientation. On clear days, the magnets had no effect. On overcast days, pigeons with brass bars (controls) showed good orientation, while those with magnets (that distorted magnetic patterns around the birds) showed random orientation. Walcott went step further to attach a "Helmholtz coil," a device inducing uniform magnetic fields, to pigeons' heads. Once more, the birds showed disorientation when there was interference with their magnetic fields—but only on overcast days. Evidently, there is a redundancy built into the navigation system of pigeons; they use the sun when available but switch to magnetic cues when the sun is not available.

Olfactory Cues As discussed earlier, pigeons are able to detect and discriminate different odors. A group of Italian workers (e.g., Papi, Fiore, Fiaschi, & Benvenuti, 1973) has suggested that pigeons are able to home on the basis of olfactory cues as the result of a kind of "olfactory imprinting" similar to that of salmon. Although Papi and his associates presented some evidence to support the olfactory hypothesis, there is still some question as to its replicability (e.g., Keeton & Brown, 1976) and the utility of olfactory stimuli over long distances.

Auditory Cues Griffin and Hopkins (1974) measured sound intensities of frog choruses at heights of 500 to 1,000 meters and suggested that natural sounds (e.g., frog choruses, insect sounds, whitecaps, vegetation movements) could provide cues for migrating birds.

Conclusion There is much yet to be learned about the sensory-perceptual mechanisms underlying long-distance migratory orientation in birds. The cues we have discussed, and others that have been proposed, all have their adherents. It is clear that there is no single cue for bird orientation but rather a complex of cues, the relative importance of which probably varies across different species. Orientation during long-distance bird migrations certainly provides some of the more spectacular examples of the sensory-perceptual control of animal behavior.

SUMMARY

To fully understand an animal's behavior, one must study it with a consideration for the animal's *Umwelt*, or unique sensory-perceptual world. Methods available for the study of sensory-perceptual function in animals include anatomical methods, electrophysiological methods, measurements of species-characteristic responses, and measurement of learned responses. The effective stimuli in the regulation of particular behavior patterns can be studied using correlational methods, experimental manipulation of stimuli, and the interruption of sensory systems.

A brief survey of the nature and functioning of vision, audition, chemical senses, active sensory systems, and other sensory systems is presented.

Although many sensory-perceptual capacities appear present in young animals (e.g., at birth, hatching, or eye opening), experience plays an important role in the sensory-perceptual development of many systems.

There are many potential cues for orientation in the spectacular long-distance migrations of birds. The relative importance of the stars, the sun, magnetic cues, olfactory cues, auditory cues, and other cues appears to vary with both species and situation.

Chapter 11

Neural Mechanisms and Behavior

The study of neural mechanisms constitutes the second of the three mechanism-related topics that we shall consider. Because (1) many studies of neural mechanisms deal with topics outside of the usual coverage of courses in animal behavior and (2) most universities offer whole courses in physiological psychology or its equivalent, there will be no attempt at exhaustive coverage. Rather, we shall consider some interesting research that is related to the study of the types of naturally occurring behavior that are emphasized throughout this book. After a brief discussion of nervous systems and the methods used in their study, we shall consider two major topics: neuroethology and the neural regulation of mammalian behavior.

NERVOUS SYSTEMS

That a complex nervous system is not necessary for evolutionary success is evident from a consideration of acellular animals, the Protozoa. Even without structural differentiation into a cellular pattern of organization, protozoans are able to carry out the functions necessary for survival and reproduction. However, the behavioral capacities of protozoans are obviously quite limited; complex behavior requires a cellular organization.

Nerve Cells

The nervous systems of multicellular animals are "interrupted" systems, composed of individual, self-contained units or cells. They are brick buildings rather than buildings of poured concrete (Dethier & Stellar, 1961). The unit of the nervous system is the nerve cell, or *neuron*. The neurons of different parts of the nervous system are quite different in both structure and function, but they typically have three major parts: *dendrites*, which receive excitation and conduct it to the cell body; a *cell body*, which contains a nucleus; and an *axon*, which carries excitation away from the cell body. The conduction of excitation within a cell is primarily a result of *action potentials*, electrical activity progressing along the cell and resulting from a complex of membrane, chemical, and electrical changes. Action potentials are of an either-or, all-or-none character, with all action potentials in a given cell occurring at the same voltage.

The places where cells come together are called *synapses*. One cell communicates with and stimulates another across a synapse, as when the axon of one cell synapses on the dendrites of another. Whereas transmission within cells is generally all or none, that across synapses is graded; that is, the degree of stimulation of the second cell is variable. While conduction of excitation across some synapses may be electrical, it more commonly has a chemical basis. Chemicals called *neurotransmitters*, stored in "packets" in the presynaptic membrane (first cell), are released on the arrival of an action potential. The chemicals cross the synaptic gap and affect the generation of an action potential in the postsynaptic membrane (second cell). Whereas some cells stimulate activity in others, some inhibit other cells by releasing inhibitory neurotransmitters rather than excitatory neurotransmitters.

Invertebrate Nervous Systems

Through evolution, nervous systems have become increasingly *differentiated* (specialized) and *centralized*. Whereas the cells of the nervous systems of phylogenetically old species all perform very similar functions, those of more recently evolved species are specialized. The most obvious specialization is that into (1) sensory cells, which are sensitive to environmental changes, are altered by them, and transmit such changes; (2) conducting and processing cells, which transmit and process information (sometimes imposing their own rhythms upon it); and (3) effectors, muscle cells which contract in response to stimulation by neurons. Even within each of these broad categories—sensory cells, conducting and processing cells, and effectors—there is considerable specialization of function.

Centralization of function is a second important evolutionary trend. Whereas the nervous systems of phylogenetically old species tend to be diffusely located about the body, those of more recent species tend to be progressively more clumped together into ganglia or brains. Most complex organisms generally locomote in one direction—forward—and the major sensory organs and brain tend to be located at the forward-facing portion of the body.

The nervous systems of the phylum Coelenterata (hydras, jellyfishes, and so on) are diffuse *nerve nets*. The neurons in nerve nets are anatomically dispersed in an apparently random form, somewhat like a game of "pick-up sticks" scattered about a living room floor. Impulses in nerve nets are conducted in a diffuse manner, out in all direction from the point of stimulation. The diffuse conduction pattern through a decentralized nervous system contrasts with the more precise conduction patterns through organized tracts and nuclei that is characteristic of vertebrate nervous systems. The nervous system of the jellyfish is more centralized than that of the hydra, with two nerve rings organized in the bell of the animal.

The nervous systems of echinoderms, such as starfish, are characterized by increased specialization and centralization. Nerve tracts, association neurons, and reflex arcs appear. The nerve ring is moved away from the body surface. With increased anatomical and physiological differentiation of the echinoderm nervous system comes increased differentiation of behavior, as in the organized movements of the starfish's five feet.

With flatworms, we find the first nervous systems that are organized about a plan of bilateral symmetry (where the two halves of the system are virtual mirror images) rather than radial symmetry (symmetry in a circular pattern about a single central point). This bilateral symmetry was critical to the evolution of advanced levels of centralization in the control of behavior. The nervous systems of flatworms are made up of rather small brains, which give rise to prominent nerve cords. A series of ganglia is located along the nerve cord. This plan is further developed in the segmented worms, where the ganglia are even more prominent. In cephalopod molluscs, such as octopuses, the frontmost ganglia have differentiated into a true brain. Estimates of the number of discrete sections, or lobes, in the octopus brain range up to 64 (Sanders, 1975). Octopuses are capable of a great diversity of social, orientational, and learned behavior associated with their complex nervous systems (e.g., Young, 1961; Sutherland, 1962).

The brains of arthropods represent a "pinnacle" of invertebrate evolution in that they are highly differentiated and capable of mediating quite complex behavior (see Figure 11-1). The complexity of behavior associated with the arthropod nervous system has been documented throughout the book.

The differences between insect and vertebrate brains are important. The insect nervous system is a segmented string of ganglia located near the ventral (lower) surface of the animal; the nervous systems of vertebrates develop from neural tubes which lie near the dorsal (upper) surface. Whereas the nervous systems of vertebrates are dominated by their brains, the ganglia located in different segments along the nerve cords of arthropods possess a functional autonomy. The ganglia of nerve cords in arthropods have retained "states rights," in contrast to the "centralized government" of the vertebrates (Roeder, 1970).

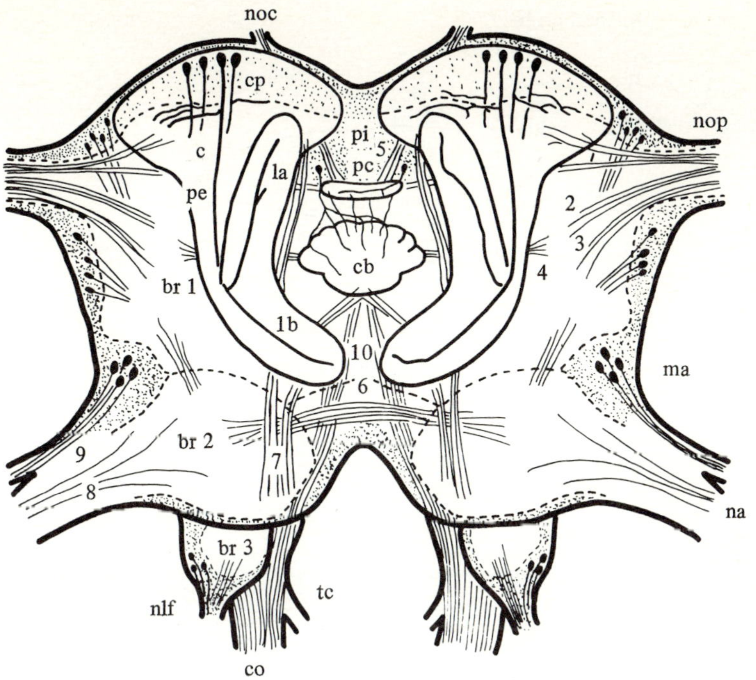

Figure 11-1 Diagram of a frontal section through the brain (supraesophageal ganglion) of a cricket. *(From Huber, 1960, and Roeder, 1967.)*

Vertebrate Nervous Systems

A diagram of a prototypical vertebrate brain is presented in Figure 11-2. Vertebrate brains contain five basic divisions:

Myelencephalon: The hindmost part of the brain, including the medulla.

Metencephalon: The next most caudal part of the brain, which includes the cerebellum and pons.

Mesencephalon: Or midbrain, which includes the optic tectum and midbrain "reticular activating system."

Diencephalon: A part of the forebrain that includes the thalamus and hypothalamus.

Telencephalon: The most rostral part of the brain, characterized by the cerebral hemispheres, olfactory bulbs, and numerous other structures.

The brains of six species of vertebrates are pictured in Figure 11-3. It is apparent that the brains differ in total size. Note the extreme development of the human brain. Part of the reason for differences in brain size is that different species have different body sizes—the larger the body size, the larger the brain size. However, even when body size is taken into consideration, there remain substantial species differences in relative brain size. Jerison (1973, 1976) has argued persuasively that within different groups of vertebrates there has been

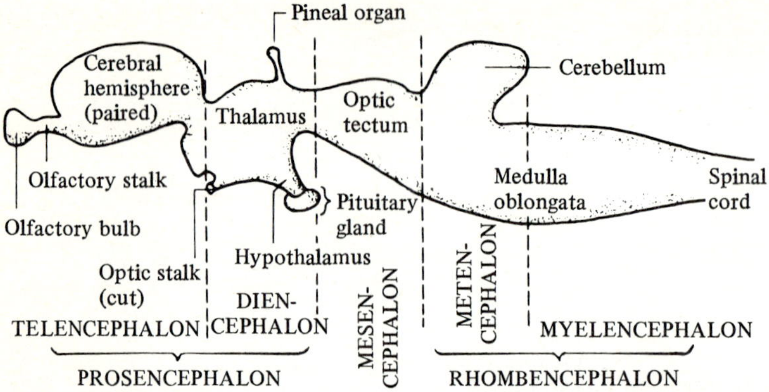

Figure 11-2 The basic divisions of the vertebrate brain. This diagram represents a fairly typical mammalian brain relatively early in development and is useful in that it indicates the regions in which some major structures are located. *(From Romer, 1962, and Manning, 1967.)*

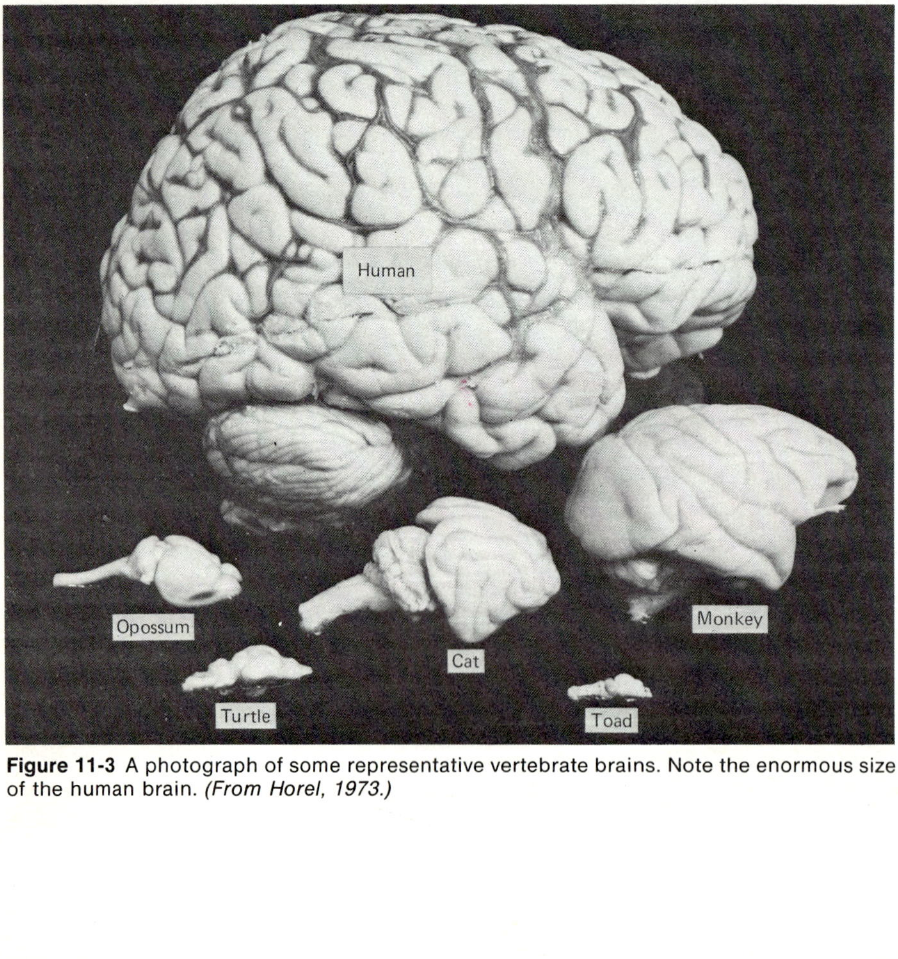

Figure 11-3 A photograph of some representative vertebrate brains. Note the enormous size of the human brain. *(From Horel, 1973.)*

an evolutionary trend toward larger-brained animals. Among both reptiles and mammals, as well as within different mammalian taxa, comparison of archaic brains (based on endocasts taken from fossils) with those of more recently evolved species reveals a consistent trend toward large brains.

While all vertebrate brains are characterized by the same basic structural divisions, the component structures and relative development thereof varies drastically across different species. The telencephalons of fishes and amphibians are quite small; true neocortex is first found in alligators and becomes so enlarged in some mammals as to envelop the rest of the brain. Some teleost fishes have prominent inferior lobes in the midbrain; midbrain optic lobes are prominent in fishes, amphibians, reptiles, and birds. Within the line of mammals leading from insectivores to primates, there have been changes in the relative sizes of structures in the telencephalon, with a dramatic increase in size of the neocortex, some increase in the size of the hippocampus, and a general decrease in the size of the olfactory bulbs (Stephan & Andy, 1969).

METHODS IN THE STUDY OF NEURAL MECHANISMS

The most complete understanding of neural mechanisms and behavior is gained when several different methods are used and produce converging results. Several methods will be surveyed.

Lesion or Ablation

A first method of studying the relationship between a particular neural structure and behavior is simply to remove that structure and look for changes in the behavior of interest. Such removal of neural tissue is referred to as a *lesion* or *ablation*. The availability of brain atlases (maps of brains) for many mammalian species and of stereotaxic machines, with which electrodes can be inserted precisely, allows for accurate placement of small lesions into the deep structures of the brain. Larger lesions in more accessible regions can be made using aspiration, a suction procedure.

Electrical Recording

As one manifestation of its functioning, the brain generates electrical activity. Insight into the control of behavior can be gained by recording that activity and correlating it with overt behavior. There are many levels at which electrical recordings can be made, and the methods differ at different levels. The recording of brain waves by means of the *EEG* is a rather gross method and was described in Chapter 4 (Figure 4-2). In another gross procedure, one can record the *evoked potentials* that are generated in large populations of neurons as a result of sensory input. At the opposite extreme, microelectrodes can be placed either into or next to individual neurons in order to record their electrical activity apart from that of the rest of the brain.

Electrical Stimulation

Behavioral patterns can be initiated as the result of electrical stimulation delivered to discrete brain regions. Typically, electrodes are implanted with the animal under anesthesia. Electrodes are placed with the use of a stereotaxic machine and the guidance of an atlas. Later, electrical stimulation is delivered while the animal is awake. As the brain contains no pain receptors, the procedure is painless. Quite complex behavioral patterns have been elicited with such electrical stimulation.

Chemical Techniques

As discussed above, transmission across synapses is generally accomplished by means of chemicals, or neurotransmitters. One approach to the study of neural function is to alter the functioning of these transmitters by various chemical methods.

While a number of chemicals are known or suspected to be neurotransmitters, much research has focused on *acetylcholine*, the transmitter in so-called "cholinergic" systems, and *serotonin, norepinephrine,* and *dopamine* (termed "biogenic amines" or "brain monoamines"), and their associated "monoaminergic" systems.

Drugs that alter neurotransmitter activity can be administered to the organism either systemically, so that they affect the entire body; directly into the total brain; or directly into specific structures in the brain. In the latter procedure, a *cannula* is implanted into the brain using much the same techniques as those for implanting electrodes. Later, various chemicals can be introduced via the cannula. Various anatomical systems have been shown to be rich in particular transmitters; lesions of such areas can be used as a "chemical" method. Biochemical assays, measurement of "turnover rates," and other techniques can be used as chemical techniques in studying brain function.

Interaction of Methods

Each method has its advantages and disadvantages; the method of choice will vary with the problem under study. For example, measurement of the activity of single neurons produces too restricted a picture of neural activity if one is interested in sleep. On the other hand, the EEG is of little value if one is interested in effects of experience on the functioning of individual cells. As behavior, chemical changes, and electrical changes occur simultaneously, interference with any of them will alter the others. The effect of a lesion of a particular brain region might not be mediated as a result of the loss of that particular nerve tissue but rather as a result of a chemical change occurring elsewhere in the brain as a secondary result of the lesion (e.g., Lints & Harvey, 1969). A comprehensive and balanced understanding of neural function can be gained from the integration of results obtained with various techniques.

NEUROETHOLOGY

Ethologists have long sought to incorporate physiological approaches into their domain. The early writings of the classical ethologists relied heavily on the work of Weiss, Gray, Lissmann, von Holst, Adrian, and W. R. Hess. With recent advances in both physiological methods and behavioral techniques, the marriage between ethology and neuroscience into neuroethology is being continually strengthened and consummated.

Endogenous Activity and Behavior

The classical ethological model of behavior, as depicted in Lorenz's hydraulic model, represented an important departure from the prevailing neurophysiology of the day. It was common at the time to regard the nervous system as a rather passive switchboard. Stimuli were conceptualized as acting on an essentially passive organism and "stimulating" it to activity. The brain functioned to route activity along appropriate connections between input from sensory systems and output to motor systems, much as a switchboard routes telephone calls. The simple reflex served as a model for nervous system function. The ethologists held quite a different view. To the ethologist, the nervous system was organized by the genes and was spontaneously active. Overt behavior failed to occur because inhibitory mechanisms held the organized and spontaneously active systems in check. Rather than viewing stimuli as *initiating* activity, the ethologists considered sign stimuli or releasers to *release* fixed action patterns by acting on the IRM to remove inhibitions to behavior. The nervous system was viewed as capable of generating patterned output by itself.

The classical ethological models were widely criticized because they appeared to contradict the prevailing neurology of the day (e.g., Lehrman, 1953). More recent information has revealed a picture of the neural regulation of at least some behavioral patterns in some species that is remarkably congruent with that proposed by the classical ethologists. Much of the patterning of animal behavior appears to result from endogenous neural organization and activity.

Copulation in Praying Mantids Mantids are relatively inactive, predatory insects that are hidden by cryptic coloration. They respond to slight movement of an insect, be it of another species or its own, with a rapid strike and often with ingestion. This presents a considerable problem for the male, who is smaller than the female and must approach her if he is to mate. The male approaches slowly, stopping whenever the female moves. Finally, he takes a short, flying leap and lands on the female. If he does not land oriented head to head, the male turns rapidly and clasps the female with his forelegs. The male then initiates copulatory movements. Coupling occurs after 5 to 30 minutes, and copulation lasts for several hours while a spermatophore forms in the female's genitalia (Roeder, 1967).

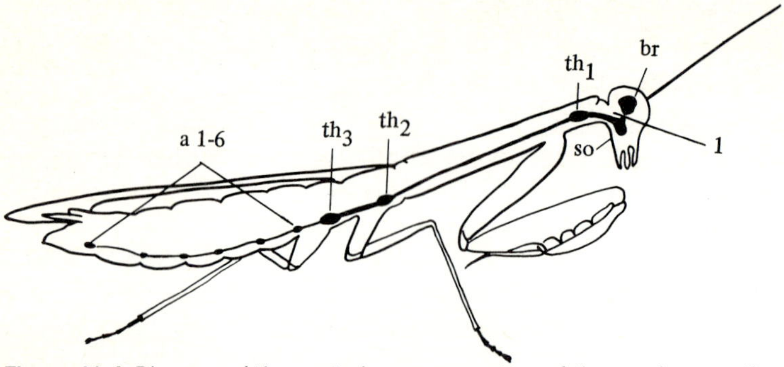

Figure 11-4 Diagram of the central nervous system of the praying mantis. *a 1-6,* abdominal ganglia; *br,* brain or supraesophageal ganglion; *so,* subesophageal ganglion; *th 1-3,* thoracic ganglia. *[From Roeder, K. D.* Nerve cells and insect behavior. *(Rev. Ed.). Cambridge, Mass.: Harvard University Press, copyright © 1963 and 1967 by the President and Fellows of Harvard College.]*

Cannibalistic attack by the female may take place during the male's approach, directly after mounting, or as the couple separate. Typically, the female bites off the male's head and prothorax (see Figure 11-4), thus initiating a rather gross natural ablation experiment. But decapitation, far from eliminating copulatory behavior, appears to enhance it. Decapitation is followed by normal copulatory movements and lateral or rotary walking movements that lead to orientation in the head-to-head position. Copulation continues for several hours and a normal spermatophore is formed. So effective is the copulation of headless mantids that when Roeder experienced difficulty in breeding males of an inbred strain that was deficient in courtship behavior, he decapitated them and placed them on females. The headless male will display these behaviors, clasping objects of appropriate size, for days.

Roeder interprets this enhancement of sexual behavior as due to the removal of neural structures in the head that normally inhibit copulatory movements. The ablation technique was supplemented with information from electrical recording to strengthen the case for this interpretation. Severing of the nerve cord just above the terminal ganglion produced an increase in organized electrical activity in the phallic nerve in the absence of any sensory input (Roeder, 1967). Such an increase is exactly what one would expect according to Roeder's interpretation.

Walking in Praying Mantids The head of the praying mantid contains two ganglia, the supraesophageal (brain), or frontmost, and the subesophageal, or second ganglion (see Figure 11-4). If the supraesophageal ganglion is surgically removed, the normally inactive mantid walks continuously, unable to climb, turn, or avoid obstacles. Removal of both ganglia, as in decapitation, results only in the copulatory movements previously described; the mantid never walks spontaneously. These data suggest that there are mechanisms in the subesophageal ganglion that excite walking and that are inhibited by continuous activity from the supraesophageal ganglion (brain). Thus, the brain must be active in order for the mantid to be inactive (Roeder, 1967).

Flying in Locusts According to a pure reflexology position, complex rhythmic behavioral patterns, such as the flight of insects, should be regulated as a reflex chain, with feedback from the periphery triggering changes in behavior, affecting the periphery, and so on. If the rhythms underlying flight were endogenously generated, however, feedback from the periphery should be relatively unimportant. D. M. Wilson (1966, 1968) studied the flight of desert locusts in wind tunnels so arranged that the animals would fly but remain in approximately the same position. Thus, Wilson was able to make electrical recordings of muscle firing patterns. Fixing the wings in a constant position or removing the wings entirely had little effect on the muscle firing rhythm. Cutting sensory nerves also had little effect. Wilson viewed feedback from sensory systems as important not in generating rhythmic behavioral patterns but in superimposing error corrections on an already active system. Flight movements are patterned spontaneously in the nervous system.

Cricket Song The courtship songs of crickets are a familiar sound on summer nights in many parts of the United States. The cricket makes these sounds by rubbing its wings together, thus rubbing the file on one wing across the scraper of the other. The song patterns within a species tend to be quite stereotyped. Several investigators have sought to determine how the highly stereotyped movements that produce cricket song *(stridulation)* are produced and regulated.

Alexander (1968) and Bentley and Hoy (1974) have summarized research on the neurobiology of cricket stridulation. Only two thoracic ganglia, the two nearest to the head, are necessary for singing. (The location of thoracic ganglia in mantids is shown in Figure 11-4.) Bentley completely isolated these two ganglia from peripheral sensory cues and found that the motor patterns produced by the ganglia were identical to those of the intact animal. Thus the pattern is generated endogenously. Through the implantation of electrodes in a number of neurons and muscles, it was shown that there are two groups of cells that fire in synchrony within groups to alternately open and close the wings.

The pattern of neural organization should be familiar by now. Cricket song patterns are generated in the lower ganglia in the absence of sensory input. The next step is to look for higher ganglia involved in the regulation of these patterned movements. Earlier work by Franz Huber and his students showed that singing could be elicited either by electrical stimulation or by making small lesions in the brain (see Figure 11-1). We thus see the complete picture of behavior organized in lower ganglia and regulated, in relation to sensory input, by higher ganglia. Patterning is generated endogenously from the lower, thoracic ganglia.

Alexander (1968) related the organization of cricket stridulation to the classical ethological model. He noted four important characteristics:

1 All the form of some cricket stridulations and some of the form of all of them is independent of external stimuli in adults and is very minimally affected by feedback.

2 There is a strong genetic component to stridulation patterns, and effects of environmental manipulations studied thus far have been minimal with regard to the development of species-characteristic songs.

3 The patterning of cricket songs is dependent on pacemakers that are unaffected by most environmental stimuli.

4 When a male is inhibited from calling, the intensity and specificity of stimuli necessary to elicit calling is greatly reduced.

Alexander concluded: "I can think of no significant way that cricket stridulation deviates from the criteria of Lorenz's original 'hydraulic' model, constructed in a time of almost complete ignorance of CNS function and apparently largely discarded now as a useful construct, perhaps even by Lorenz himself" (p. 212).

Conclusion In attempting to illustrate a point, one always faces the danger of overemphasis. Reflexes and external stimuli are important to the organization of many behavioral patterns. The important point behind the examples we have cited, and others we could have cited, is that there exist behavioral patterns that gain their patterning as the result of organized neural activity that is generated endogenously and is minimally altered by external stimuli. Typically these behavioral patterns are organized in lower regions and regulated by higher regions of the nervous system. The resemblance of this organizational pattern to the classical ethological model provides a dramatic illustration of the value and importance of the model, even though its applicability today appears quite limited.

Electrical Stimulation and Integrated Behavioral Patterns

According to Tinbergen's (1951) hierarchical model of behavior, instinctive behaviors are organized into a hierarchical structure, with instinctive centers constituting the highest levels in the hierarchy and individual appetitive behavior, consummatory acts, muscles, and muscle units subordinated to the higher instinctive centers. On the basis of a simple physiological interpretation of Tinbergen's model, it ought to be possible to find loci in the brain at which these successively more inclusive organizational levels are located.

Early attempts to locate neural "centers" mediating the hierarchical organization of behavior appeared quite successful (e.g., von Holst and von St. Paul, 1962, 1963). Electrical stimulation delivered to the brains of chickens produced either simple elements of behavioral patterns or complex integrated patterns. Von Holst and von St. Paul reported that they were able to evoke "almost all of the forms of activity and vocalization familiar to those acquainted with chickens" (1962, p. 59). Some stimulation sites produced a complete, integrated pattern leading to sleep and including a cessation of eating, looking around and walking, fluttering the eyelids, yawning, sitting down, fluffing the plumage, retracting the head, and closing the eyes. Von Holst and von St. Paul attempted to study conflict situations by simultaneously stimulating loci for two different behavioral patterns. While the result was often a combination of two compatible behavioral patterns or the suppression of one of two incompatible patterns, some results were more spectacular. In the most extreme few cases, a completely new behavioral pattern arose. When loci yielding (1) aggressive pecking and (2) flight with smoothed feathers were simultaneously stimulated, the chicken displayed a new pattern—(3) a frantic bout of running and vocalization, characteristic of an animal near its nest confronted with a superior opponent.

The early reports of von Holst and von St. Paul led later investigators to use more sophisticated techniques, including histological verification of the location of electrodes, to continue investigation of the organization of behavior using electrical stimulation. These more recent studies have rarely produced the kind of dramatic results reported in the earlier work. Phillips and Youngren (1971) concluded, after examining results from more than 2,600 brain points in ducks and chickens, that most of the behavioral patterns they observed were simple ones, like turning the head. The only complex natural sequences elicited in their studies were agonistic behaviors. Simultaneous stimulation of different loci in the brains of juvenile gulls resulted only in the suppression of one of the two behavioral patterns, even with careful adjustment of current levels (Delius, 1973). Complex behavioral patterns, including mating behavior, biting attack, defensive threat, eating, grooming, yawning, investigation, escape, and gnawing have been elicited with electrical stimulation to mammalian brains (e.g., Roberts, Steinberg, & Means, 1967; Waldbillig, 1975). However, the organization of loci producing these behavioral patterns appears much more diffuse than that suggested by von Holst and von St. Paul.

It now appears that complex behavioral patterns are mediated by neural circuits that are located diffusely about the brain rather than in tightly organized centers. While the hierarchical model may be of value as a conceptual tool in dealing with behavior, it appears to lack close correspondence with neural organizational patterns.

NEURAL MECHANISMS AND MAMMALIAN BEHAVIOR

Most of the efforts that have characterized the dramatic growth of physiological psychology and the related neurosciences in the 1960s and 1970s have been directed at the study of mammalian behavior. We shall sample results as related to the control of several functional categories of behavior. This survey will be brief and will not do justice to the research area. If you lack background in physiological psychology and neuroanatomy, you probably will not be able to keep differentiated all the brain regions discussed. Reference to Figure 11-2 should help. However, it is general principles rather than particular structures that are important for the present.

Copulatory Behavior

The control of behavior by a system which is organized in lower neural areas and regulated by higher ones is apparently not limited to insect nervous systems but applies to the regulation of copulatory behavior in mammals as well. For example, Beach (1967) proposed four hypotheses relating to the cerebral and hormonal control of mammalian copulatory behavior:

1 Species-characteristic copulatory patterns consist in part of reflexes organized in the spinal cord and myelencephalon (see Figure 11-2) and capable of functioning after separation from the more rostral (higher) portions of the central nervous system (CNS).

2 In the intact adult organism, these lower regions are under varying degrees of control by higher brain regions and, in some cases, this control is inhibitory.

3 Hormones secreted by the ovaries and testes have little effect on the lower regions.

4 One of the ways in which hormones influence copulatory behavior is by modifying the control of lower regions by higher regions.

With the possible exception of hypothesis 3, all of Beach's proposals have been supported by research on copulatory behavior. Research on hormones will be considered in Chapter 12. Here, we shall consider some of the evidence that supports the view that the reflexes mediating copulatory behavior are organized in lower regions of the CNS and regulated by the activity of higher regions.

The Spinal Cord The spinal cord, which runs downward from the myelencephalon as shown in Figure 11-2, is often considered as a mere conduit for neural messages. However, there is considerable evidence that reflexes, including those which together constitute copulatory behavior, are organized in the cord.

Perhaps the clearest example of spinal organization lies in the study by Hart (1967) of "spinal" male dogs, dogs whose spinal cord connections with the brain had been completely interrupted by lesions. Such animals are unable to maintain normal posture or to walk about. However, their reflexes could be tested when they were maintained in special apparatus. The results were remarkable. By applying tactile stimulation to different parts of the spinal male's penis, Hart was able to elicit a set of four different reflexes that could easily be seen to be related to the normal copulatory pattern of dogs. These patterns included penile erection, shallow pelvic thrusting, intense thrusting, and expulsion of seminal fluid. It was clear from Hart's work that elements of the copulatory pattern of dogs, including the intense ejaculatory reaction (IER) and copulatory lock, are mediated by spinal mechanisms. Similarly complex reflexes were found in a related study of male rats, although in rats the relationship of the reflexes to normal behavior is less clear (Hart, 1968a).

Similar organizational patterns have been found in the spinal cords of females. Tactile stimulation to the genital region of spinal female cats produced several clearly defined reflexes which could be related to the normal mating pattern (Bard, 1940). Lordosis, the female copulatory posture in rats, was elicited, although somewhat inconsistently, by tactile stimulation or mounting by a male in spinal female rats (Hart, 1969).

The Clasp Reflex of Frogs A particularly clear example of the organization of reflexes mediating vertebrate copulatory behavior can be found in the clasp reflex of frogs, *Xenopus laevis* (Hutchison & Poynton, 1963). During mating, frogs remain in *amplexus*, wherein the male clasps the female tightly about the waist and remains in that position for 10 to 30 hours. Hutchison and Poynton studied the male's clasp reflex after a number of different transections of the brain. In such transections, a complete cut is made through the brain, interrupting all rostral-caudal connections. Results indicated that clasping is organized in the upper medulla and normally inhibited by neural structures higher than the medulla (see Figure 11-5). When a cut was made below the upper medulla (A), no clasping was elicited because the connections from the organizing region to the muscles were interrupted. When a cut was made above the level of the inhibitory region (E and F), no clasping was observed because the inhibitory mechanisms remained functional. When intermediate cuts were

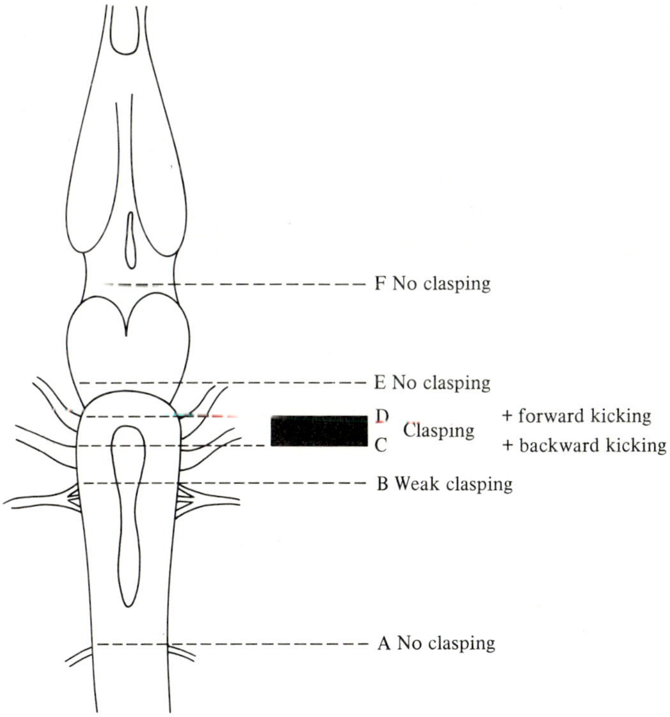

Figure 11-5 The brain of *Xenopus laevis*, showing the effects of transections at various levels. Cuts below the excitatory region (A) or above the inhibitory region (E and F) resulted in no clasping. Cuts between the inhibitory and excitatory centers (C and D) produced strong clasping. *(From Hutchison & Poynton, 1963.)*

made, at C and D in Figure 11-5, the inhibitory action of higher regions was removed and clasping was observed.

Mammalian Brains If the reflexes underlying copulatory behavior that are organized in the spinal cord are regulated by the brain, the occurrence of copulatory behavior ought to be altered as the result of manipulations of the brain. At least some of these manipulations ought to indicate an inhibition of lower regions by higher ones. Thus, once more, we should find evidence of the remarkable phenomenon wherein the *removal* of part of the brain *improves* performance. As the literature on neural control of copulatory behavior is voluminous (Bermant & Davidson, 1974; Malsbury & Pfaff, 1973; Montagna & Sadler, 1974; Diakow, 1974), we shall cite but a few representative studies.

Various brain lesions have been found to facilitate male copulatory behavior in rats. Destruction of a region at the junction of the diencephalon and mesencephalon (see Figure 11-2) results in such a facilitation (Heimer & Larsson, 1974; Lisk, 1966). Destruction of a smaller region in the rostral midbrain produces a dramatic reduction in the length of the refractory periods that follow ejaculation (Clark, Caggiula, McConnell, & Antelman, 1975; Barfield, Wilson, & McDonald, 1975). While postejaculatory "singing" by the males is eliminated or reduced, other aspects of the copulatory pattern are normal.

Beach (1944) found that the lordotic responses of some female rats were intensified and longer lasting after the cerebral cortex was removed. A different approach was taken by Clemens, Wallen, and Gorski (1967), who use a technique called *spreading depression*, in which potassium chloride (KCl) is applied to the cortex. KCl produced a substantial reduction in cortical activity, or "functional decortication." Lordotic responding in estrogen-primed females was *enhanced*.

Not all brain regions are inhibitory. The hypothalamic area, located at the base of the diencephalon (Figure 11-2), is particularly important and generally facilitative. Lesions in either the medial forebrain bundle or medial preoptic area of the hypothalamic region of male rats produce severe deficits in copulatory behavior that are typically not reversed with hormonal or other therapy (see Malsbury & Pfaff, 1973). Electrical stimulation in the region of the hypothalamus can facilitate copulatory behavior. Ablations of the cerebral cortex of male rats interfere with copulatory behavior (Larsson, 1962, 1964).

The available literature suggests that the regions we have discussed, as well as other regions, interact in subtle ways and in relation to hormonal and sensory input to play upon the spinal reflexes and govern overt copulatory behavior.

Neurochemical Regulation The results of many studies indicate that the neurotransmitters classed as biogenic amines (serotonin, norepinephrine, and dopamine) are involved in the control of copulatory behavior. One or more of these neurochemicals appears to act to inhibit copulatory behavior in both males and females.

Administration to estrogen-primed female rats of drugs that lower levels of brain serotonin produces a clear facilitation of lordotic responding (Meyerson, 1964; Zemlan, Ward, Crowley, & Margules, 1973). Drugs that elevate brain biogenic amine levels generally interfere with lordosis, whereas those that lower biogenic amine levels facilitate its occurrence. Kow, Malsbury, and Pfaff (1974) suggested that in its normal action, the hormone progesterone facilitates female copulatory behavior by turning off serotonergic brain systems that otherwise inhibit copulatory behavior.

Analogous mechanisms have been observed in male rats (e.g., Dewsbury, 1975b). For example, administration of *p*CPA (*p*-Chlorophenylalanine), which lowers brain serotonin levels, facilitates male copulatory behavior (Tagliamonte, Tagliamonte, Gessa, & Brodie, 1969; Salis & Dewsbury, 1973).

More sophisticated techniques are now being utilized in studying exactly which biogenic amines are involved in the control of copulatory behavior and the precise mechanisms through which this control is accomplished. For example, different pathways in the brain are especially rich in particular neurotransmitters. Lesions or other manipulations of such small regions provide more precise information than that gathered from systemic administration of drugs. This is because systemically administered drugs produce many effects throughout the body. The results discussed above, in which small lesions of the rostral midbrain decrease postejaculatory refractory periods in male rats, are interpretable as resulting from an interruption of the "dorsal norepinephrine bundle."

Cannulas can be used to permit implantation of drugs into particular structures. Thus, when Zemlen et al. (1973) placed drugs which block receptors of serotonin directly into hypothalamic structures known to contain serotonin terminals, they found a facilitation of lordosis.

It thus appears that much of the subtle interplay of different brain structures regulating the spinal reflexes of copulation may involve interactions of different neurotransmitters. Future progress will be made as neural and neurochemical methods are combined to develop a comprehensive view of brain function.

Sleep

It should be apparent from the discussion of copulatory behavior that the brain structures important in regulating particular behavioral patterns are dispersed throughout a large portion of the brain. Early hope that there might be a "sex center" or an "aggression center" have proved far from correct. Never is the necessity of dealing with circuits in the total brain more apparent than when the neural bases of sleep are considered (e.g., Webb, 1968; Jouvet, 1969; Williams, Holloway, & Griffiths, 1973). Relevant structures are scattered throughout a wide region of the brain.

A region important for waking is the *reticular activating system*, a region in the central core of the brainstem that begins in the medulla and courses up through the midbrain, to the thalamus and the diencephalon (see Figure 11-2). Electrical stimulation of this area will arouse a sleeping animal; lesions to this area produce an animal that sleeps incessantly. However, the reticular activating system is not the only brain area important to waking; a number of other structures, particularly the posterior hypothalamus, seem important.

While it once was thought that sleep was simply a *passive* result of the removal of sensory stimulation, it is now apparent that sleep is an *active* process, the result of increased neural activity in certain regions. Sleep can be induced by electrical stimulation delivered to a great range of brain areas including the cortex, thalamus, subthalamus, hypothalamus, mesencephalon, pons, cerebellum, medulla, and spinal cord. The Raphé nuclei, an elongated group of nuclei in the center of the brainstem, together with some structures in the base of the forebrain appear most critical to the initiation and maintenance of slow-wave sleep (SWS; see page 58). Lesions in the Raphé nuclei of cats eliminate sleep for a period of several days. The Raphé nuclei are an especially interesting region because they are extremely rich in serotonin. Indeed, it appears that serotonin is critical to SWS. *p*CPA, which inhibits the synthesis of serotonin, interferes with normal sleep. Drugs which increase levels of brain biogenic amines (including serotonin), such as monoamine oxidase inhibitors, generally increase slow-wave sleep.

A group of nuclei in the caudal pons appear particularly important to the regulation of paradoxical sleep. Lesions in this region eliminate paradoxical sleep, although the normal sleep-wake cycle remains.

A region of the midbrain and pons called the *locus coeruleus* is particularly rich in norepinephrine-containing fibers. Lesions in this region interfere with paradoxical sleep. A number of drugs that affect norepinephrine levels generally alter paradoxical sleep. The control of paradoxical sleep is not limited to norepinephrine, however. Jouvet hypothesized a "three-key" mechanism for the initiation of paradoxical sleep that involves (1) serotonergic regions of the Raphé nuclei, (2) a cholinergic mediation, and (3) a final triggering of the norepinephrine-rich *locus coeruleus*.

The nature of the mechanisms regulating sleep is poorly understood. This brief presentation represents a considerable oversimplification of the control of sleep. The regulation of sleep is being actively studied by a large group of scientists using anatomical, physiological, and biochemical techniques.

Maternal Behavior

A number of neural structures have been implicated in the regulation of maternal behavior (Lamb, 1975). Effects of lesions generally have been of two kinds: they either eliminate maternal behavior or interfere with its fine integration.

Lesions in the medial preoptic area of the rat hypothalamus produce severe deficits in maternal behavior; maternal females do not build nests, do not retrieve pups, and do not nurse (Numan, 1974). There is some evidence that stimulation of the hypothalamus can produce maternal behavior. Changes in hypothalamic norepinephrine levels have been correlated with birth and the onset of maternal behavior (Moltz, Rowland, Steele, & Halaris, 1975).

A number of studies of maternal behavior have focused on the *limbic system* of the forebrain, a group of structures implicated in the regulation of species-characteristic behavior. The deficits produced by lesions of the cingulate cortex in female rats appear quite different from those created with hypothalamic lesions. While the females appeared "motivated" toward the display of maternal behavior, they were generally unable to engage in the complete, organized behavioral sequences necessary for efficient behavior. Thus, rather than retrieve a pup to the nest, they might pick it up, wander about, and deposit the pup outside the nest (Slotnick, 1967). Further, deficits found in mice with lesions of the septal formation appeared to result from an inability of the mice to inhibit irrelevant responses rather than from the loss of maternal behavior (Carlson & Thomas, 1968).

Interpretation of these and other experiments is complicated by a persistent problem—that of species differences. For example, lesions of the cingulate cortex produce substantial deficits in rats but very minimal effects in mice. As emphasized earlier, the nature of the organism must be considered in any attempts to formulate "general laws."

Eating

As the reader should anticipate by now, the control of eating involves a considerable number of neural structures. However, interest has centered on two areas of the hypothalamus, the lateral hypothalamus (LH) and ventromedial hypothalamus (VMH) (reviewed by Grossman, 1967; Wayner & Carey, 1973). Both are intimately involved in eating behavior, but in very different ways. LH lesions produce a "lateral hypothalamic syndrome," the time course of which was described by Teitelbaum and Epstein (1962). Immediately after the lesion is made, the rat will neither eat nor drink. Gradually, the animal passes through successive stages in which it will first eat palatable foods, such as chocolate; then regulate food intake; eat dry food; and later drink water. Electrical stimulation of the LH in intact rats increases food consumption, while the injection of an anesthetic into the LH decreases eating. That there is a chemical coding of both eating and drinking the LH was indicated by research by Grossman (1960). Cannulation of a cholinergic transmitter into the LH produced drinking, whereas cannulation of norepinephrine resulted in eating.

In contrast to the *aphagia*, or lack of eating, produced by LH lesions, VMH lesions produce *hyperphagia*, or overeating. The rats at first gain weight very rapidly, but they later regulate body weight in such a way as to remain at their new level of obesity. Electrical stimulation of the VMH decreases eating, while cannulation of an anesthetic produces eating in satiated rats. While the rat with a VMH lesion overeats, it displays some interesting patterns of eating. Dilution of the caloric content of the food, decreasing palatability as by adding quinine, or increasing the amount of work required by rats as a prerequisite to eating all decrease the amount of eating by VMH-lesioned rats as compared with normal animals.

While most workers agree on the basic empirical phenomena of manipulations of the LH and VMH regions and the control of eating, interpretations vary greatly. The classical view was that the LH is a "feeding center" and the VMH a "satiety center." Chemical messages from the bloodstream and cerebral fluids were thought to affect the VMH and alter its inhibition of the LH. Feeding would occur when the inhibition of the LH by the VMH was removed as the result of responses to chemical or other stimuli acting on the VMH. While this picture is a neat one, results of a number of more recent lines of research suggest that it may be too simple; thus other interpretations have been proposed. On one view, the LH and VMH are involved in the determination of "set points" that are important in regulating body weight (Keesey and Powley, 1975). According to this view, body weight is regulated in relation to a predetermined value, analogous to the setting of a desired temperature on the thermostat of your home. LH and VMH lesions are viewed as resetting the set point.

According to another view, LH and VMH lesions do not actually produce the observed deficits they are generally believed to produce. Rather, in making such lesions, experimenters damage nearby bundles of fibers rich in biogenic amines, and it is this damage that produces the observed lesion effects (Gold, 1973; Wayner & Carey, 1973). A counterposition was presented by Ahlskog, Randall, & Hoebel (1975).

The ultimate resolution and intergration of these and other views of hypothalamic regulation of eating is not yet in sight. It is clear, however, that the classical view of discrete centers controlling behavior is an oversimplification.

Zeigler (1976) has conducted an extensive program of research on the regulation of feeding in pigeons. Zeigler's work on pigeons has not only explored an interesting control system but has forced a reevaluation of much mammalian research, especially in relation to sensory bases. This work contains an excellent argument for increased emphasis on comparative research in studies of neural mechanisms associated with behavior.

Aggressive Behaviors

It will be recalled from Chapter 6 that Moyer (1968, 1976) proposed that aggressive behavior be viewed not as a unitary category but as a conglomerate of very different behavioral patterns grouped together for convenience of organization. As a first step in developing a useful taxonomy of aggressive behavior, Moyer proposed that eight categories of aggressive behavior be considered: predatory, intermale, fear-induced, irritable, territorial, maternal, instrumental, and sex-oriented aggression. Moyer's main interest in proposing such a list was in attempting to find order in the mass of diverse findings from research on the physiological determinants and correlates of aggressive behavior. Moyer's proposals have had some impact, and several researchers have used such a classification in reviewing research on aggressive behavioral patterns (e.g., Moyer, 1968; Conner, 1972; Eichelman & Thoa, 1973; Avis, 1974). Little is known of the physiological bases of fear-induced, territorial, maternal, instrumental, or sex-related aggression. Attention has focused on studies of (1) predatory aggression, (2) irritable aggression, and (3) a pattern of behavior variously termed "intermale," "spontaneous," or "isolation-induced" aggression. We shall consider each in turn.

Predatory Aggression It should be recognized at the outset that many animal behaviorists would not consider predation to be a form of aggression but a functionally separate category of behavior. However it is classified, much research has focused on the interspecific attack behavior that often appears to be of a predatory nature. Emphasis has been placed on study of convenient laboratory preparations, including rat killing by cats, mouse killing by rats (termed "muricide"), and frog-killing by rats (termed "ranacide").

Areas of the brain that have been implicated in the regulation of predatory aggression include the hypothalamus, the frontal lobes of the cerebral cortex, and the amygdala. Wasman and Flynn (1962) noticed two types of behavior elicited by brain stimulation delivered to the hypothalamus of cats. Stimulation of the medial hypothalamus produced an "affective attack," accompanied by symptoms of rage, which could be directed at a wide range of objects. By contrast, stimulation of the lateral hypothalamus (LH) elicited a quiet biting attack. Given our discussion of the involvement of LH with feeding, the fact that it is implicated in the control of predatory behavior should not be surprising. However, there is considerable evidence to suggest that predatory attack is not based primarily on hunger, although it is often related to feeding (see O'Boyle, 1974).

In contrast to their actions on most other forms of aggressive behavior, gonadal hormones have little effect on predatory behavior. Muricide is affected by the administration of a wide variety of drugs, including drugs affecting both the cholinergic system and the biogenic amines. For example, *p*CPA, which lowers brain serotonin levels, increases mouse killing by rats (Eichelman & Thoa, 1973).

Irritable Aggression The most typical paradigm for the study of irritable aggression involves use of electric shock. Two animals, such as rats, that normally display little aggression when placed together, will often begin fighting if an unavoidable shock is delivered. The ventromedial nucleus of the hypothalamus, among other structures, is important in the control of irritable aggression (Grossman, 1972). Gonadal hormones are also important to irritable aggression (see Chapter 12).

There is evidence to suggest that both cholinergic systems and norepinephrine are involved in the regulation of shock-elicited aggression. Serotonin appears not to be involved. Indeed, shock-elicited aggression may be one of the few species-characteristic behavioral patterns that is not affected by *p*CPA (Conner, Stolk, Barchas, Dement, & Levine, 1970).

Intermale Aggression Intermale aggression is often studied by isolating individuals and later pairing them for testing. Isolation is often critical to later aggressive behavior. As males are isolated, their tendency to fight increases and levels of norepinephrine and serotonin in the brain decrease. There is some indication that the increased tendency to fight may be related to decreased levels of biogenic amines (Welch & Welch, 1970). A wide range of pharmacological manipulations can alter probabilities of intermale aggression. Anticholinergic drugs (those that interfere with acetylcholine function) inhibit isolation-induced aggression. Interpretations of drug effects are complicated by the fact that isolation-induced aggression can be inhibited by drugs that either increase or decrease levels of acetylcholine, serotonin, or norepinephrine.

A program of research on the neural bases of isolation-induced, dominance-related aggressive behavior in hamsters has been conducted by Bunnell and his associates (e.g., Sodetz & Bunnell, 1970; Bunnell, Sodetz, & Shalloway, 1970). Males are housed individually except for periodic tests for dominance and aggression. This group used a 28-category behavioral inventory to provide a complete description of the behavioral patterns of their hamsters. By constructing diagrams of the sequential probabilities of transitions from one behavioral pattern to another, the precise behavioral effects of a lesion could be clearly delineated. The combination of sophisticated behavioral measurement used in conjunction with physiological manipulations is important. Lesions of the amygdala were found to alter the total amount of social interaction. Preoperatively dominant males decreased their frequency of aggressive patterns, while preoperatively submissive males decreased the frequency of their submissive behavioral patterns.

Sodetz and Bunnell (1970) found that lesions of the septal formation increased the aggressiveness of both socially naïve and preoperatively dominant hamsters. However, if a male experienced defeat before the operation was done, there was no increase in aggressive behavior. Indeed, there was some tendency for such hamsters to become even more submissive. The lesson to be learned from this study is that there is an interaction between a subject's experience and the effectiveness of a neural intervention in that subject. Removal of the same neural tissue produces a different effect in animals with differing past histories.

An extensive program of research using electrical stimulation of the brain has been conducted by Delgado (e.g., Delgado, 1966). Various loci have been found to elicit aggressive behavior when stimulated. Equally interesting, some nuclei have been found to inhibit aggressive behavior. In Delgado's most dramatic demonstration, electrical stimulation inhibited the charge of a bull in a bullring. Stimulation of the caudate nucleus of the forebrain inhibited aggressive behavior in rhesus monkeys. When several rhesus monkeys were housed in a group, the subordinate animals learned to press a lever that delivered electrical stimulation to the caudate nucleus of the dominant male and thus inhibited his aggression (Delgado, 1963).

SUMMARY

The evolution of more recent species has seen a general trend toward nervous systems that are increasingly differentiated and centralized. Among vertebrates, increases and decreases in the relative sizes of particular brain structures have been observed in different evolutionary paths. There is a general trend toward the evolution of larger brains.

The major methods used in studying brain function include lesion or ablation, electrical recording, electrical stimulation, and chemical techniques.

The proposal of classical ethologists that behavioral patterns can be organized in lower brain regions and inhibited by the activity of higher regions has been confirmed. Patterned behavior can be produced from lower structures, even in the absence of input from higher structures or from the senses. Examples of this general organizational pattern for the neural control of behavior can be found in copulation in praying mantids, walking in praying mantids, flying in locusts, and cricket song.

Electrical stimulation studies have revealed no simple hierarchical organization of behavior. Rather, structures regulating behavior tend to be located diffusely about the brain.

Studies of the neural mechanisms of copulatory behavior, sleep, maternal behavior, eating, and aggressive behaviors were briefly reviewed. Reflexes underlying the copulatory patterns of mammals appear to be organized in lower brain regions and regulated by the action of higher regions. Species-characteristic behavioral patterns generally appear to be affected by a wide range of regions located diffusely about the brain. Although particular neural circuits may be critical to a given behavioral pattern, it appears that there are no "centers" uniquely and completely mediating complex, species-characteristic behavioral patterns. Understanding of the control of mammalian behavior is increasing with the combined use of a variety of techniques, including neurochemical analyses.

Chapter 12

Hormones and Behavior

A hormone is a chemical that is secreted by a specific gland of the body and that passes into the bloodstream to produce a specific effect on some other organ in the body. Thus, hormones are chemical messengers which function in "communication" *within* the organism and which are transported by the bloodstream. In 1849, A. A. Berthold noted that cockerels, if castrated, showed a reduction in the size of comb and spurs, a cessation of crowing, and a loss of sexual behavior. Replacement of a single gonad, at a location in the bird where a functional connection to the bloodstream could be established, prevented these losses. Berthold concluded that the testes must secrete something into the bloodstream that maintains secondary sex characteristics and behavior. Since the time of Berthold's observations, a whole interdisciplinary research area—called "behavioral endocrinology" or "psychoneuroendocrinology"—has developed. We shall review some of the major principles and findings of this, the third of the three mechanism-related research areas that is to be considered. Although hormones are important in a number of invertebrates, especially the insects, we shall concentrate on vertebrates, especially mammals.

HORMONES AND THE REGULATION OF HORMONE SECRETION

Organs that secrete hormones and thus function as *endocrine glands* are scattered about the body. Hormones secreted by the thyroid glands, parathyroids, pancreas, and gastrointestinal mucosa are without obvious direct behavioral function and will not be considered further; hormones secreted by other endocrine glands (see Turner and Bagnara, 1971) have pervasive and clearly defined behavioral effects.

Hormones Important to Behavioral Regulation

The primary producers of behavior-related hormones are the gonads and placenta (located in the groin), the adrenal glands (located on the kidneys), and the pituitary gland (located at the base of the brain, just beneath the hypothalamus).

Hormones of the Gonads and Placenta The primary hormones secreted by the male's testes are androgens (e.g., testosterone). Testosterone influences the development and maintenance of the male reproductive tract, secondary sex characteristics, and behavior. The female's ovaries produce both *estrogens* (e.g., estradiol and estrone) and *progestins* (e.g., progesterone). Estrogens function in the development and maintenance of the female reproductive tract, in the development of secondary sex characteristics, in stimulating the mammary glands, and in regulating behavior. Progestins often work together with estrogens, as in the regulation of the development of the female reproductive tract, mammary glands, and behavior. Progestins are important in preparing the uterus for the implantation of fertilized ova and in maintaining pregnancy.

Hormones of the Adrenal Gland The central core, or *medulla*, and the outer region, or *cortex*, of the adrenal gland secrete different hormones with different functions. The adrenal medulla secretes *epinephrine and norepinephrine.* Both function in emergency situations. Epinephrine has primary effects on heart muscle, smooth muscle, and arterioles; thus it increases pulse rate and blood pressure. Norepinephrine functions as a transmitter in the autonomic nervous system, causes contraction of arterioles, increases peripheral resistance, and helps maintain blood pressure.

The adrenal cortex secretes at least 28 different steroid hormones which function in metabolism and resistance to infection.

Hormones of the Pituitary The pituitary gland (see Figure 12-1) can be divided into an anterior lobe, an intermediate lobe, and a posterior lobe. The posterior lobe, or neurohypophysis, secretes *vasopressin* and *oxytocin*, both of which are actually synthesized in the hypothalamus. Vasopressin, or antidiu-

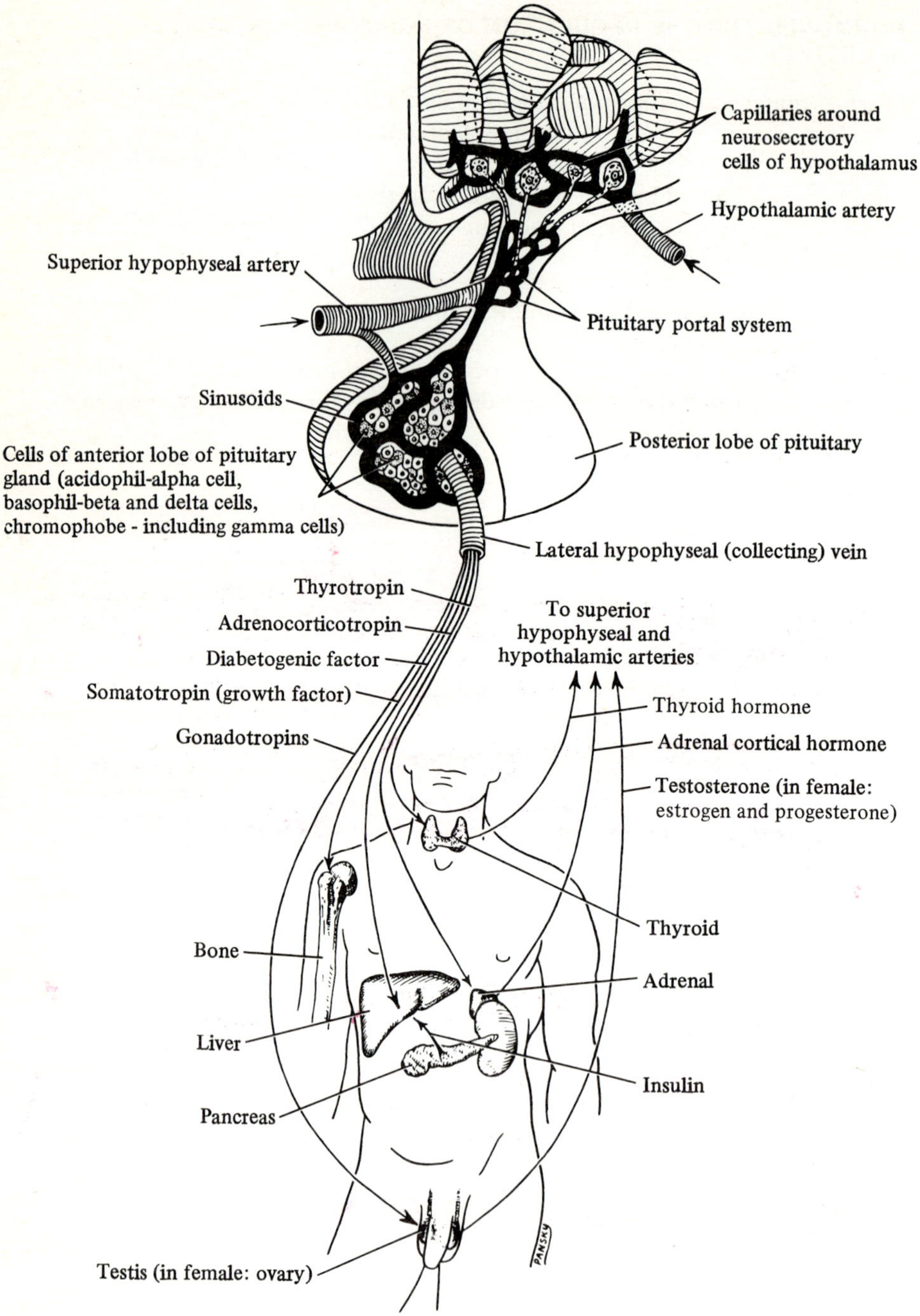

Figure 12-1 A diagram showing the relationship of the hypothalamus to the anterior pituitary, other endocrine glands, and target glands. *(From House & Pansky, 1967.)*

retic hormone, elevates blood pressure and promotes reabsorption of water by the kidney. Oxytocin promotes the contraction of smooth muscle, as in stimulating the secretion of milk in the mammary glands, the contraction of the uterus at birth, and transport of sperm.

The intermediate lobe secretes *melanocyte-stimulating hormone* (MSH), which functions in the dispersal of granules of pigment.

The anterior pituitary secretes an array of hormones that have both direct effects on target organs and effects that are mediated via the modulation of the secretions of other endocrine glands. These include:

Somatotropin (STH): Stimulates growth of bone and muscle and functions in metabolism

Adrenocorticotropin (ACTH): Stimulates the secretion of steroid hormones by the adrenal cortex

Thyrotropin (TSH): Stimulates the thyroid gland to release growth-related hormones

Prolactin: Functions in the mammary gland to initiate secretion of milk and in the ovary to maintain the corpus luteum and modulate secretion of progesterone

Follicle-stimulating hormone (FSH): A gonadotrophin which acts on the ovary to promote growth of ovarian follicles and secretion of estrogen; also acts in the testes to promote the formation of sperm

Luteinizing hormone (LH) or interstitial-cell-stimulating hormone (ICHS): A gonadotropin that acts in the ovary to trigger ovulation and the formation of corpora lutea and to modulate progesterone secretion; it also acts in the testes to promote the secretion of androgens (male hormones).

Although different hormones produce different effects, many of them are chemically very similar; indeed, some hormones lie along the pathway of the synthesis of others. Thus, norepinephrine lies along the synthetic pathway leading to epinephrine; testosterone is an intermediate step in the synthesis of estradiol.

Hormonal output is not rigidly limited by organ or sex. Small amounts of the sex hormones are secreted in the adrenals. Males typically have measureable amounts of "female hormones" and females have measureable amounts of "male hormones."

It is important to note that the hormones have changed little throughout the course of evolution. Testosterone injected into a bird elicits bird behavior; injected into a rat, it elicits rat behavior (Beach, 1958). "Endocrine evolution is not an evolution of hormones but an evolution of the uses to which they are put; an evolution not, to put it crudely, of chemical formulae but of reactivities, reaction patterns and tissue competences" (Medewar, 1953, p. 334).

Neuroendocrine Feedback Mechanisms

Some of these hormones interact in intricate feedback circuits that function in the regulation of their production and secretion. The pituitary gland is often termed the "master control gland" of the endocrine system, and with good reason. However, the pituitary, in turn, is regulated by the hypothalamus.

In Chapter 8, where we discussed the effects of early experience on the adrenal system, Levine's concept of the "hormonostat" was cited. A thermostat initiates the production of heat and is turned off by negative feedback (increased temperature). A homonostat works in a manner analogous to that of a thermostat in that hormone secretion is increased as the result of the action of a pituitary hormone and decreased as the result of negative feedback resulting from the production of the hormone. The negative feedback ultimately shuts off production of the pituitary-produced stimulating hormone.

The hypothalamus exerts its influence upon the anterior pituitary by secreting a family of regulatory hormones called *releasing factors*. The hypothalamus is linked to the anterior pituitary by a specialized system of blood vessels called the *hypophyseal portal vessel system* (see Figure 12-1). The hypothalamus dumps releasing factors into the portal system which, in turn, carries them to the pituitary. There is a whole family of releasing factors, including FSH-releasing factor (FSH-RF), LH-releasing factor (LH-RF), and prolactin-inhibiting factor. Each governs the production and release of a pituitary hormone. The pituitary hormone is secreted into the bloodstream and transported to a *target organ*. There, many pituitary hormones may have terminal effects, but they often act by stimulating the production and secretion of other hormones. These hormones not only have their specified effects on other target organs but also *feed back* to act upon the hypothalamus (and possibly directly on the pituitary) to decrease production of the releasing factor, thus completing the circuit (see Figure 12-1). Were any step of this system to fail, hormone levels would reach values much greater or lower than usual, depending on the location of the failure, just as a failure of a thermostat would result in a house that was too cold or too hot.

Female estrous cycles are governed by such negative feedback systems. The hypothalamus sends FSH-RF to the pituitary, which in turn secretes FSH into the bloodstream (see Figure 12-1). In the ovary, FSH stimulates the growth of a Graafian follicle and the secretion of estrogen into the bloodstream. Estrogen feeds back on the hypothalamus and pituitary and decreases the secretion of FSH-RF and FSH while increasing the production of LH-RF and LH. LH reaches the ovary via the bloodstream and triggers the rupture of the follicle (ovulation) and the secretion of progesterone. Progesterone feeds back on the hypothalamus and pituitary, completing the circuit (Levine, 1972).

Similar systems operate in the regulation of adrenal and other hormones.

METHODS IN THE STUDY OF HORMONES AND BEHAVIOR

With this background in the nature and regulation of hormones, we can proceed with the topic of interest: hormone-behavior interactions. A number of hormones, especially those of the adrenal and reproductive systems, have been shown to have direct effects on behavior in addition to whatever other functions they might perform in the body. We shall begin with a discussion of the methods used in studying hormone-behavior interactions.

Correlational Method

The simplest method used in relating behavior to hormones is simply to study their correlations. Hormone levels fluctuate according to circadian rhythms in both sexes, the estrous cycles of females, annual breeding cycles, and other cycles. In relating hormones to behavior using the correlational method, one simply searches for fluctuations in behavior that parallel hormonal fluctuations. Thus, Udry and Morris (1968) noted that the pattern of fluctuations in frequency of intercourse and orgasm in human females paralleled the fluctuations of estrogen levels during their menstrual cycles.

Some correlates of the 4-day estrous cycles of laboratory rats are shown in Figure 12-2. Stages of the estrous cycle can be determined by *vaginal smear*, a

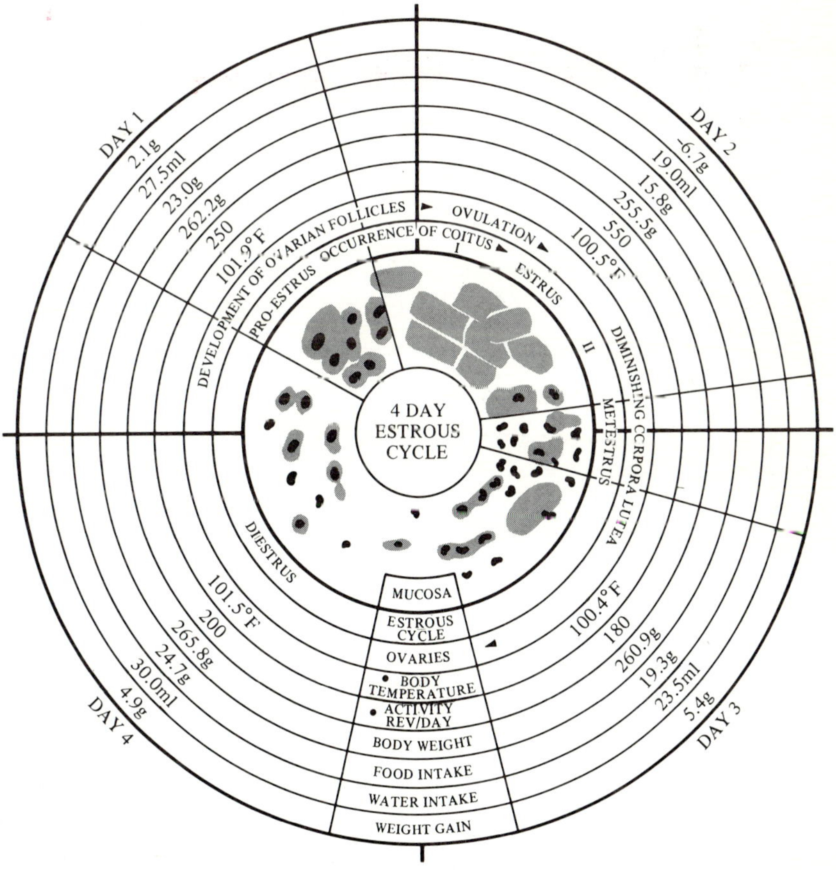

Figure 12-2 A diagram showing effects of the estrous cycle of the female rat on the cells of the vaginal mucosa, physiological function, and behavior. *(From Simpson & May, 1973.)*

procedure in which the contents of the vaginal mucosa are sampled periodically by a lavage of the vagina. Cell types characteristic of the stages of the cycle—proestrus, estrus, metestrus, and diestrus—are shown in the center of Figure 12-2. As can be seen in the figure a great variety of behavioral patterns—including eating, drinking, activity, and coitus—vary in conjunction with the estrous cycle.

With new and more sophisticated techniques available for obtaining assays of hormone levels, use of the correlational technique has become quite sophisticated.

Castration and Replacement

More direct evidence of a hormonal regulation of behavior can be gained through use of the method of castration and replacement. The method is simple and can be applied to almost any endocrine gland. With castration, one removes the source of a hormone. If a behavioral pattern is hormone-dependent, it should show a decrement with the loss of hormone. However, such a decrement could be due to a number of factors (e.g., illness). If one replaces the hormone, as by injecting synthetic hormone or transplanting a new endocrine organ, behavior should return to baseline levels. This provides more conclusive evidence of the function of the hormone. Grunt and Young (1952) conducted a castration and replacement experiment using male guinea pigs having three levels of performance of copulatory behavior before castration. As can be seen in Figure 12-3, castration produced a substantial decrement in

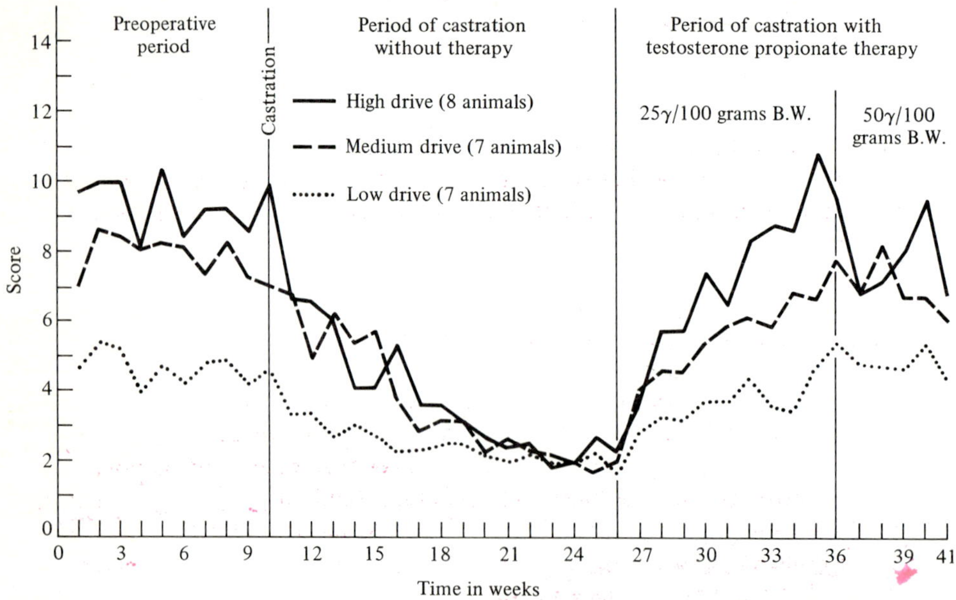

Figure 12-3 The effect of castration and replacement with testosterone propionate on sexual behavior in male guinea pigs with initially high, medium, and low scores of sexual activity. *(From Grunt & Young, 1952.)*

copulatory behavior; daily injections of a male hormone, testosterone propionate, restored copulatory behavior to its preoperative levels.

Other Methods

Various other methods have somewhat more specialized application.

Cannulation Hormones have been found to affect behavior through effects on numerous specific brain structures within and outside the hypothalamic region. Just as cannulas can be used to introduce drugs into the brain and thus probe the location of effective sites, so hormones can be introduced and studied. Barfield (1971) cannulated testosterone propionate into the brains of castrated male ring doves. He found the preoptic area of the hypothalamic region to be the most effective site for the restoration of sexual and aggressive behavior in castrate doves.

Hormone Uptake Autoradiography, a procedure in which hormones are "labeled" with radioactivity, can be used to determine the brain regions in which hormones are concentrated. If labeled hormones are injected into the animal, their locations in the brain can be detected later. Such information provides sustantial corroborative evidence or clues regarding the brain regions upon which hormones act. For example, androgen-concentrating cells were found the *nucleus intercollicularis* of the midbrain of chaffinches. This is the identical region from which vocalizations can be elicited by the use of electrical stimulation. Thus, hormones appear to affect singing in chaffinches through actions on this part of the midbrain (Zigmond, Nottebohm, & Pfaff, 1973).

Cross-Transfusion Terkel and Rosenblatt (1972) developed a technique whereby blood could be cross-transfused between two freely moving rats. In cross-transfusing blood between mother and virgin females, they found that only blood from newly paturient females was effective in inducing maternal behavior in virgins.

Differing Forms of Hormone A variation on the castration-replacement technique involves the utilization of various related forms of hormones. There is not just one female hormone, estrogen, but a family (e.g., estrone, estradiol). Recently, different forms of hormone have been found to have different behavioral effects; they have thus become a promising research tool. Whalen and Luttge (1971) compared the ability of testosterone, androstenedione, and dihydrotestosterone (all androgens) to maintain copulatory behavior in castrate rats. The results indicated that those hormones that had effects on the brain were effective in maintaining behavior, whereas those that worked primarily on peripheral structures were ineffective.

Antihormones Various drugs are available that have been shown to function as "antihormones" (i.e., to antagonize the effects of hormones). Both "antiandrogens" and "antiestrogens" have been used in the study of behavior. For example, Arai and Gorski (1968) showed that administration of an antiestrogen blocks lordosis induced by estradiol benzoate in female rats.

MECHANISMS OF HORMONE ACTION

Hormones affect behavior by altering one or more of the following: (1) sensory-perceptual function, (2) neural activity, and (3) the effector mechanisms important in the execution of overt behavioral patterns.

Effects of Hormones on Sensory-Perceptual Mechanisms

Numerous studies have demonstrated that both the ability to detect stimuli and preferences for particular stimuli are affected by hormones. The visual sensitivity of human females, as measured by their ability to detect a test light, varies in relation to the menstrual cycle. Visual sensitivity is greatest at about the time of ovulation and declines abruptly at the time of menstruation (Diamond, Diamond, & Mast, 1972). The ability of female rats to detect odors varies with varying levels of estrogen and progesterone as modified by the estrous cycle, pseudopregnancy, ovariectomy, and administration of androgens (Pietras & Moulton, 1974). Komisaruk, Adler, and Hutchison (1972) recorded neuronal activity in the pudendal nerve of female rats. The receptive field, the part of the body whose tactile stimulation affects the nerve, lies in the genital region and varies in size in relation to estrogen levels.

Intact and castrate male rats do not differ in their threshold for *detecting* the urine of estrous females (Carr, Solberg, & Pfaffmann, 1962). However, whereas intact, experienced males *prefer* the odor of estrous females to that of diestrous females, inexperienced males and castrate males show no such preference (Carr, Loeb, & Dissinger, 1965). Thus, both experience and hormones alter preferences.

It will be recalled that sticklebacks move from salt water to freshwater in order to breed. Administration of thyroxin to sticklebacks living in salt water in winter shifts their preference from salt water to freshwater—a prerequisite for the migration upstream to freshwater (Baggerman, 1962).

Effects of Hormones on Neural Mechanisms

Hormones affect the anatomy, physiological activity, neurotransmitter function, and physiological activity of neural systems. For example, a sexual dimorphism in the anatomy of synapses on dendritic spines in the preoptic area of the hypothalamus in rats was traced to the action of hormones (Raisman & Field, 1973). Estrogen, progesterone, and testosterone all have influences on neural activity in the brain, as reflected in the EEG, evoked potentials and the activity of single nerve cells (Komisaruk, 1971; Pfaff, Lewis, Diakow, & Keiner, 1973). Hormones affect levels of brain monoamines and are in turn affected by brain monoamine levels. For example, estrogen produces a reduction in the concentration of norepinephrine in the anterior hypothalamus (Donoso & Cukier, 1968). Kow et al. (1974) proposed that the effects of progesterone in the brain are mediated by its inhibition of serotonergic activity. Hormones enter particular "target" cells in the brain and function by altering the actions of DNA and RNA in the synthesis of proteins (see McEwen, 1976).

Testosterone propionate maintains the levels of responsivity of sexual reflexes in castrated, spinal male rats. Dihydrotestosterone maintains the integrity of peripheral receptors but does not maintain the presence of spinal reflexes (Hart, 1973). Thus, it appears that hormones such as testosterone propionate act directly on mechanisms in the spinal cord to affect its functioning. This is somewhat counter to Beach's (1967) third proposal regarding the control of copulatory behavior (see Chapter 11).

Effects of Hormones on Effectors

Hormones affect an array of diverse structures that are important to the execution of behavior. For example, both the behavior of red deer stags and their antler cycle are dependent upon androgens. Because a male's social interactions are partially dependent upon his antlers, it is difficult to dissociate direct effects of hormones on behavior from those resulting from the development of his antlers (Lincoln, Guinness, & Short, 1972).

The surface of a male rat's penis is covered with spines (papillae) (Figure 12-4) that appear to function in tactile sensation. Castration produces both a

Figure 12-4 Electronmicrographs of the surface of the penis of male rats showing the effect of castration on penile papillae. Shown at left is an intact male rat with prominent spines; at right is an 8-week castrate male, whose spines have disappeared and epidermal folds have flattened. *(From Phoenix, Copenhaver, & Brenner, 1976.)*

loss of copulatory behavior and a disappearance of the spines (Phoenix, Copenhaver, & Brenner, 1976). These spines are critical to the execution of copulatory behavior, possibly because of their sensory function.

Another effect of hormones that is critical to behavior is their effect on pheromone production. As discussed above, intact, experienced male rats can discriminate the odor of receptive versus nonreceptive females. Further, exposure to androgens early in life alters the nature of the pheromones that female mice produce and increases the likelihood that they will be attacked by males (Lee & Griffo, 1973).

EARLY HORMONE EFFECTS

Most of the hormone effects considered thus far have been studied in adults. These effects generally have been reversible and not strictly age-dependent. Reversibility is indicated by the ease with which a behavior can be increased and decreased in correlation with fluctuating hormone levels, as in the normal female estrous cycle or with successive periods of administration and termination of hormone treatment in castrate males. Such effects can be observed at almost any time in the adult life of the animal.

The actions of hormones on the developing organism are quite different from those on the adult. Beach (1975) notes that the actions of hormones during the perinatal (around the time of birth) period differ in that:

1 They are *permanent and irreversible.*

2 They are *limited to very clearly defined periods* in the development of the organism.

3 They affect systems that are *not manifest at the time of occurrence* of the hormone treatment but only later in life.

Perinatal hormone effects, then, have a clearly defined sensitive period in which they produce irreversible effects later in the life of the animal. The parallel with imprinting is interesting.

Male and female rodents, like many other species, differ in a number of respects. For example, males and females differ anatomically, as with respect to body weight and the presence of a penis or clitoris. While the normal male copulatory pattern includes mounts, intromissions, and ejaculations, the normal female pattern is made up of lordosis and associated soliciting behavior. Whereas the release of hormones in the female rodent is cyclical, thus producing an estrous cycle, that in the male is tonic, showing little cyclical variation. Males and females differ with respect to aggressive behavior, open-field exploratory behavior, and in many other ways. It has been demonstrated that many of these sex differences can be traced to the actions of perinatal hormones.

Two general principles have been found to hold for the actions of perinatal hormones in rodents:

1 If androgens are present during the sensitive period, the individual will develop the characteristics of a male, and female characteristics will be suppressed.

2 If androgens are absent during the sensitive period, the individual will develop the characteristics of a female, and male characteristics will fail to develop.

The 1960s and 1970s have seen a flood of research in support of these generalizations. We shall consider a typical experiment, that of Whalen and Edwards (1967), the procedures for which are summarized in Table 12-1 (page 230).

Although the logic behind experiments on early hormone effects is quite simple, there are a number of details that must be kept straight in order to understand them. The critical variable is the presence or absence of male hormone during the sensitive period, which in rats occurs in the first few days of life. Whalen and Edwards studied both males and females. The first group of males had a "sham" operation and were given no neonatal hormones. Since newborn males have functional amounts of androgen in their bloodstream (Resko, Feder, & Goy, 1968), these rats had androgens present during the sensitive period. A second group of males were castrated at birth and given no neonatal injections; thus they had no androgens present during the sensitive period. The third group of males were castrated but given an injection of testosterone propionate (TP); thus they had androgen present despite castration. This design simply represents a variation of the old castration-replacement paradigm adapted to the neonate. Groups of females that received sham operations and no injections or ovariectomy and no injections had no androgen present. Those females that were ovariectomized and injected with TP did have androgen present during the sensitive period.

The animals were to be tested under equivalent conditions, but conditions needed to be such that normal animals would display male or female mating behavior. In order to even things up across groups, the testes and ovaries were removed from the adult animals that still had them. Because even normal adult female rats rarely display lordosis unless the appropriate hormones are present at the time of testing, female behavior was tested in the presence of a normal male after adult administration to the test animal of estradiol benzoate and progesterone (EB+P). Similarly, male behavior was tested with a receptive female after all experimental animals were given TP as adults.

The results, as shown in Table 12-1, were clear. Regardless of genetic sex or the origin of the male hormone, rats that had appreciable amounts of androgen present during the first few days of life displayed much higher levels of intromission or intromission-like behavior and much lower levels of lordosis

Table 12-1 Procedures for the Demonstration of Effects of Early Hormones in Rats

Genetic sex	Neonatal operation	Neonatal injections	Neonatal androgen present?	Adult operation	Lordosis after adult EB + P*?	Intromission after adult TP†?
♂	Sham	None	Yes	Castration	Low	High
♂	Castrate	None	No	None	High	Low
♂	Castrate	TP	Yes	None	Low	High
♀	Sham	None	No	Ovariectomy	High	Low
♀	Ovariectomy	None	No	None	High	Low
♀	Ovariectomy	TP	Yes	None	Low	High

*Estradiol benzoate and progesterone.
†Testosterone propionate.
Source: Based on Whalen and Edwards, 1967.

than animals lacking early androgens. These data are completely consistent with the two general statements given above.

Because the development of peripheral structures, the penis and clitoris, are hormone-dependent, there is some difficulty in interpreting effects of early hormones on male behavior. Early androgens produce no effect on the frequency of mounts (without intromission) in some experiments. However, it is now clear that while peripheral factors often play a role, they cannot account for all of the observed differences. Early hormones also affect the central nervous system.

The timing of the sensitive period varies with different mammalian species. In laboratory rats and mice and in hamsters, the sensitive period is postnatal; in guinea pigs and rhesus monkeys, it is prenatal. Clemens (1973) related the timing of the sensitive periods in different rodent species to the timing of the embryonic differentiation of the developing gonads into testes and ovaries.

Hormones acting early in life have been shown to affect a wide range of mammalian species, including rhesus monkeys (Phoenix, Goy, & Resko, 1968). Studies of human sex differentiation were summarized by Money and Ehrhart (1972). Because experiments such as that of Whalen and Edwards cannot be performed on humans for ethical reasons, interest has focused on three clinical syndromes, two of which are characterized by the exposure of females to hormones with masculinizing effects. In the case of *progestin-induced hermaphroditism,* progestins given to mothers to help maintain pregnancy were found to masculinize the appearance of the female. Naturally, this procedure was stopped once this was discovered. In the *adrenogenital syndrome,* females have a hereditary disease in which the adrenal gland secretes androgens instead of cortisone. In *testicular feminization,* genetic males produce functional amounts of androgens, but the target tissues cannot respond to the androgens because of an enzyme deficit. The role of experience is overwhelming in the development of humans with these syndromes. Thus, if an individual is raised and treated as male or female and given appropriate hormone injections and prosthetic devices, he or she will adjust to the sex of rearing regardless of genetic sex. Humans appear to differ from most other species in this regard. However, Money and Ehrhart did detect some more subtle effects of the presence of prenatal androgen-like hormones in individuals with progestin-induced hermaphroditism and those with the adrenogenital syndrome. Such females showed increased frequencies compared to other females of a number of behavioral patterns that differentiated males from females in their culture at the time of the study (e.g., interest in outdoor activities, priority of career over marriage, preference of slacks to dresses).

The alterations of rodent behavior produced by perinatal hormones have been demonstrated to be due at least in part to alterations in the central nervous system. As discussed above, the sexual dimorphism of the preoptic area of the brains of male and female rats can be traced to the presence or absence of perinatal hormones (Raisman & Field, 1973). Implantation of androgen in the ventromedial-arcuate nucleus of the hypothalami of neonatal rats was found to have a much greater effect on sexual differentiation than implants in other regions (Nadler, 1972). Hart (1968b) found that neonatal castration produced a decrease in the sexual reflexes of spinal male rats.

COMPLEXITIES AND INTERACTIONS IN HORMONE ACTION

It should already be apparent that hormones do not act in a vacuum but interact with many other factors in producing their effects on behavior. We shall now make more explicit a few of these complexities and interactions.

Genotype-Hormone Interactions

The fact that hormones and genotype interact in their effects on behavior should come as no surprise to readers of earlier chapters. McCollom, Siegel, and Van Krey (1971), for example, studied the response to androgen administration of male chickens from two different selected lines—one selected for high and one for low numbers of completed matings. Identical hormone treatment produced different results in the two lines, returning the mating behavior of both lines to their own precastration levels. Thus, the strain difference in behavior results from the soma, not the hormone output.

Vale, Ray, and Vale (1972) studied the interaction of strain with early hormone treatment in an experiment on perinatal androgenization and aggressive behavior in inbred female mice. Neonatal androgen produced substantial increases in the aggressive behavior of females of strains wherein the males are aggressive but little change in females of strains wherein the males are nonaggressive.

McGill and Haynes (1973) found that male mice of heterozygous genotypes retained the ejaculatory reflex for a longer period following castration than did inbred animals.

Individual Differences

Individual differences may arise as the result of genotypic factors, environmental factors, or their interaction. The data in Figure 12-3 show that individual differences in copulatory behavior in guinea pigs are not a function of differential hormone output but rather result from differences in the soma, probably in the central nervous system. Individual differences were restored after castration and replacement with a constant hormone level for all animals. Similar results have been found in work with rats (Larsson, 1966).

Interactions among Individuals

The individuals of social groups interact in complex ways and can influence each other by means of their behavior. Thus, a behavioral change in one animal that is created by hormone administration can, in turn, affect the behavior of other animals in the group. The continuing interaction between male and female ring doves (e.g., Lehrman, 1965) discussed in Chapter 5, provides a case in point. An even more dramatic example can be found in rhesus monkeys. Two females from a free-ranging organized troop in Puerto Rico were brought into estrus (a state of sexual receptivity) with estradiol benzoate during the nonmating season. This treatment produced an increase in both sexual behavior and reddening of the sex skin among adult males of the troop that were untreated and in the nonmating season (Vandenbergh & Drickamer, 1974).

Interaction with Season

The same treatment can produce very different effects when administered in different seasons. For example, implantation of testosterone in male red deer in December produced all elements of the rutting pattern within a few weeks. The same treatment given in April or June resulted in no such behavioral modification until the normal autumn rutting time (Lincoln et al., 1972). When any hormone treatment is being administered, the possibility of seasonal effects must be considered. Seasonal fluctuations in behavior and hormone levels are common in many species, as, for example, in rhesus monkeys (Gordon, Rose, & Bernstein, 1976). Indeed, such cycles can persist even in the controlled environment of the laboratory (Michael & Keverne, 1971).

Maintenance versus Restoration

The difference between maintenance and restoration lies in the difference between needing only to provide exogenous hormone that will keep a functional behavioral pattern active as opposed to needing to restore a system that has been permitted to remain inactive for some period of time. It has long been known that a maintenance dose for behavior is much less than the dose required to restore behavior once it has been permitted to decline. Hutchison (1976) showed that such effects can be mediated by the brain. Hypothalamic implants of androgen were less effective in restoring mating behavior in doves that had gone some time since castration than in doves that were tested after a relatively short time.

Interaction of Hormones and Experience

Endocrine manipulations can produce differential effects that are a function of the experiential history of the animal. Thus, the postcastration maintenance of copulatory behavior in male cats is more prolonged if they have had previous copulatory experience than if they are naïve (Rosenblatt, 1965). The induction of sexual receptivity by the administration of estradiol benzoate and progesterone to female mice becomes more effective as a function of the experience of the mice with such procedures. Both repeated hormonal primings and repeated testing appear important (Thompson & Edwards, 1971).

Heterotypical Hormone Effects

Thus far, we have emphasized the behavioral effects of the administration of hormones appropriate to the sex—androgens to males and estrogens and progestins to females. What of heterotypical hormonal effects—the effects of estrogens on males and progestins on females? Pfaff (1970) conducted a systematic exploration of homotypical and heterotypical hormone administration. He found that estrogen injections reproduced part of the testosterone effect on male sexual behavior, and testosterone, when given with progesterone, reproduced part of the effect of estrogen on female behavior. In general, estrogen mimicked testosterone better than testosterone mimicked estrogen. Females appeared more bisexual than males in that they performed more elements of the male pattern than males did of the female pattern.

In the interest of simplicity of presentation, two experimental groups were deliberately omitted from the discussion of the study of effects of perinatal hormone administration by Whalen and Edwards (1967) presented in Table 12-1. The additional groups were males and females that were gonadectomized and treated with estradiol benzoate. Estradiol benzoate, a female hormone, produced defeminization in exactly the same way as did testosterone propionate. This raises an interesting problem. If females were masculinized by estrogens, why do their endogenous estrogens not defeminize them even when no unscrupulous behavioral endocrinologists are conducting experiments? The answer appears to be that the females have a specialized chemical system (an estrogen-binding protein) that prevents this from happening (Weisz & Gunsalus, 1973). Endogenous ovarian secretions may enhance female sexual behavior (Dunlap, Gerall, and McLean, 1973).

As was mentioned earlier, some androgens are converted to estrogens as a normal step in biosynthesis. This process is termed *aromatization*. It has generally been found that only those androgens that are aromatized to estrogens are effective in restoring copulatory behavior in castrate male rats. It will be recalled that estrogens by themselves can have masculinizing effects. According to one current theory, the aromatization theory, androgens *normally* exert their effects on behavior after they are aromatized to estrogens (see Luttge, 1975).

Species Differences

As both genotype and individual differences within species interact with hormones, it will be no surprise that substantial species differences can be found in hormonal effects, even among closely related species. For example, androstenedione has very different effects in laboratory rats and house mice (Luttge & Hall, 1973). Whereas progesterone facilitates sexual receptivity in rodents, it inhibits receptivity in rhesus monkeys (Michael, Saayman, & Zumpe, 1967).

Even more substantial differences appear when hormonal effects on birds are compared with those on mammals. Injection of androgens or estrogens into the eggs of Japanese quail had no effect on adult female behavior, but it did demasculinize males (Adkins, 1975). The latter effect, of course, is opposite to the early hormone effects found in mammals. Administration of estrogen to adult female quail produced normal female but no male behavior. The same treatment given males produced both male and female behavior. Testosterone had little effect on females, while it produced male behavior when given to males (Adkins & Adler, 1972). Thus, it appears that, among birds, males are more "bisexual" than females, while the reverse is generally true in mammals. Adkins and Adler proposed that these differences may be related to two factors: (1) the difference between birds and mammals with respect to the homogametic (XX) sex (female in mammals and male in birds) and (2) the difference in the hormonal determination of sex in birds and mammals (in the absence of hormone, development in birds is masculine; in mammals it is feminine).

HORMONES AS DEPENDENT VARIABLES

Most of the studies discussed thus far have used hormones as an independent variable and examined their effects on behavior, the dependent variable. The reverse relationship is just as important.

Reproductive Systems

A classification system for different patterns of reproductive cycles in female mammals is shown in Table 12-2 (see Conaway, 1971). Two pivotal events are considered: (1) *ovulation*, the release of an ovum, and (2) the initiation of a *functional luteal phase*, in which the corpora lutea of the ovary are stimulated to secrete functional amounts of progesterone that act to prepare the uterus for the implantation of the fertilized ovum. Each of these pivotal events can be either spontaneous (i.e., occurring every cycle, whether or not the female mates) or induced (i.e., occurring only if the female mates). Many species, such as humans, display a complete cycle whether or not they mate. Species with induced ovulation, such as cats and rabbits, do not ovulate unless they mate. Laboratory rats and mice ovulate spontaneously but have very short estrous cycles (4 days) because they skip the long luteal phase if they do not mate. In species with induced ovulation or induced luteal phases, it is copulation that is critical to triggering a set of neuroendocrine reflexes in the female that result in initiation of a whole sequence of events rendering pregnancy possible. Increased hormone output is a critical step in this process. Thus, copulation leads to increased plasma levels of progestins in species with induced luteal phases such as laboratory rats (Adler, Resko, & Goy, 1970) and increased levels of

Table 12-2 Patterns of Female Reproductive Cycles

Type	Ovulation	Luteal phase	Example
I	Spontaneous	Spontaneous	Humans
II	Induced	Spontaneous	Prairie voles
III	Spontaneous	Induced	Laboratory rats
IV	Induced	Induced	Montane voles

Source: Based on Conaway (1971).

both LH and progesterone in induced ovulators such as montane voles (Gray, Davis, Kenney, & Dewsbury, 1976). The amount of copulatory behavior essential to the initiation of pregnancy varies across species. The mere occurrence of the ejaculatory reflex triggers a luteal phase in house mice (McGill, 1970a); one complete ejaculatory series is sufficient for the initiation of pregnancy in young female rats (Wilson, Adler, & LeBoeuf, 1965); four to seven ejaculatory series are necessary for maximal initiation of pregnancy in hamsters (Lanier, Estep, & Dewsbury, 1975).

Copulatory activity produces not only a complex chain of neuroendocrine events in the female but also dramatic alterations in the hormonal milieu of the male (see Davidson & Levine, 1972). One anonymous human reported stimulative effects of sexual activity and anticipation of such activity on his beard growth, presumably indicating increased hormonal activity (Anonymous, 1970).

Crowding

As was discussed in Chapter 6, social organization changes as a function of population density. In some rodent species, for example, a territorial organizational pattern may prevail at low population densities, but this pattern may shift to a dominance hierarchy as population density increases. As population density increases further, social organization turns to disorganization and the health of the animals tends to decline. Reproduction ceases.

Endocrine events appear to play a critical role in behavioral and other alterations that are found when mammals live at very high population densities. According to the dominant theory (e.g., Christian & Davis, 1964; Davis, 1964; Thiessen & Rodgers, 1961), increased "social pressure," fighting, or both result in increased production of pituitary ACTH and thus of the steroid hormones of the adrenal cortex. Increased adrenal activity, in turn, results in an impairment in reproductive efficiency as revealed in a number of indexes.

There are many lines of evidence in support of this view of the effects of crowding. As population size is increased, adrenal weights increase and gonadal weights decrease. These effects are particularly notable in low-ranking animals. The latter finding is consistent with observations to the effect that in individual interactions, adrenocortical function is increased in defeated mice but not in trained fighters (Archer, 1970).

Thus, hormone levels in both the gonadal and adrenal systems, in interaction with pituitary mechanisms, are altered as a result of the experience and behavior of animals at varying population densities and act, in turn, to alter the behavior of the animals.

A BRIEF SURVEY OF HORMONAL EFFECTS ON DIFFERENT BEHAVIORAL PATTERNS

We shall now briefly survey some effects of hormones on a variety of different behavioral patterns. As some hormonal effects on behavior have already been discussed in the earlier sections, only remaining details will be considered.

Sexual Behavior

The role of hormones in the control of sexual behavior is well known and has been repeatedly demonstrated throughout this chapter. A few additional points merit mention. Most studies of female behavior in animals have considered the adoption of the lordotic posture to the be sole dependent variable. Beach (1976) has emphasized that female behavioral patterns are much more complex than this and that the full range of female behavioral patterns ought to be considered. Beach divides female behavior into three phases: (1) *attractivity*, the female's stimulus value in evoking sexual responses from the male; (2) *proceptivity*, reactions by the female that constitute her initiative in establishing and maintaining the sexual interaction (e.g., soliciting); and (3) *receptivity*, the female's postural responses that are necessary and sufficient for the achievement of ejaculation by the male. In a study of female rhesus monkeys, Johnson and Phoenix (1976) found different effects of hormones on the different behavioral patterns. Female attractiveness was enhanced by estrogen, while effects of androgen varied with the dose; female proceptivity was heightened by both estrogen and androgen; and female receptivity was stimulated by estrogen but not androgen.

Hormones other than those produced by the gonads play a role in the control of sexual behavior. Pituitary hormones, ACTH, and related peptides have been reported to affect male copulatory behavior, especially erection and ejaculation (Bertolini, Gessa, & Ferrari, 1975). Even the hypothalamic releasing factors, such as LH-RF, appear to have direct effects on copulatory behavior (Moss & McCann, 1973).

Maternal Behavior

Maternal behavior can be elicited in the apparent absence of hormonal modifications. Exposure of adult rats to rat pups for 10 to 15 days produces in the adults a tendency to display a full complement of maternal patterns. This nonhormonal maternal behavior can be elicited in intact, ovariectomized, and hypophysectomized females and in intact and castrated males (Rosenblatt, 1967). Whereas cross-transfusion of blood from new mothers stimulates maternal behavior in virgin females, cross-transfusion of blood from animals that became maternal as a result of exposure to pups does not induce maternal behavior in recipients (Terkel & Rosenblatt, 1971). This further demonstrates the nonhormonal basis of maternal behavior.

The presence of androgens during the sensitive period decreases the ease of elicitation of maternal behavior under some test conditions (Quadagno & Rockwell, 1972).

The critical role of hormones in the display of maternal behavior in the adult lies in its *timing* (as opposed to its occurrence). If pups had to wait 10 to 15 days for their mothers to become responsive, none would survive. The unique hormonal milieu of the new mother assures that she is ready with maternal behavior at the time when her pups are ready to receive it. Hormonal elicitation of maternal behavior has proven a challenge to behavioral endocrinologists, but several have succeeded. Moltz, Lubin, Leon, and Numan (1970) used an 11-day treatment program in which estradiol, progesterone, and prolactin were administered according to a carefully designed schedule. All females showed maternal behavior, although a latency of 35 to 40 hours was required for its appearance. Zarrow, Gandelman, and Denenberg (1971) used a 27-day administration program for the same three hormones to produce maternal behavior. Estrogen and progesterone work in synergy in stimulating maternal nest building (Lisk, 1971). Obviously, the hormonal milieu of the new mother is critical to the survival of her young and very difficult to duplicate experimentally.

Aggressive Behaviors

Aggressive behaviors are affected by a considerable variety of hormones from both the pituitary-gonadal system and the pituitary-adrenal system. Gonadal hormones function both in the neonatal development of the potential for aggressive behavior (see above) and in the maintenance of aggressive behavior in the adult, as in male hamsters (e.g., Vandenbergh, 1971b). In monkeys, the levels of both testosterone and adrenal hormones have been shown to vary as a function of the dominance rank of individual males (Rose, Holaday, & Bernstein, 1971; Leshner & Candland, 1972). Luteinizing hormone (LH) from the pituitary gland has been demonstrated to play a role in the control of aggressive behavior in red-billed weaver birds (Lazarus & Crook, 1973).

As with all studies of aggressive behavior, it is important to differentiate the forms of aggression under study. Kostowski, Rewerski, and Piechocki (1970) compared the effectiveness of different adrenal steroid hormones in altering both isolation-induced aggression in mice and muricide in rats. Hydrocortisone increased aggressiveness and hydroxydione decreased it in both tests. Desoxycorticosterone decreased muricide but failed to alter isolation-induced aggression. By contrast, G-stophanthin decreased isolation-induced aggression and had no effect on muricide. Castration and androgen replacement affect shock-elicited aggression, but to a lesser degree than they generally affect other forms of aggression (Milligan, Powell, & Borasio, 1973). Neither castration nor testosterone injections was found to exert an influence on frog killing by rats (Bernard, 1974).

Leshner (1975) has reviewed a considerable number of studies of hormones and aggressive behavior and proposed a theoretical model that incorporates hormonal states, environmental stimulation, experience, and the animal's own quality as a stimulus for aggression.

Territorial Marking

Thiessen and his associates (e.g., Thiessen, 1973) have conducted an extensive research program on the control of the territorial marking behavior of Mongolian gerbils. This program provides a model for the systematic, programmatic investigation of hormonal and other determinants of a clearly defined and easily measured behavioral pattern. Both males and females have a midventral sebaceous gland that secretes pheromones used in marking. Marking is accomplished with a brief, stereotyped behavioral pattern in which a gerbil approaches a prominent object, lowers its abdomen and scent gland to it, and moves forward, thus depositing glandular secretions. Thiessen measured marking behavior in an open field that had been modified by the inclusion of some prominent pegs appropriate for marking.

Scent marking is clearly hormone-dependent in males. Castration leads to a decline in marking that is reversed by androgen administration. Restoration is dose-dependent—the greater the dose, the more marking results. Some males with high doses become "supermarkers." When the cannulation technique was used to probe the regions of the brain where implants of testosterone were most effective in restoring marking in castrated males, the preoptic area of the hypothalamic region was found especially important. This information was important in demonstrating that castration and hormones affect the brain and not just the sebaceous gland, the size and activity of which are also hormone-dependent. By using 11 different steroid hormones in their cannulae, Yahr and Thiessen (1972) were able to determine the precise molecular configurations of hormones that were effective in restoring behavior.

Ovariectomy decreases marking in females, while injections of either testosterone or estrogen plus progesterone restore it. Both estrogen and testosterone implanted into the preoptic area restore marking in females. Lactating females show exceptionally high frequencies of marking.

Other Behavioral Patterns

A few of the many additional behavioral patterns that have been shown to be affected by hormones will be discussed.

Wheel Running A classical phenomenon in the literature on hormones and behavior is the association between hormones and wheel running (e.g., Wang, 1923). The nonpregnant female rat shows a clear-cut 4-day cycle of wheel-running activity that is correlated with her estrous cycle (see Figure 12-2). Recent evidence indicates that hormonal fluctuations, rather than intrinsic neural rhythmicity, are important in controlling this cyclical wheel-running pattern (Gerall, Napoli, & Cooper, 1973). Manipulations of thyroid and adrenal activity also affect wheel running (Stern & Murphy, 1972; Leshner, 1971).

Open-Field Behavior Open-field behavior has been shown to be affected by a number of hormonal manipulations, including neonatal presence or absence of androgen and the stage of the estrous cycle of adult females at the time of testing (Swanson, 1967; Quadagno, Shryne, Anderson, & Gorski, 1972; Stewart, Skvarenina, & Pottier, 1975).

Hoarding Food hoarding varies as a function of hormonal milieu. Hoarding by male gerbils is increased by castration and decreased by testosterone treatment, indicating that testosterone inhibits hoarding (Nyby, Wallace,

Owen, & Thiessen, 1973). Hoarding by female rats varies in association with the estrous cycle (Herberg, Pye, & Blundell, 1972).

Urinary Patterns The urinary postures used by dogs vary as a function of the hormonal milieu early in life, although castration of the juvenile or adult has little effect on urinary posture (Beach, 1974).

Eating and Body Weight Effects of hormones on eating and the regulation of body weight are considerable (see Wade, 1976). Both estrogens and progesterone are important in the regulation of body weight in female rats, although estrogens, probably acting on the ventromedial hypothalamus, appear particularly important. Sex differences in eating and body weight appear due both to organizing effects of perinatal hormones and activational effects of hormones in adults. Ovarian secretions generally reduce body weight, whereas androgens increase body weight. The adrenal system interacts with gonadal hormones in affecting body weight.

Persistence Andrew (1972) has proposed that testosterone increases the persistence of animals, particularly young chicks, in a variety of tasks, such as searching, discrimination, and open-field tasks. Such effects may be related to various effects of androgens on cognitive performance in humans (see Andrew, 1972).

Play An important program on the role of perinatal hormonal conditions in the development of behavior in rhesus monkeys has been conducted at the Oregon Regional Primate Research Center and the University of Wisconsin. One of the more interesting findings of this program is the effect of early hormonal makeup on active play behavior. Normal males display much more active play than females. "Pseudohermaphroditic" females, those that were exposed to androgens prenatally, showed levels of play that were intermediate between those of the normal males and females (Phoenix, Goy, & Resko, 1968).

PSYCHONEUROENDOCRINE INTEGRATION

While it is necessary to fractionate behavior in order to provide an orderly discussion of the role of hormones in its regulation, such fractionation can present a misleading picture. Behavior is integrated, and natural selection works not in favor of the perfection of particular "ideal" mechanisms but on integrated organisms that survive and reproduce. A major role of hormones lies in the coordination and integration of behavior into functional patterns that produce viable, adapted organisms. If this important point is not understood, then the functioning of hormones in psychoneuroendocrine integration is not understood.

While the role of hormones in behavioral integration should be apparent from many mammalian examples, it has been made particularly explicit in work on the breeding cycles of birds, especially in that of Daniel Lehrman and Robert Hinde. Lehrman's description of the reproductive cycle in ring doves was discussed on pages 86–87 and might be reviewed by the reader at this time. A long and distinguished research program has shown that the dramatic shifts in behavior that are apparent in ring doves are mediated primarily by hormones as affected by and affecting the behavior of the organism and its partner. Injection of doves with progesterone results in the immediate initiation of incubation (i.e., sitting on eggs), whereas injection of an estrogenic substance initiates intensive nest-building, followed several days later by incubation. Injections of testosterone have little effect on behavior. While prolactin is critical to the development of the crop and its functioning in feeding the young, injections of prolactin produce little effect on incubation. Thus, estrogen appears to play a major role in stimulating nest building, while progesterone functions in stimulating incubation (Lehrman, 1965).

Komisaruk (1967) studied the effects of the cannulation of progesterone into specific sites in the brain. Incubation was stimulated in both sexes by implantation of progesterone in the preoptic region and the lateral forebrain system. Progesterone implants suppressed male courtship behavior. While progesterone implants stimulate incubation in both sexes, progesterone appears to be functional in normal cycles only in the female. Assays of progesterone levels show them to rise at about the time of incubation in females (Silver, Reboulleau, Lehrman, & Feder, 1974). By contrast, levels of estradiol in females reach their peak during courtship and nest-building, and estradiol is not detectable at the time of incubation (Korenbrot, Schomberg, & Erickson, 1974).

The basis for the pervasive hormonal changes that so control behavior and produce integrated cycles in ring doves lies in stimulation from the environment and from conspecifics. Both visual and auditory stimuli appear particularly important. Exposing isolated females to either their own visual image (by providing them with a mirror) or to colony sounds by means of a speaker produces little egg laying. Exposure to both visual input and the colony sounds elicits egg laying in most females (Lott & Brody, 1966). Exposing isolated females to a male from whom she is separated by a glass barrier can also stimulate reproductive activity. Auditory input facilitates this phenomenon, and intact males are more effective than castrate males (Lott, Scholz, & Lehrman, 1967). The latter finding means that female endocrine development must be dependent upon stimulation from particular displays that are shown by intact males but not by castrated males. The degree of ovarian development in females varies as a function of the vigor of androgen-induced courtship displayed by castrate males; not all androgen-dependent male courtship behaviors are equally effective (Erickson, 1970).

It should be apparent that there is an intricate, continuing interaction of the behavior and the endocrine systems of both male and female that results in an integrated and biologically functional reproductive sequence. A research program parallel to that of Lehrman was conducted by Hinde (e.g., Hinde, 1965) on the reproductive cycles of canaries. The complex interactions among behavior, environmental stimuli, and hormones is made dramatically apparent in Hinde's summary diagram of the determinants of the reproductive cycles of canaries (Figure 12-5).

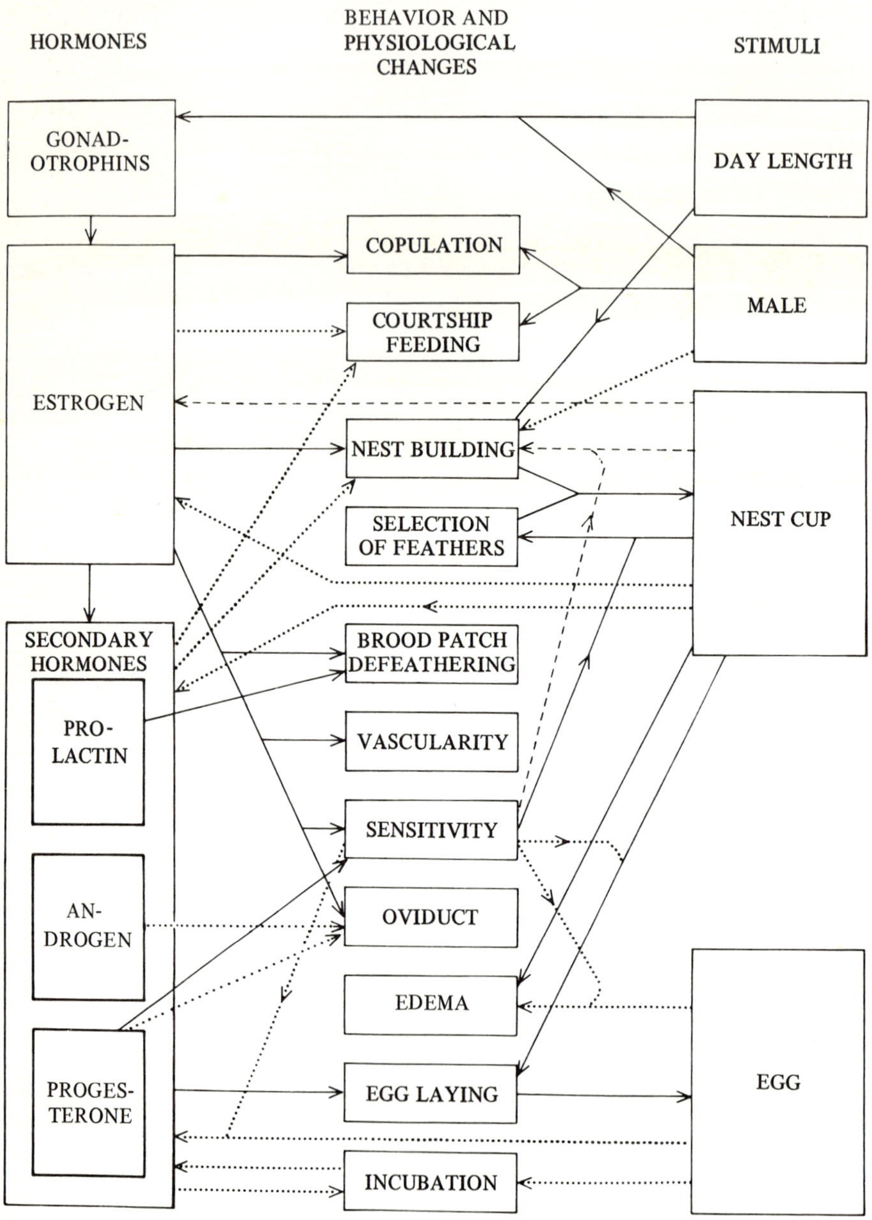

Figure 12-5 Summary of relationships among hormones, external stimuli, physiological function, and behavior in the reproductive cycle of canaries. Continuous lines indicate positive effects; discontinuous lines indicate negative effects. Dotted lines represent probable positive relationships. *(After Hinde, 1965.)*

SUMMARY

Hormones are glandular secretions that function as chemical messengers among the organs of the body. A considerable array of different hormones is produced by endocrine glands that are located about the body. Production and release of hormones are controlled by a series of intricate neuroendocrine feedback systems.

Hormone-behavior interactions are usually studied using correlational methods, castration and replacement, or a variety of more specialized methods. Hormones can produce behavioral effects by acting upon sensory-perceptual systems, neural mechanisms, or effectors.

Whereas the actions of hormones in the adult are short-term and reversible, actions of perinatal hormones tend to be permanent, irreversible, and produced only during a limited sensitive period.

In affecting behavior, hormones have been found to interact with genotype, individual differences, season, interindividual interactions, maintenance versus restoration, experience, sex of recipient, and species.

Hormones are affected by behavior, just as behavior is affected by hormones. Hormones and behavior interact in a large number of functional systems, including sexual behavior, maternal behavior, aggressive behaviors, territorial marking, and numerous other behavioral patterns. An accurate picture of hormone effects can be gained only when their functioning is viewed in relation to a variety of other important determinants to produce functional psychoneuroendocrine integration.

Part Five

Evolution of Behavior

Part Five deals with the third of the four sets of questions around which this book is organized—questions related to the evolution of behavior. Most psychologists and biologists agree that, like morphology, behavior evolves. When dealing with questions of the evolution of behavior, we consider the ways in which behavioral patterns have evolved as well as the nature of the processes underlying that evolution. Parts Three and Four dealt with what have been termed "how" questions or questions of "proximate causation." In Parts Five and Six, the emphasis is shifted to questions that are alternatively called "why" questions or questions of "ultimate causation." With this shift, we shift time spans as well, as questions of evolution and adaptive significance generally deal with sequences of events that transcend the lives of individual organisms. Associated with these changes must come changes in methodology. Because of the long time spans involved, it is not possible to make the kind of repeated observations, empirical tests, and controlled manipulations that characterize the consideration of "how" questions. However, "why" questions are of such importance that some tolerance of the necessary ambiguity inherent in their study is appropriate. Intolerance of unnecessary ambiguity remains appropriate.

Chapter 13

Evolution of Behavior

Two important characteristics of behavior should be apparent from the discussions that have appeared thus far in this book—its diversity and its adaptiveness. The tremendous diversity of the behavior of different species, not only among distantly related species but often among closely related species as well, has been apparent in virtually every chapter. However, this diversity is by no means random variation. Animals of each species have become adapted to live a particular "life-style" that is appropriate for a given environment; that is, they have become adapted. The most parsimonious explanation for the diversity and adaptedness of behavior, and indeed of all biological phenomena, comes from the theory of evolution. Seldom has the development of a theory so revolutionized and synthesized a scientific endeavor as the theory of evolution has the life sciences.

The driving force of evolution has been natural selection. It will be recalled from Chapter 3 that natural selection was defined as a process that resulted in the differential reproduction by individuals of different genotypes. It will be recalled from Chapter 7 that there is good evidence in support of the view that behavior is, at least in part, dependent upon genotype. It follows that as different genotypes reproduce differentially, the behavioral patterns that are dependent upon those differentially reproducing genes will become differentially represented; that is, behavior evolves.

Prerequisite to an understanding of the evolution of behavior is an understanding of evolutionary processes in general. Some of these processes will be discussed in this chapter; others were mentioned in Chapter 3. For the reader without extensive background in the evolutionary approach, it might be appropriate to review the material in Chapter 3 before continuing with the present chapter.

METHODS IN THE STUDY OF THE EVOLUTION OF BEHAVIOR

The comparative method lies at the heart of the study of the evolution of behavior. Relatively little can be learned about the evolution of behavior from the study of a single species as opposed to comparing a group of different species. King (1963, 1970) has distinguished two levels at which comparisons among species can be made—the *species level* and the *phyletic level*. When making comparisons at the species level, one compares the behavior of species that are relatively closely related, often falling within a single genus or a few closely related genera. Comparisons at the phyletic level are made across much broader taxonomic classifications—often across families, orders, classes, or phyla. Comparisons at both levels are complicated by the problem that behavior does not fossilize. Thus, we lack evidence of many of the intermediate steps in evolutionary history that are available to the student of structure, especially bone structure. In dealing with behavior, most inferences must be based on extrapolations from the behavior of contemporary species.

The Species Level

Statements about behavioral evolution are most precise when related to a clearly defined and measureable behavioral pattern in a restricted group of species (i.e., at the species level). Lineages are most clear when a graded sequence of species is available for study, so that probable graded steps in the evolution of the behavior can be inferred. Inferences based on groups of species where such graded steps are available have been termed *"strong" comparative inference*, as opposed to inferences where such graded sequences are lacking, which may be termed *"weak" comparative inferences* (Klopfer, 1974). The graded sequences of evolutionary trends seen among closely related species have been termed *ethoclines* (Johnsgard, 1972).

An excellent example of study at the species level is provided by Dilger's (1960, 1962) research on the behavior of lovebirds of the genus *Agapornis*. Dilger was able to study eight of the nine forms that make up this genus of birds. All display stereotyped and easily elicited behavioral patterns in the laboratory. On the basis of morphological and behavioral studies, a reasonable reconstruction of the course of evolution of the forms of this genus can be proposed, moving from the more primitive Madagascar, Abyssinian, and red-faced lovebirds through the peach-faced lovebird to forms with distinct white rings about their eyes. Some of the differences between the more primitive and more recently evolved forms of lovebirds are summarized in Table 13-1. As is evident from the table, the more recently evolved forms are more colonial, display more complex social interactions, and generally have

Table 13-1 Differences in Behavior between Primitive and Recent Species of Lovebirds (*Agapornis*)

Primitive forms	Recently evolved forms
Sexual dichromatism (males and females differently colored).	No sexual dichromatism.
Nest as individual pairs.	Nest in colonies.
Form pairs at about 4 months of age.	Form pairs earlier—at about 2 months of age.
"Squeak-twitter" call by male during courtship is variable in pitch and purity, has no recognizable rhythm, and occurs only when female retreats to nest cavity.	Squeak-twitter is rhythmic, purer in tone, less variable in pitch, and occurs even when the female is present.
"Displacement scratching" in male courtship resembles ordinary head scratching except for context and fact that it is always done with foot nearest female; occurs only when thwarted by female.	Displacement scratching is a ritualized display—more rapid and perfunctory than ordinary scratching; male can use either foot; occurs even when male is not thwarted.
"Courtship feeding" is reciprocal (i.e., female may offer regurgitated food to male); associated "head-bobbing" pattern is rapid and related to actual food transfer.	Courtship feeding done only by male; head-bobbing pattern is slower and more ritualized (i.e., may not always be related to actual food transfer).
Nest is a simple pad of soft material deposited on the floor of a nest cavity in a tree.	Elaborate nest consists of a roofed chamber at the end of a tunnel.
Elaborate individual nest-cavity defense displays; no colonial "mobbing."	No individual nest cavity display; colonial mobbing of predators.
Carry small bits of nesting material tucked in feathers about body.	Use longer strips of material or twigs and carry either in restricted group of feathers or in bill.
Despite elaborate threat and appeasement displays, occasionally have serious fights.	Ritualized display fighting (e.g., "bill-fencing" in which only the toes get bitten).

Source: Based on Dilger, 1960, 1962.

more ritualized and stereotyped behavioral patterns than the less highly evolved species.

Lindauer (1961) used the species level of comparative analysis to study the evolution of the complex dances that signal the location of food sources in honeybees. A plausible sequence was recreated from the stingless bees, in which the guidance of a "leader" and odor markers are necessary for the transmission of information, through bees of different species of the genus *Apis*, in which the direction is coded first directly with respect to the sun and later indirectly relative to gravity.

The Phyletic Level

Comparisons at the phyletic level are often more difficult than those at the species level because the species under study differ in so many ways that (1) it is often impossible to be sure that they are studied under comparable conditions and (2) once differences are observed, it is difficult to know which of the many factors that differentiate the species are associated with the particular difference of interest. The major advantage of comparisons at the phyletic level is that they allow one to reconstruct the broad sweep of evolutionary change occurring in a particular group of species. Thus, while comparisons at the phyletic level may be of little value for elucidating the nature of evolutionary processes, they can be of great interest in reconstructing the broad evolutionary history of behavior.

There are many examples of the application of comparison at the phyletic level. Alexander (1964) traced the history of the evolution of mating behavior in arthropods and found it related to the evolution of spermatophores, adoption of aquatic or terrestrial habitats, and more specific environmental conditions within particular groups. Glickman and Sroges (1966) found object exploration to be increased from reptiles to "primitive mammals" and primates and carnivores. Allison and VanTwyver (1970) proposed that sleep must have evolved independently among birds and mammals, with paradoxical sleep in mammals developing after and slow-wave sleep before, a proposed divergence of therian from nontherian mammals. Heffner et al. (1969) proposed that selective pressures for responsivity to sound waves of low frequencies must have characterized the evolutionary lineage leading to human beings. Nottebohm (1970, 1972) proposed that there is a general trend in the evolution of patterns of the ontogeny of birdsong from "self-centered strategies" to more environmentally dependent patterns. The trend toward environmental dependence appears to have evolved independently at least three times. Electric organs appear to have evolved independently in fishes at least six different times (Bennett, 1970).

THE CONCEPT OF HOMOLOGY

"A *homology* can be defined as any resemblence between two species that can be explained by their common descent from an ancestor possessing the character in which they are similar to each other" (Lorenz, 1974, p. 230). The concept of homology has played a pivotal role in the development of ethology and the study of behavioral evolution from the time of Darwin, Whitman, and Heinroth to the present. Indeed, Lorenz (1974) regards the elaboration of the usefulness of the concepts of homology and analogy in the study of behavior as his most important contribution to science.

Homology and Analogy

The concept of homology originated in studies of the evolution of morphological structures. Thus, the wings of bats and birds may be regarded as homologous *as forelimbs* in that they developed from a common reptile ancestor possessing forelimbs. By contrast, the wings of insects and birds evolved as appendages along independent lines and are thus not homologous even as appendages.

It is important that homologous structures and behaviors be differentiated from analogous structures and behaviors. "Analogy" may be defined as a correspondence between structures or behavioral patterns that is due to similarity in function and that is not necessarily derived from a common ancestor. Analogous structures or behavioral patterns derive their similarity from evolution along similar paths, but paths which occur in different lineages. Thus, the wings of bats and birds are analogous *as wings*; they are not homologous as wings. The evolution of forelimbs into wings must have occurred quite independently in the evolutionary lines leading to bats and birds. Although the wings of birds and bats are quite similar in structure, and their evolution presumably occurred in response to similar selective pressures, they are analogous as wings, not homologous. The addition of the conditional phrase (e.g., "as wings") is important in delineating homologies clearly (Bock, 1969). Analogies are very important in the study of behavioral evolution (see Lorenz, 1974) but must be kept distinct from homologies.

Criteria for the Recognition of Homologies

While homologies are defined in relation to patterns of ancestry, such ancestry often is not immediately apparent, so that various criteria are necessary for the recognition of homologies. Among the criteria that have been proposed (see Atz, 1970; Eibl-Eiblesfeldt, 1975; Campbell & Hodos, 1970) are:

1 *Position:* A specific structure may occupy the same location in a structural system, or a behavioral pattern may occupy a similar temporal position in a complex behavioral chain in two species.

2 *Minuteness of similarities:* The more minute and specific the similarities among characters between species, the more confident we may be that they are homologous.

3 *Multiplicity of similarities:* The greater the number of similarities, the more confident we may be of homology.

4 *Linkage by intermediate forms:* Even quite dissimilar behavioral patterns may be homologized when they can be traced along a path of plausible linkage by way of intermediate steps.

5 *Large number of related forms:* If there is a large number of closely related species that display similar behavior, we can be more confident of the homology than with a smaller number.

6 *Inheritance in hybrids:* The appearance of the particular behavioral pattern in a hybrid between two species increases the likelihood that behavioral patterns are homologous (e.g., Hinde, 1956).

Sources of Confusion in Attributing Homology

It should be apparent that the determination of homologies is a difficult task. The complex ontogenetic pathways leading from genes to behavior provide one important reason for this difficulty. It is particularly difficult to delineate homologies when species under consideration are not closely related and the behavioral pattern under study is not clearly defined. It thus comes as no surprise that the most convincing demonstrations of homology are in the display patterns of closely related species of insects and birds, such as the Anatidae (ducks and geese) of Lorenz's study (1941).

Because of the lack of paleontological evidence to support proposed homologies, there is a danger that similarities due to (1) analogy arising from convergent evolution, (2) mimicry, and (3) transmission by nongenetic means may be mistaken from homologies. The problem of confusing similarity derived through homology with that derived through analogy has already been discussed.

Similarities achieved as the result of mimicry might sometimes be confused with homology. In *Batesian mimicry,* a palatable insect comes to resemble in appearance an unpalatable one (Brower, 1969, Chapter 4). In *Muellerian mimicry,* two unpalatable species may come to resemble each other and thus derive benefit in relation to protection from predators. Both forms of mimicry appear in butterflies (Brower, 1969) and produce similarity that might be mistaken for homology.

Some stereotypy of behavior within a population can be maintained by nongenetic, "cultural" transmission. This stereotypy is potentially confused with homology (see Galef, 1976). Certain aspects of birdsong fall within this class. Among Japanese macaques, patterns of eating candy, washing sweet potatoes, and eating wheat have been propagated by cultural transmission (Kawamura, 1963). Such behavioral patterns transmitted across species might be confused with homology.

Applicability of Homology to Behavior

Animal behaviorists disagree regarding the applicability of the concept of homology to the study of behavior. For those whose work remains close to the roots of classical ethology (e.g., Wickler, 1973; Eibl-Eibesfeldt, 1975), the concept of homology remains a pivotal one. By contrast, Atz (1970) was a particularly harsh critic of attempts to homologize behavioral patterns, arguing that the concept of homology is essentially a morphological concept. An intermediate position has been adopted by Beer (1973).

Few animal behaviorists would deny the applicability of the concept of homology to cases where there is a series of closely related species and a clear ethocline with respect to a clearly defined behavioral pattern. The research of Dilger on lovebirds provides a good example. Such studies are of considerable importance in understanding behavioral evolution. However, as one deals with behavioral patterns that are less clearly defined and species that are less closely related, many animal behaviorists would caution against facile homologizing. Dewsbury (1973) and Mason and Lott (1976) have expressed the view that while behavioral homologies appear valid, the study of analogies is likely to play a more important role in the future development of animal behavior (see Chapter 14).

PRIME MOVERS OF BEHAVIORAL EVOLUTION

Evolutionary rates differ greatly for different behavioral patterns and among different groups of species. The rate at which evolutionary change occurs is determined by the interaction of forces that can be considered under the rubrics of *phylogenetic inertia* and *ecological pressure* (Wilson, 1975). These factors are considered "prime movers" in that they appear to be the ultimate determinants of the degree to which behavior (or other characteristics) will remain phylogenetically unchanged or show substantial evolutionary change.

Phylogenetic Inertia

Great phylogenetic inertia (of rest) implies considerable resistance to evolutionary change, whereas weak inertia implies a high degree of lability. There are numerous cases where species have failed to evolve behavioral changes in situations where such change would seem appropriate. For example, gray seals typically breed in pairs or small groups on ice floes in the North Atlantic. No advantage is gained from discriminating their own from other young, as there are few other young. When they breed in crowded rookeries in more southward locations, however, gray seals fail to adopt the more typical pinniped pattern of selectively nursing only their own young (E. A. Smith, 1968; Wilson, 1975). The failure to show evolutionary change may be understood in relation to phylogenetic inertia and ecological pressure.

There are a number of factors that promote conservatism in evolution. One is limited genetic variability. As evolutionary change is dependent upon changing gene frequencies, it is obviously limited in situations where genetic variability is minimal. Inbred populations should show slower evolutionary change than those that are outbred. Asexually reproducing forms should be slow to change.

A second major conservative force is the tendency of many species to resist change, as, for example in avoiding exposure to new habitats. As most species have become quite well adapted to particular habitats, there are obvious disadvantages in forsaking that habitat in favor of an alternative one that may be less hospitable. Often the dominant or territory-holding animals are those that are most likely to resist migration to new environments. Thus, Christian (1970a) has argued that it is the subordinate animals that are driven from the colony and are the primary agents of movement into new areas and of evolutionary change (see also Udvary, 1970; Christian, 1970b). By analogy, social change in humans is often perceived as instituted by the "have nots" rather than the established power structure. By way of example, the strong tendency of mountain sheep to follow traditional routes within known portions of their habitat is a major factor limiting exploitation of new habitats and may be threatening their survival (Geist, 1967).

Another factor limiting evolutionary change may be lack of appropriate preadaptations. A *preadaptation* is "any previously existing anatomical structure, physiological process, or behavioral pattern that makes new forms of evolutionary adaptation more likely" (Wilson, 1975, p. 592). Natural selection does not act on a *tabula rasa* but on existing organisms. There must exist appropriate "raw material" if a particular character is to be evolved. The source of the raw material is unimportant and often unrelated to the selective pressures acting toward evolution of new behavior. An obvious feature of natural selection is its opportunism. Like the good cook who uses every part of the chicken, the forces of natural selection appear to utilize all available material in creating organisms that are fit. Even urine and feces are utilized, as in communication by means of pheromones.

Various authors have speculated about preadaptations for particular behavioral patterns and structures. For example, Alexander and Brown (1963) considered the problem of the evolution of insect wings. How could wings evolve through a gradual evolutionary process if tiny wing buds, an essential intermediate step, would be nonfunctional in flying? All intermediate stages must be functional. They proposed that the first phase of the evolution of wings was functional in courtship displays. Once functional structures evolved for courtship, they served as a preadaptation for the evolution of gliding and eventually became wings. Huber (1962) speculated that cricket stridulation may have developed from preadapted intention movements for flying. Alcock (1972) speculated about the preadaptations and evolution of numerous examples of use of tools. For example, the tendency of woodpecker finches to use twigs to dislodge insects from tree bark may have evolved from displacement activities. Finches unable to secure insects from bark may have performed displacement nest gathering (see Chapter 2) and returned to the task of hunting insects with twigs still in their mouths. Once this step was reached, reinforcement by other processes was possible.

The lack of appropriate preadaptations can be a factor severely limiting evolutionary change. Other potential factors include limitations due to detrimental effects of change upon other adapted characteristics of the organism, limitations due to interspecific competition, and limits imposed by the complexity of the trait under consideration (more complex traits are more difficult to change than simple ones).

Ecological Pressure

Wilson (1975) defined ecological pressure as "the set of all environmental influences that constitute the agents of natural selection" (p. 592). Thus, ecological pressure is simply the summation of all selective pressures that act to produce evolution and increase fitness. These selective forces act to increase survival—as, for example, through increased efficiency in feeding, protection from predators, and resistance to disease—and to optimize reproductive processes so as to maximize the number of viable, fertile offspring.

It is the continuing tug-of-war between the conservative forces of phylogenetic inertia and the progressive forces of ecological pressure that determines rates of evolutionary change.

MICROEVOLUTION OF BEHAVIOR

The study of *microevolution* is concerned with small amounts of evolutionary change, as, for example, in the earliest stages of the differentiation of behavioral patterns. The end product of evolutionary action—that is, macroevolutionary change—is a clearly differentiated, qualitative difference. However, it is axiomatic in the study of evolution that such changes do not generally occur in single steps. Rather, they often represent the accumulation of a continuing series of many very small changes, all in the same direction. Natural selection acts on a preadaptation to produce a series of gradual, small changes that ultimately result in large qualitative differences. Each step along the way must confer some selective advantage. Remember that natural selection does not work in order to perfect specific systems. Natural selection works through differential reproduction by individuals of differing genotypes. Thus, the accumulation of small changes into qualitative differences and the apparent perfection of particular mechanisms is not an end in itself but the by-product of natural selection acting through differential reproduction.

Quantitative Changes in Microevolution

If natural selection produces qualitative change as the result of an accumulation of quantitative changes, it ought to be possible to locate and study such small quantitative changes and thus view the actions of evolutionary processes at their most subtle level. This is what microevolutionary study is all about. The importance of examining small, quantitative microevolutionary changes as a step in understanding evolutionary processes has been emphasized by Manning (1964; 1971). At least five different kinds of quantitative changes among species can be observed in microevolutionary study (Blest, 1961; Manning, 1964, 1971).

1 *Changes in threshold for the elicitation of homologous behavioral patterns:* Microevolutionary changes can produce differences in the threshold for elicitation of particular homologous behavioral patterns. For example, moths of different species differ with respect to the intensity of stimulus required to elicit defensive displays. The more vividly colored, unpalatable species respond to the slightest touch with movement and a display. By contrast, the more cryptically colored species remain motionless and only display in response to a hard prod (Blest, 1957).

2 *Changes in the frequency with which homologous behavioral patterns are performed:* With changes in frequency there is a common repertoire of behavioral patterns between closely related species, but the species differ in the relative frequency with which they display the behavioral pattern. Thus, for example, two closely related species of fruit flies, *Drosophila melanogaster* and *D. simulans*, each show two patterns of wing display during courtship: "scissoring" (in which both wings are slowly opened) and "vibration" (in which one wing is extended and rapidly vibrated in the vertical plane). Vibration is the dominant display among *D. melanogaster*, whereas scissoring is more common in *D. simulans*. Indeed, the casual observer may never observe scissoring in *D. melanogaster*; it is revealed only with prolonged observations (Manning, 1959).

3 *Changes in the speed with which homologous behavioral patterns are performed:* Differences in the speed of homologous behavioral patterns imply that while the form of the movement is unchanged, the movement occurs at a faster or slower pace. Examples have been reported in studies of the protective displays of moths (Blest, 1957) and threat displays of gulls (Tinbergen, 1959).

4 *Changes in the "emphasis" given various parts of homologous patterns:* While two closely related species may show similar movements, those of one species may be more exaggerated or emphasized than those of another. Thus, a similar movement may be carried further, so that the pattern appears more exaggerated or extreme. The movements associated with the "long call" in two closely related species of gulls, herring gulls and lesser black-backed gulls, provide an example (Figure 13-1). Differences can be seen in the emphasis of movements at both extremes of the motor pattern in the display. The lesser black-backs throw their heads down further and up higher than do the herring gulls.

5 *Changes in the degree of rhythmic repetition of homologous behavioral patterns:* A movement that is but a single discrete act in one species may become elaborated into a rhythmically repeated pattern in another species. Such a process may have been important in the evolution of courtship in some bird species, such as zebra finches (Morris, 1954).

These five, in addition to other forms of quantitative change in behavior occurring at the microevolutionary level, are summed over time and generations to produce the qualitative differences that so characterize evolutionary change across species.

An example of the kind of differences that can result from quantitative evolutionary changes is provided by the courtship displays of lizards of the genus *Sceloporus* studied by Hunsaker (1962) (Figure 13-2). Head-bobbing movements differing in amplitude, speed, and pattern are sufficient to maintain reproductive isolation.

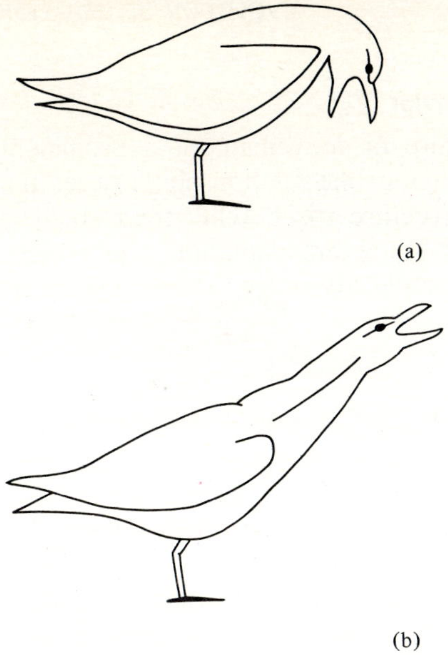

Figure 13-1 Variations in the form of homologous displays in herring gulls *(Larus argentatus)* and lesser black-backed gulls *(L. fuscus)*. In each species, birds begin in the head-down position *(a)* and progress to the head-up position *(b)*. The two species differ in the emphasis given to these patterns. *(After Brown, 1967.)*

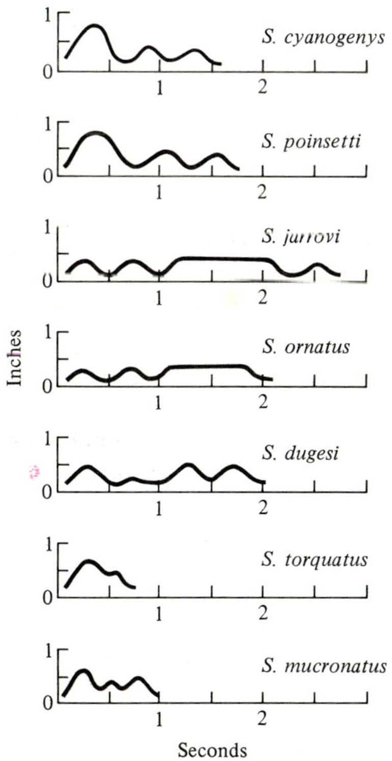

Figure 13-2 Diagrams illustrating the head-bobbing movements in the courtship of seven species of lizards of the genus *Sceloporus*. The elevation of the head is portrayed on the ordinate as a function of time, which is shown on the abscissa. *(After Hunsaker, 1962.)*

Changes in Structure Associated with Behavior

Where a display pattern involves exposure or movement of a prominent structural adaptation, the interesting problem of the relationship between the evolution of the behavior and that of the structure arises. While there could be no such structural modification without some preadaptation, there often appears to be no function and hence no selective advantage to the exaggeration of the structure other than in the behavioral context. Thus, it appears that the evolution of structural changes may often follow behavioral changes.

A case can be made for the view that behavior is among the most progessive characteristics of organisms. Wickler (1972) and Wilson (1975) went so far as to refer to behavior as an "evolutionary pacemaker." By that, they meant that in situations such as that just described, where behavior and structure are changing simultaneously, it is often behavior that constitutes the leading edge or pacemaker of evolutionary change, while structural change follows. For example, the elaborate morphological adaptations displayed during courtship in peacocks are likely to have developed after display behavior was already established.

There are numerous other examples of morphological changes trailing behavioral evolution. In developing a plausible model for the evolution of horns and horn-like organs, Geist (1966) proposed numerous steps in which horn-like organs change following various behavioral modifications. Different behavioral patterns used in combat provided impetus for structural modifications in skull thickness and the development of horn-like organs. The largest antlers ever known are those of the now extinct "Irish elk," which are known to have ranged up to 4 meters in total span. Gould (1973) proposed that these enormous antlers developed as symbols of dominance rank that permitted differential reproductive advantages typical of those accruing to dominant males. A final example can be found in the evolution of rump patches in numerous species of mammals. Many species, particularly ungulates, have prominent patches of white and other colors in the rump region. Guthrie (1971) argued that such rump patches have evolved as organs that support subordination behavior. Guthrie proposed that rump patches permit unambiguous subordination signals and that they evolved in conjunction with appeasement gestures in which the anogenital region is exposed.

SPECIATION AND REPRODUCTIVE ISOLATION

An important question in the study of the evolution of behavior is that of how species are formed. In Chapter 3 we defined a species as a group of organisms that are capable of interbreeding freely under natural conditions but which do not interbreed with organisms from other groups. In the study of the evolution of behavior, it is differences of at least the species level that are generally of interest. There are many quantitative and even qualitative differences among members of different populations of the same species. In considering the problem of the formation of species (*speciation*), we consider how such differences come to produce reproductive isolation and the formation of species.

Allopatric Speciation

Two groups of animals are said to be *sympatric* if they occur together in the same geographical region and *allopatric* if they are geographically separated. The dominant view among biologists is that speciation requires geographic isolation; that is, that speciation must take place in allopatry. However, some biologists maintain that sympatric speciation occurs. Thus, while Mayr (1963) maintains that allopatry is necessary for speciation, Scudder (1974) maintains that alternative forms, including sympatric speciation, are possible. We shall limit our discussion to allopatric speciation.

The process of allopatric speciation begins with a single population that clearly represents but one species. Somehow this population must become divided by a geographical and ecological barrier, so that animals from what are now subpopulations are no longer able to interbreed freely. Populations may be separated by the development of a mountain range during a time of glaciation, by being swept up on different islands, or by being isolated in different lakes. During the period that gene exchange between the two subpopulations is prevented, there will be changes in the gene pool within each subpopulation. Many of these changes will be related to natural selection as the subpopulations adapt to slightly different habitats. The extent and rate of these changes will be a function of phylogenetic inertia and ecological pressure. Still other genetic changes may result from random processes, such as genetic drift. The latter factors may be particularly important in small subpopulations.

At some point the geographical barrier is crossed or removed and the two subpopulations interact. What happens then depends on the degrees and kinds of changes within the two subpopulations that have occurred during their period of allopatry. If changes in the gene pool have been minimal, the two subpopulations will freely interbreed and eventually merge to become a single population once more. Alternatively, the two subpopulations may have been separated for so long that no breeding can occur; they have already formed species. However, divergence may be less extreme than this. The two subpopulations may have diverged to an intermediate degree, so that hybrids could occur but would be at a competitive disadvantage. If hybrid animals are not as healthy or as fertile as those conceived from mating within subpopulations, the fitness of animals breeding with partners from outside their subpopulation would be lower than that of animals choosing mates from within the subpopulation. In that case, there would be selection against matings between animals from different subpopulations.

It is at this point, after two divergent subpopulations have come together, that reproductive isolating mechanisms confer selective advantages. Reproductive isolating mechanisms are those biological properties of individuals that prevent the interbreeding of sympatric groups (Chapter 3). They develop under the present circumstances because of the competitive disadvantage of hybrids between the two subpopulations. Individuals that breed with members of the foreign subpopulation would be at a reproductive disadvantage relative to those that breed with members of their own subpopulation. Thus the reproductive isolating mechanisms are reinforced and mating between the two subpopulations becomes infrequent or impossible. They now have become two different species.

Typically, sympatry will occur among previously isolated subpopulations only in part of their geographical range. It is in the areas of sympatry that selection will act against interpopulation breeding and foster the development of isolating mechanisms. The animals from the two subpopulations that occur in the region of sympatry should diverge more from each other than those living in areas of allopatry. This is the phenomenon of *character displacement*. Character displacement is the phenomenon whereby animals from two newly evolved species interact so that further divergence between the species results in the area of sympatry.

Darwin's Finches

An excellent example of the manner in which a presumed single population can diversify in adapting to new habitats (called *adaptive radiation*) is provided by a group of finches living in the Galapagos Islands off the coast of Ecuador (Lack, 1947, 1953). Because Darwin studied these birds during his voyage on the *Beagle*, they are called "Darwin's finches." There are currently 14 species of finches on the Galapagos Islands. They have radiated into a number of very diverse forms. This diversification can be seen most easily in an examination of their feeding habits and bills. The bills of some species are adapted for crushing seeds of various sizes; some are adapted for grasping insects and buds; and others are adapted for probing after insects. The appearance of the bill may also function in displays and hence foster reproductive isolation.

Lack proposed that a flock of ancestral finches may have arrived as the first passerine birds on the Galapagos. They found great potential for exploitation of resources because of an absence of both competition from other species and of predators. The result was an adaptive radiation in which the original animals became isolated on different islands, adapted to local conditions, and formed the total of 14 species from the single founding population. Many of these species have evolved into distinctly "unfinchlike ecological niches" as a result of the unique conditions on the Galapagos. The geographical isolation from both mainland birds and from each other, together with the opportunity for rapid diversification in adapting to new habitats, probably was critical to the adaptive radiation of Darwin's finches.

Experimental Studies of Reproductive Isolation

A number of researchers have attempted to recreate artificially the kinds of interactions that must occur when two previously sympatric subpopulations are reunited (Koopman, 1950; Knight, Robertson, & Waddington, 1956; Kessler, 1966; Crossley, 1974). All studies used fruit flies of the genus

Drosophila, the "geneticist's white rat." Flies of two closely related species or of two similar genotypes differing in some prominent marker gene are intermixed and permitted to mate. Such flies can either mate with partners of their own genotypes or form hybrid matings. The offspring of hybrid matings are removed and not permitted to reproduce. If this procedure is followed over several generations, reproductive isolation between the populations can be seen to develop. Whereas a relatively large proportion of matings in the first few generations and in control lines may be hybrid matings, the percent of matings in the selected lines that produce hybrids decreases over 20 to 40 generations.

Crossley (1974) analyzed the details of the reproductive interactions between her ebony and vestigial flies after selection against hybrids. She found that the courtship interactions of her two populations had been changed over the course of her experiment. There occurred (1) an increase in the tendency of females to resist courtship from males of the strange (i.e., heterogametic) strain and (2) a greater tendency of males to respond to the repelling movements from females of the foreign strain by terminating courtship than to respond similarly to the repelling movements of females of their own strain.

These studies differ from the selection studies discussed in Chapter 7 in that the experimenters did not select for particular behavioral characteristics. Rather, as natural selection appears to do on occasion, the experimenters selected against hybrids. As a result of selecting against hybrids, increased reproductive isolation and evolutionary changes in behavior developed. Events that occur when two previously allopatric subpopulations first interact may not be too much different from those that occur under experimental conditions with these *Drosophila*.

RITUALIZATION AND THE EVOLUTION OF DISPLAYS

The evolution of the displays that function in the context of communication in agonistic and sexual encounters has provided a focal area for research by ethologists. Such displays tend to be quite elaborate and quite diverse among different species. The question of the nature of the evolution of displays immediately arises. The processes underlying the evolution of displays are not fundamentally different from those underlying the evolution of any other behavioral patterns. However, displays provide many of the clearest examples of the actions of evolution on behavior. Therefore, we shall consider some of the processes underlying the evolution of displays as an illuminating special case of the evolution of behavior in general.

Evolutionary Origins of Displays

As is characteristic of evolution in general, natural selection has proved to be quite opportunistic in seizing upon behavioral patterns as sources for the evolution of displays. Classically, three primary sources of displays were recognized by ethologists (e.g., Tinbergen, 1952b; Hinde & Tinbergen, 1958):

1 *Intention movements:* Many displays appear to have evolved from intention movements, the preparatory or incomplete movements that often appear at the beginning of an activity (see Chapter 2). Such movements appear to have been an important source of "preadaptations" for the evolution of display patterns in birds (Daanje, 1950). Components of displays such as flexing of the tail are incipient movements that are preparatory to flight, yet they often occur when a bird is alarmed but does not take flight. They appear to have been a source of many displays. The full-forward display of green herons (Figure 2-1) provides an example.

2 *Displacement activities:* Other displays have apparently evolved from displacement activities, the "out of context" behavioral patterns that often appear in conflict situations. Examples include "displacement scratching" in lovebirds (Table 13-1) and the snap display of green herons (Figure 2-1), which appears to be derived from displacement activities related to the collection of nest material.

3 *Redirection activities:* The third classical source of material for the evolution of displays is redirected activities, that is, the direction of a behavioral pattern, such as attack, toward an object other than that eliciting it. Certain displays of terns and related species may have been derived from redirected attack (Cullen, 1960).

More recently, it has become apparent that signals can be derived from almost any convenient behavioral pattern, anatomical structure, or physiological process (Andrew, 1963; Wilson, 1975). Given the opportunistic nature of natural selection, this should come as no surprise. Some additional sources of displays include:

1 *Food exchange:* Displays can evolve from food exchange, as in the ritualization of courtship feeding in lovebirds (Table 13-1).

2 *Protective responses:* Greeting displays of primates appear to be derived from protective movements given by mammals to startling or unpleasant stimuli (Andrew, 1963).

3 *Comfort movements:* Many precopulatory signals in ducks and geese appear to be derived from comfort movements such as preening, shaking, stretching, and washing (McKinney, 1965).

4 *Thermoregulatory patterns:* Avian displays involving feather erection may have been derived from thermoregulatory patterns (Morris, 1956).

Ritualization

Ritualization is the evolutionary process whereby a behavioral pattern is modified in such a way that it either comes to serve as a signal in communication or becomes more efficient as such a signal. Virtually all the quantitative changes of microevolution discussed above (e.g., changes in threshold, frequency, speed, emphasis, and repetition) occur during the transition from the original function of the behavioral pattern into its new signal function. Hinde and Tinbergen (1958) emphasized three characteristics of the ritualization of displays:

1 *Development of conspicuous structures:* As discussed above, various structures associated with signaling behavior become modified in conjunction with behavioral evolution.

2 *Schematization of the movement:* The nature of the movement becomes changed in accordance with the general nature of microevolutionary changes (e.g., see Table 13-1).

3 *Emancipation:* As ritualization occurs and the behavioral pattern comes to function in a new context, it becomes "emancipated" from its original motivational context. Thus, a display that originated from a displacement activity no longer occurs in association with conflict but is associated with courtship, threat, or whatever might be the appropriate situation for its appearance.

Selective Pressures on Displays

A number of selective pressures act on behavioral patterns that serve as signals, and these pressures are responsible for the refinement and change of the signals through evolution (Cullen, 1966). Typically, evolutionary changes act to make signals more conspicuous and to decrease ambiguity. For example, the evolution of display patterns of fiddler crabs (*Uca*) entails changes in the speed, rhythm, and form of the pattern with which the chelipeds (large claws) are waved. An evolutionary trend toward increased conspicuousness has been achieved through such alterations (Crane, 1957).

Three fundamental classes of selective pressures were distinguished by Cullen (1966):

1 *Between species:* Reduced ambiguity of signals functions to foster reproductive isolation and prevent hybridization; it also acts to reduce aggressive encounters in closely related species not competing for identical ecological resources.

2 *Within species*: Within species, it is advantageous for signals to be as distinctive as possible so as to minimize confusion among different signals. According to Darwin's (1873) *principle of antithesis*, two displays with opposite meaning should look as different as possible, often entailing movements of opposite orientation. Thus, the stretch display and the full-forward display of green herons (Figure 2-1) function respectively in threat and appeasement and entail quite opposite postures.

3 *Individual differences:* Signals can serve to identify the individual emitting the display.

Receptor Systems and Compromise

Two interesting problems in the study of ritualization remain. One lies in the correlation between evolutionary changes in the sender and those in receiver. As display patterns evolve, so must patterns of responsivity on the part of the responders to displays (Blest, 1961). Thus, if a male courtship pattern changes through evolution, so must the preferences of the female. Little work has been done to determine whether there are common sets of genes controlling both displays and responsivity patterns or whether the two systems are controlled by different sets of genes capable of showing parallel evolutionary change.

Another problem is that of perfectionism and compromise in the evolution of displays (see Manning, 1971). Because natural selection acts on whole organisms, many of their characteristics represent a compromise among opposing evolutionary forces. Thus, the extent to which behavioral and morphological change can take place in response to one set of selective pressures is limited by other sets of pressures. While large antlers may aid in courtship, there are obvious limits to the strength and metabolic capacities of the animals that must bear them. Although long, narrow bills are more effective releasers of pecking by young gulls than normal bills (Tinbergen & Perdeck, 1950), bill shape and size obviously are constrained by other functions, such as feeding habits.

Evolution of Courtship in Balloon Flies

We close this section on ritualization and the evolution of displays with a particularly interesting example, the evolution of courtship in balloon flies (Kessel, 1955), that illustrates many of the principles and phenomena discussed in this chapter. The empid fly, *Hilara sartor*, displays an interesting courtship pattern. Males construct a silk "balloon" about as large as they are. They then join other males in a swarm, from which a female selects a mate. The female accepts her mate's balloon and the pair depart to mate. The obvious problem is that of how such a pattern could evolve. Kessel studied a number of species from among the thousands of the family of empid flies and found a graded eight-stage sequence in different species corresponding with the probable evolution of this unusual courtship pattern:

Stage 1: Males search for a female and court her in isolation from other pairs.

Stage 2: Males capture a prey, present it to a female, and she consumes it before mating.

Stage 3: Males capture prey, form groups, and present prey to females, who receive the prey and mate.

Stage 4: Males mate as in stage 3 except that a few strands of silk are applied to the prey prior to swarming.

Stage 5: Males mate as in stage 4 except that males wrap a complete silk covering about the prey.

Stage 6: Mating is the same as in stage 5 except that males consume the edible portion of the prey before presenting it, so that the female receives a nonnutritious husk.

Stage 7: These species are not carnivorous, unlike those of the previous six stages. A dried insect fragment is collected and used as the base for the weaving of a balloon.

Stage 8: Males omit the insect fragment in the formation of a balloon, as in *H. sartor.*

The processes of ritualization and of progressive changes among closely related forms are readily apparent in this example. The advantages, for the study of behavioral evolution, of a taxon with many closely related species and of a behavioral pattern that is both clearly defined and species-characteristic should be apparent.

BEHAVIOR AND SYSTEMATICS

The area of systematics consists of the endeavor to develop a useful system for the classification of living forms. There is a tendency among some students to regard classification as completed, and to think that all living species must have been identified and classified by now. This is clearly not the case. Biologists are continually finding new species and shifting the classifications of some forms. Much debate still abounds with respect to the classification of certain groups. Recently, there has been an increased use of behavior in determining taxonomic relationships. As has been demonstrated repeatedly in this book, many behavioral patterns are species-characteristic, have strong genetic determination, and show little variability within a species; often, they are coupled with substantial variation among species. Such behavioral patterns have proved useful to systematists in refining the classificatory system of animal species (Mayr, 1958).

In some cases, behavioral patterns provide the only means of discriminating animals of different species. An example appears among the fireflies (Barber, 1951; Lloyd, 1969). It will be recalled from Figure 5-2 that the males of different species display species-characteristic flash patterns as they fly about, while females are selectively responsive to the flash patterns of males of their own species. It will be recalled further that species are defined in relation to their breeding patterns and reproductive isolation. There are a number of species of fireflies that cannot be discriminated on the basis of any known morphological characteristic. However, they display different courtship patterns and do not interbreed. These represent "good" species even though the only basis for separating them is their behavior.

Behavioral data can be useful in determining the appropriate taxonomic classification for ambiguous species. Bekoff, Hill, and Mitton (1975) quantified this approach, using "discriminant function analysis" to assess the appropriate location of the controversial New England canid relative to other canids by using behavioral data. Their results agreed with anatomical data suggesting that the New England canid is more closely related to coyotes than to wolves. Similarly, Rovner (1973) used detailed data on the copulatory behavior of the wolf spider, *Schizocosa avida*, to support its reclassification into a new genus.

Behavioral data have been useful in the classification of whole groups of species. An excellent example can be found in the classification of ducks and geese (Lorenz, 1941; Johnsgard, 1961). As can be seen in Figure 13-3, different species of ducks and geese display different behavioral patterns. Construction

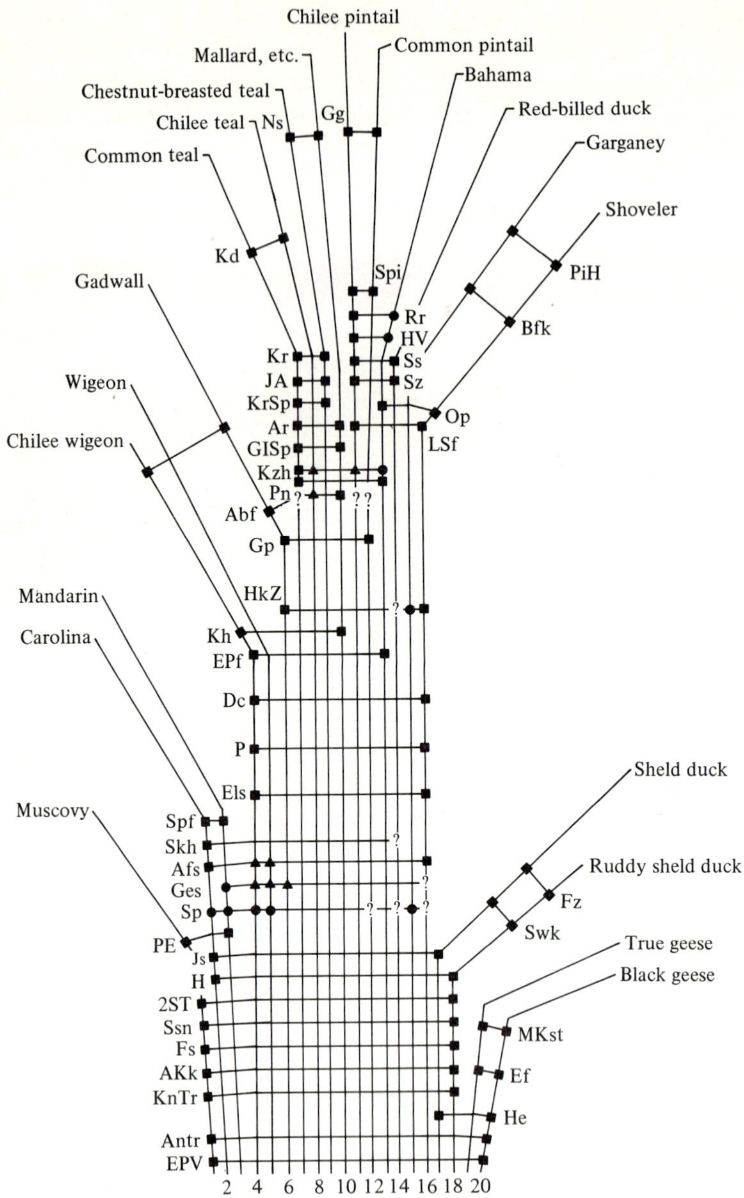

Figure 13-3 A taxonomy of Anatidae based on behavioral characteristics. Vertical and oblique lines represent species; each horizontal line signifies the presence of a characteristic in the species whose lines it crosses. *(After Lorenz, 1941.)*

of a "shaving brush" model based on these behavioral characteristics provides valuable data in affirming and revising taxonomic classifications.

Spider wasps, Pompilidae, are morphologically so distinct from other wasps that they are not placed in any major family. By considering a set of five major behavioral characteristics, Evans (1953) was able to make reasonable inferences concerning their derivation and appropriate classification.

All these examples should serve to reinforce the view that, while many behavioral patterns are quite labile and difficult to study in an evolutionary framework, others are quite stable and can be studied in such a framework in a manner quite similar to that used for anatomical characteristics.

DOMESTICATION

Domestication "is defined as that condition wherein the breeding, care, and feeding of animals are more or less controlled by man. . . . Domestication involves some biological change (morphological, physiological, or behavioral) in the animal. Thus, domestication differs from *taming*, which is defined as the elimination of tendencies to flee from man" (Hale, 1969, p. 22).

The issue of the effects of domestication on behavior has been a prominent one in the study of animal behavior. Many animal behaviorists have used domesticated animals—including birds, rodents, ungulates, and carnivores—in their research. Other behaviorists have been highly critical of such usage, as were the classical ethologists when they criticized the reliance of classical comparative psychologists upon laboratory rats. Opinions regarding the effects of domestication on behavior have often been expressed with extremist statements. The only proper way to settle the question of the effects of domestication on behavior is through careful research.

Origins of Domesticated Species

Hale (1969) provided a list of the major domestic species, all birds and mammals, together with a discussion of their origins. It is likely that domestication began with the utilization of animals in ceremonious or religious contexts, as pets, and for economic reasons. It appears that domestic dogs were derived exclusively from wolves and that the process began in Central Europe (Scott & Fuller, 1965). Richter (1954) and Lockard (1968) traced the domestication of rats and found the first laboratory use of albino rats to have occurred in Paris in 1856. The waltzing mouse, a genetic mutant, can be traced to China as early as 80 B.C. (Linzey & Thiessen, 1970). The origins of different strains of inbred mice are quite diverse, with many derived from pet fanciers (Staats, 1966). Van Oortmerssen (1970) went so far as to suggest that many of the observed behavioral differences among inbred strains may represent remnants of adaptations to different habitats that occurred before domestication rather than effects of domestication and inbreeding per se.

Not all species are easily domesticated. Hale (1969) attempted to list characteristics of species that are easily domesticated. Included in Hale's list were large social groupings, dominance patterns of social organization, promiscuous mating patterns, precocial young, short flight distance to humans, and omnivorousness. Traits that are unfavorable for domestication include territorial structure, pair bonding, altricial young, specialized dietary habits, and extreme agility.

Effects of Domestication on Behavior

While the effects of domestication on numerous avian and mammalian species have been analyzed (e.g., Hale, 1969; Eibl-Eibesfeldt, 1975; Desforges & Wood-Gush, 1975), the most substantial research has concentrated on rodents.

Methods The two major means of studying effects of domestication are through reenactment and the comparison of domesticated and wild forms. The more satisfactory method is reenactment, in which wild animals are trapped and changes in the population are tracked over years in captivity. King and Donaldson (1929) and King (1939) reported on such a reenactment in laboratory rats, while Connor (e.g., 1975) is conducting a reenactment of domestication in house mice. Such experiments require much time, are expensive, and are difficult to control. Thus, they have been relatively rare.

More commonly, existing domestic strains are compared with animals from wild-trapped populations. A major problem with this method lies in sampling error. Many of the differences observed between wild-trapped and domesticated animals are quantitative. It is known that there are quantitative differences in behavior among both different domesticated strains and naturally occurring populations. The proper experiment, in which an array of domesticated strains is compared with wild animals from an array of different populations, is yet to be conducted.

Norway Rats While it is true that wild Norway rats are more difficult to work with than domesticated (i.e., laboratory) rats, these difficulties are often overestimated. Methods are available with which wild Norway rats can reasonably be studied in the laboratory (Boice, 1971).

A considerable number of anatomical, physiological, and behavioral characteristics have been found to differ between wild and laboratory rats. During their reenactment of domestication, King and Donaldson (1929) noted that rats became heavier, showed decreased weights of their brains and certain endocrine glands, showed increased fertility, and lost much of their viciousness.

As discussed in Chapter 4, laboratory rats show less of a neophobic reaction than wild animals (Barnett, 1963; Mitchell, 1976; Barnett & Cowan, 1976). Further, in comparison with wild rats, laboratory animals have been shown to display less open-field activity (Price & Huck, 1976; Hughes, 1975), a greater tendency to display spontaneous gnawing (Price, 1973), a lesser tendency to jump from a platform (Price, 1973), a lesser tendency to kill mice (Karli, 1956), less aggression directed at both other rats and humans (Richter, 1949; Galef, 1970), and generally better performance in learning tasks (Boice, 1972, 1973). Wild and laboratory rats differ quantitatively in a number of measures of copulatory behavior (Adler, 1974; McClintock, 1974).

The sources of these differences and others that have been reported are not completely clear. Cross-fostering between strains has generally been found to produce little change in behavior (e.g., Galef, 1970; Price and Loomis, 1973). It is likely that at least some of the differences are genetically based. On the other hand, early experience does play some role in the development of aggression. Thus, wild rats showed a reduction in aggression-like behavior directed at mice if they were reared with mice or directed at humans if given periodic handling (Galef, 1970). Galef suggested that the high aggression levels in wild rats may be related to their tendency to avoid novel objects.

House Mice In his reenactment of the domestication of house mice (carefully ensuring minimal genetic change in the wild, random-bred descendants of wild mice), Connor (1975) found no change in any of nine behavioral measures over ten generations in the laboratory. However, inbreeding reduced both intermale aggression and resistance to handling, although it failed to affect seven other measures. In comparing inbred strains with feral mice, Smith (1972) found wild mice to be characterized by poor avoidance learning, high activity, and extensive freezing behavior. Estep, Lanier, and Dewsbury (1975) found wild mice generally to require less stimulation prior to ejaculating than inbred mice. Wild mice also used less cotton in building nests than did inbred animals. The differences between wild and domestic mice are complex and cannot be understood in relation to genetic "wildness" conceived as a unitary trait (Smith & Connor, 1974).

Deer Mice Although they have been bred by humans for a shorter period of time than laboratory rats and house mice, deer mice (*Peromyscus maniculatus*) already show signs of the effects of domestication. Effects have been noted on reproductive performance (Price, 1967), novelty-induced self-food deprivation (Price, 1972), and open-field behavior (Price, 1970). Changes in habitat preference resulting from domestication were described by Wecker (1964) (see Chapter 9).

Implications It should be clear even from this cursory review that there are many behavioral differences between wild and domesticated forms of the same species of rodents. However, different investigators have drawn different conclusions from these observations. Among the most extreme statements are those of Lockard (1968), who variously referred to laboratory rats as "freaks," "unnatural animals," and "degenerates." Lockard concluded that laboratory rats are a "bad habit" rather than a "defensible choice" for research in animal behavior. Other investigators have adopted the opposite extreme position, utilized laboratory forms exclusively, and attempted to generalize freely to behavior and adaptive significance in the natural habitat.

An intermediate position has been suggested by Dewsbury (1973) and Boice (1973). The laboratory environment differs from the natural environment. Selection works in the laboratory just as it does in nature, although the selective pressures are quite different. Laboratory rats and wild rats differ in a number of ways. These differences do not represent a degeneration in the laboratory but rather an adaptation to a different set of conditions. Because of these differences, one must exercise care in generalizing from studies of domesticated forms to behavior in the natural habitat, as in studies of adaptive significance. On the other hand, laboratory rats may be an ideal species for many studies of development and mechanism. The animal of choice will vary according to the question to be asked and the generalizations to be drawn.

SUMMARY

Studies of the evolution of behavior can be conducted at either the species level or the phyletic level. The species level provides more precise information regarding evolutionary processes, while the phyletic level is appropriate for tracing the broad sweep of evolutionary history.

A homology is defined as a resemblance between species that can be traced to their common ancestry. Conclusions regarding homologies among the behavioral patterns of different species appear legitimate under some circumstances.

Microevolutionary changes often entail quantitative changes in behavior, often with morphological changes following behavioral alterations.

Allopatric speciation occurs when a population is divided into geographically and ecologically isolated subpopulations which may diverge due to actions of natural selection and genetic drift. Reproductive isolation is fostered when sympatry is restored.

The evolution of displays, or ritualization, provides a particularly clear locus for the study of behavioral evolution.

Behavior can be usefully employed in taxonomic classification.

Through domestication, many behavioral, anatomical, and physiological changes have occurred. The appropriate usage of domesticated species is considered.

Part Six

Functions of Behavior

With this section, we complete our survey of the four classes of questions to be asked of animal behavior—development, mechanism, evolution, and function. In the study of the function of behavior, or teleonomy, we inquire into the ways in which behavior contributes to the survival and reproduction of organisms. The discussion of function, Part Six, is divided into two chapters. The first deals with methods that are useful in the study of adaptation and with the adaptive significance of behavioral patterns that are generally displayed by individuals. The second chapter of Part Six is devoted to social behavior and the area of "sociobiology."

A caveat already expressed several times in this book bears repeating here, for it is of great importance. When we study the adaptive significance of behavior, we do not in any way imply that the individual organism necessarily comprehends or understands the relationship between its behavior and the ultimate consequences of the behavior. Rather, we propose that natural selection has acted to create organisms that display particular behavioral patterns under given sets of circumstances. While shorthand statements by reputable students of animal behavior may sometimes imply such comprehension of ultimate consequences by the animals, all valid statements can be translated into more precise terminology, though often requiring more space and less vivid language or analogy.

Chapter 14

Behavior and Adaptive Significance

In the study of the adaptive significance of behavior, one attempts to determine the function or functions that behavior serves in fostering survival and reproduction. Such an approach is essentially identical to the study of "teleonomy" (Pittendrigh, 1958), the "analysis of adaptation" (Hodos & Campbell, 1969), and the "ecological method" (Lockard, 1971). Emphasis is placed upon the study of ecological factors that are important in the evolution and maintenance of behavior in a population. Whereas similarities based on analogy are generally to be avoided in the study of behavioral evolution, they can often be of considerable usefulness in the study of adaptive significance.

METHODS

The problem to be addressed in this chapter is that of how one goes about making valid inferences about the adaptive significance of behavioral patterns. Much writing about adaptive significance consists of little more than armchair speculation. A behavioral pattern is observed and the writer sits back, lights a cigarette, gazes into space, and attempts to think of a reasonable function that might be served by the behavioral pattern. The explanation is formulated after the fact, or *post hoc*. Such armchair speculation can be quite valuable; it provides an excellent source of hypotheses.

However, convincing arguments concerning adaptive significance require more rigorous methods (see Hailman, 1965). While such methods do not often match the degree of rigor obtained in studies of development and mechanism, they do permit reasonable inferences about adaptive significance. Further, such inferences are generally open to further testing, so that hypotheses can be validated, refined, or eliminated. Four methods will be considered: the experimental method, the method of within-species correlation, the method of adaptive correlation, and the behavior-genetic method.

Experimental Method

When using the experimental method in the study of adaptive significance, one reverses the traditional role of behavior in experimental studies. Rather than being the dependent variable, upon which the effects of independent variables are observed, behavior becomes the independent variable. One manipulates behavior and determines the effect of such manipulation on a consequence or consequences that are presumed to be related to survival, reproduction, or both.

An example of the application of the experimental method can be found in de Ruiter's (1956) study of the orientation behavior of countershaded caterpillars. The animals of many species are *countershaded.* The body surface that is typically exposed to light is more darkly colored than the body surface that faces away from light. If the dorsal (upper) surface of the body is the one that is dark, the animal positions itself so that it faces up. If the ventral (lower) surface is dark, the animal may orient itself upside down. When one observes such animals, it is immediately apparent that countershaded animals are more difficult to locate from almost any angle than are animals colored in different patterns. Using armchair speculation, we can formulate the hypothesis that countershading functions in camouflage and thus confers a selective advantage through protection from predators. However, this hypothesis needs testing. de Ruiter killed caterpillars of several species, and then fixed them to twigs that were distributed in the shrubbery of an aviary containing several hungry birds. Some caterpillars were placed in the normal orientation; others were placed upside down, so that the pattern of countershading was reversed (light side up). The birds found and ate significantly more of the caterpillars that were upside down than those that were normally oriented. When countershading was eliminated by painting the caterpillars a uniform color, no difference due to orientation was observed. Thus, countershading per se was the relevant factor, not orientation. Thus de Ruiter was able to use the experimental method to go beyond armchair speculation in determining the adaptive significance of the orientation behavior of caterpillars.

Method of Within-Species Correlation

The method of within-species correlation resembles the experimental method in that one examines the consequences of variations in behavior. However, rather than contriving an experimenter-designed independent variable, one takes advantage of natural variation. Essentially, this is a correlational method applied within a species and capitalizing on "experiments of nature."

Patterson (1965) used the method of within-species correlation in a study of the adaptive significance of synchronized colonial breeding in black-headed gulls. Most pairs of gulls nest inside the colony and lay their eggs during a peak laying period. Gulls breeding just outside the colony or on the colony fringe had lower breeding success, measured by the percent of eggs fledging, than animals breeding within the colony. Birds laying their eggs before or after the period of peak laying also suffered reduced breeding success. Such differences appeared to be due to excessive predation on eggs laid outside the colony or outside the peak breeding period. Thus, Patterson correlated naturally occurring variations in behavior (i.e., time and place of breeding) with reproductive success and was able to demonstrate a function for the species-characteristic breeding pattern.

Method of Adaptive Correlation

With the method of adaptive correlation, one also attempts to find correlations between naturally occuring variations in behavior and some other factor, one that is presumably related to survival and reproduction. However, rather than focusing on within-species variation, the study of variability among different species provides the material for analysis. Thus, one attempts to find correlates of differences in species-characteristic behavioral patterns. The hope is that such correlations will lead to inferences regarding adaptive significance.

The classical example of a study of adaptive correlation is Cullen's (1957) study of the behavior patterns of kittiwakes. Whereas most species of gulls nest in sand dunes, marshes, or other relatively flat habitats, kittiwakes nest on small ledges in steep cliffs. Cullen used the method of adaptive correlation to find behavioral correlates of the atypical nest location of kittiwakes. She found over 30 ways in which the behavior of the cliff-nesting kittiwakes differed from that of ground-nesting gulls. Whereas ground-nesting gulls disperse droppings and eggshells away from the nest, kittiwakes do not. In contrast to the young of ground-nesting birds, young kittiwakes do not run when attacked. In contrast to the relatively simple nests of ground-nesting gulls, kittiwakes build more elaborate nests, using mud in their construction and stealing nest material from neighbors. The young of kittiwakes are less mobile, and individual recognition of chicks by parents is slower to develop. All these adaptations in breeding behavior can be seen to be related to cliff nesting—either to the protection against predation afforded by the cliffs or to the dangerous height, cramped space, and scarcity of nesting sites on cliff ledges.

While Cullen's demonstration of correlates of the cliff-nesting habit are convincing, the correlations are *post hoc*. The beauty of the method of adaptive correlation is that one can treat such correlations as hypotheses and examine other, previously unstudied species as a means of testing that correlation. For example, if the behavioral patterns studied by Cullen are indeed related to cliff nesting, then other cliff-nesting species ought to have evolved similar adaptations by convergent evolution. One can thus use data on analogous adaptations to make inferences concerning adaptive significance.

Hailman (1965) studied the breeding behavior of the cliff-breeding swallow-tailed gulls of the Galapagos Islands. The habitat of this species is intermediate between that of the kittiwakes and that of ground-nesting gulls. Hailman found that swallow-tailed gulls shared many of the breeding characteristics of kittiwakes but, overall, were intermediate between kittiwakes and ground-nesting birds just as their habitat is intermediate. Swallow-tailed gulls were found to display eleven characteristics of kittiwakes, six characteristics of ground-nesting gulls, and one intermediate characteristic related to reduced living space and danger to the eggs or of falling from the nest. Nelson (1967) studied the breeding behavior of cliff-dwelling gannets and found that, although they were not closely related, these gannets shared many behavioral patterns with kittiwakes.

As was true of the studies of the evolution of behavior, the method of adaptive correlation can be used at either the phyletic or species level (see King, 1963, 1970; also Chapter 13).

Behavior-Genetic Method

The rationale for the behavior-genetic method was introduced in Chapter 7. The diallel-cross method is used. As will be recalled from Table 7-3, a diallel cross is a design for a breeding experiment in which one studies all possible F_1 crosses among three or more inbred strains. There are three possible outcomes. The F_1s may fall near the mean of their parents' scores, thus representing intermediate inheritance. Alternatively, the F_1s may generally fall nearer to the high- (or low-) scoring parent than to the other—directional dominance. Directional dominance may be either partial or complete. Finally, as in Table 7-3, the F_1s may generally fall in one direction outside the range defined by the mean parental scores, to represent overdominance (heterosis, hybrid vigor). Where significant directional dominance or overdominance is observed, it is likely that the trait under study is related to fitness. Further, it is likely that it is adaptive for animals to score in the same direction as do the F_1s relative to their parental strains. Thus, for example, one would use the data from Table 7-3 to infer that mount frequency in laboratory rats is related to fitness and that it is adaptive for rats to have low mount frequencies.

The theory underlying the use of the diallel cross to make inferences about adaptive significance was summarized by Falconer (1960) and Bruell (1964, 1967). In brief, natural selection works to alter the characteristics of the pools of genes related to adaptive characteristics in populations. Selection for extremes produces an asymmetrical gene pool, in which there is a correlation between (1) whether a gene is dominant or recessive and (2) whether, for traits that are related to fitness, the gene acts to increase or decrease the trait under study. Inbreeding results in a deterioration of scores (inbreeding depression).

Directional dominance (or overdominance) is simply the inverse of inbreeding depression (e.g., hybrid vigor). Thus, when directional dominance is found in a diallel-cross experiment, we can infer that the population in question has an asymmetrical gene pool related to the behavioral pattern under study and that the behavioral pattern probably is related to fitness.

An example of the use of the diallel cross can be found in the research of Henderson (1970 a, b). Henderson used the diallel-cross method to study the effects of environmental enrichment on brain weight and on performance in a food-getting task. After weaning, inbred and F_1 mice were reared either under standard laboratory conditions or in enriched environments containing a variety of small objects for climbing and exploring. As can be seen in Figure 14-1, the F_1 mice had heavier brains than the inbred mice. Thus, we would infer that heavy brains are adaptive. Further, the F_1 mice responded to environmental enrichment with a greater increase in brain weight than did the inbred mice. Thus, we would infer that it is adaptive for mice to respond to environmental enrichment with an increased brain weight. Genetically heterogenous mice produced by way of a "four-way cross" showed patterns of inheritance and responsivity to environmental enrichment similar to those of the F_1s.

In the food-getting task, mice were required to climb and tightrope walk along a circuitous route in order to reach a small platform containing food. Mice that were either F_1 hybrids or reared in enriched environments found the food faster than inbred animals or animals reared in standard environments. As in the study of brain weight, F_1 mice showed greater improvement with environmental enrichment than did inbred animals. It would seem that it is adaptive for mice to be able to negotiate circuitous routes to find food. While

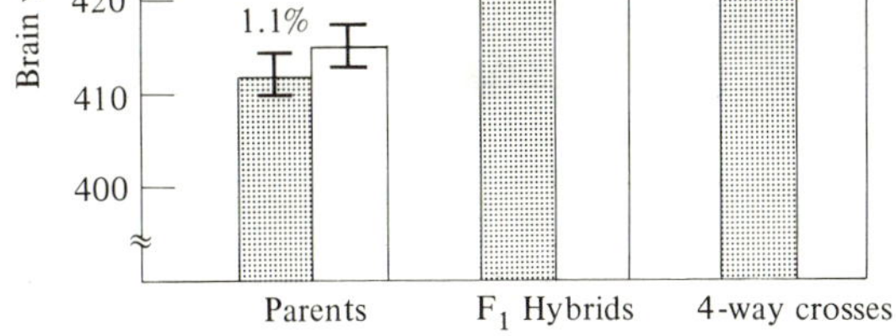

Figure 14-1 Average brain weights of six inbred strains, their F_1 hybrids, and four-way crosses when reared either in standard laboratory cages or in "enriched" environments. Small bars indicate standard errors of the mean. Both brain weight and response to environmental enrichment are greater in mice with genetic variability than in inbred mice. *(From Henderson, 1970b. Copyright 1970 by the American Association for the Advancement of Science.)*

this result may not seem earth shaking, it does increase one's confidence in the validity of the behavior-genetic method in studies of adaptive significance.

STUDIES OF ADAPTIVE SIGNIFICANCE

Animal behaviorists have utilized the methods summarized in the last section to study the adaptive significance of a variety of behavioral patterns. We shall review some of this research as it relates to a few functional categories of behavior.

Sexual Behavior

Each of the four major methods has been applied to the study of the adaptive significance of patterns of sexual behavior.

Method of Adaptive Correlation Ewer (1968) used the method of adaptive correlation at the phyletic level to generate hypotheses regarding the adaptive significance of copulatory patterns. She noted great variability in the duration of copulations in different species (e.g., whereas the copulations of many rodents last less than 1 second, those of some marsupials and carnivores last for hours). Ewer proposed that two sorts of factors were important sources of selection pressures in the evolution of patterns with short copulations—predation pressure and anatomical characteristics. As animals *in copula* would appear likely to be more exposed to predation than when not copulating, it seems reasonable that animals under great predation pressure would display brief copulations. This generally appears to be the case. Anatomical adaptations of the penis also appear related to the evolution of copulation durations of varying lengths (Ewer, 1968). Hediger (1965) also used the phyletic level in comparing the copulation durations of the predator-prone ungulates with the more predator-free large carnivores. As expected, the large carnivores have much longer copulations than plains-dwelling ungulates such as antelopes.

Dewsbury (1975c) used the method of adaptive correlation at the species level in summarizing research on 31 species of muroid rodents. The pattern displayed by each species was described using the classification system presented in Figure 5-3. Ecological correlations were consistent with the hypotheses of Ewer and Hediger. Species with copulatory patterns that entail locking (establishing a mechanical tie between penis and vagina) generally have relatively protected nest sites, whereas species that are able to ejaculate on a single, brief insertion typically live in more exposed habitats.

The study of the correlation between the dimensions of the glans penis and copulatory behavior among simple-baculum muroid rodents (those species whose penes contain a single bone) illustrates the use of the method of adaptive correlation to test proposed hypotheses. On the basis of data from 10 species, it was proposed that species that either lock or display intravaginal thrusting have glandes that are relatively thicker than those of species that neither lock nor thrust. After that correlation was proposed, an additional 10 species were studied; data were consistent with the proposed correlation. Data collection and testing of this hypothesis continues as of the time of this writing. The ability to test proposed correlations of this type constitutes an important advantage of the method of adaptive correlation.

Behavior-Genetic Method Fulker (1966) conducted a diallel-cross analysis of male mating speed, as measured by the number of successful copulations by a male in a 12-hour period, in *Drosophila melanogaster.* Male mating speed is an important component of fitness, as can be demonstrated with other data (e.g., Parsons, 1974). Consistent with this fact, substantial directional dominance—indeed, slight overdominance—was found for rapid mating speeds.

In a diallel-cross analysis of the copulatory behavior of four strains of laboratory rats, substantial overdominance was found for sets of two measures: the number of mounts and the number of intromissions preceding ejaculation (e.g., Table 7-3). Substantial directional dominance was found for a variety of latency measure of copulatory behavior (Dewsbury, 1975a). These data would suggest that it is adaptive for rats to copulate relatively rapidly and for males to attain ejaculation following relatively low numbers of mounts and intromissions.

Vale and Ray (1972) conducted a diallel-cross analysis of copulatory behavior in male house mice. Although a trend toward directional dominance of rapid mating patterns appeared, the authors, due to a number of complexities, cautioned against facile interpretations of the data. In comparing the pattern of inheritance of the F_1s produced by crossing individual pairs of inbred mouse strains, McGill and his associates (McGill & Blight, 1963a; McGill & Ransom, 1968; McGill, 1970b) have emphasized the degree to which different patterns of inheritance can be seen in the crosses of different strains and in different measures of copulatory behavior.

Wild house mice ejaculate after fewer mounts and intromissions than those of inbred strains (Estep, Lanier, & Dewsbury, 1975).

Some hybrid mice display copulatory behavior for a longer period following castration than inbred mice do (McGill & Blight, 1963b; McGill, 1969; McGill & Manning, 1976). Does this mean that it is adaptive for males to copulate after castration in the natural habitat? Obviously not! It is more likely that this result is a secondary consequence of the development of some mechanism that does confer selective advantage. Tendencies toward or away from seasonal breeding and social living may be related to the tendency to retain the ability to copulate after castration (see Hart, 1974).

A variety of patterns of inheritance, including some with substantial directional dominance, were found in crosses produced from two inbred strains of guinea pigs (Jakway, 1959; Goy & Jakway, 1959, 1962).

Experimental Method A number of laboratory experiments have been conducted in order to study the adaptive significance of various elements of rodent copulatory patterns. While one must exercise caution in generalizing to behavior in the field, such experiments provide important clues regarding male-female interactions and their relationship to successful reproduction. It will be recalled from Chapter 5 that many rodent species display a multiple intromission–multiple ejaculation copulatory pattern. The male ejaculates only after attaining several intromissions (vaginal insertions) without ejaculation; several such "ejaculatory series" precede the occurrence of sexual satiety. We shall consider the adaptive significance of both the multiple intromissions preceding ejaculation and the multiple ejaculatory series.

It will be recalled from Chapter 12 (Table 12-2) that in two prominent types of rodent estrous cycles, copulation is necessary for completion of the female cycle. Some species ovulate only in response to copulation, while in others a functional luteal phase (preparation of the uterus for implantation of the fertilized ovum) occurs only after copulation. Laboratory rats, house mice, and hamsters display the latter pattern. The preejaculatory intromissions of rats function in triggering the neuroendocrine reflexes that mediate the functional luteal phase. If a female rat receives an ejaculation after very few intromissions, she will probably not become pregnant. This is because (1) her neuroendocrine reflexes are not triggered and (2) sperm transport is deficient (Wilson, Adler, LeBoeuf, 1965; Adler, 1969; Chester & Zucker, 1970). By contrast, the preejaculatory insertions of house mice do not appear necessary for the initiation of pregnancy; the mere occurrence of the ejaculatory reflex is sufficient (Land & McGill, 1967; McGill, 1970a). In golden hamsters, by contrast, four to seven complete ejaculatory series seem essential in ensuring a high probability of pregnancy (Diamond & Yanagimachi, 1968; Lanier, Estep, & Dewsbury, 1975).

It is apparent that the stimulus requirements of the females vary in different species. (In fact, they also appear to vary in different circumstances; Davis, Gray, & Dewsbury, 1977.) It has been proposed that the stimulus requirements of the females of different rodent species are coadapted with the amount of copulatory behavior displayed by the males of the species (Dewsbury, 1977). Thus, in species where males display much copulatory behavior, females have high thresholds; the reverse is the case in species in which males display relatively little copulatory activity. This hypothesis represents a utilization of experimental data in conjunction with the method of adaptive correlation and, like other correlations, is subject to further testing.

A female rat requires a period of quiescence after receiving an ejaculation; if she receives intromission too soon, sperm transport often is suboptimal (Adler & Zoloth, 1970). Thus the postejaculatory refractory periods of males appear to have a function in the insurance of pregnancy. The interference with such quiescent periods may play a role in the breakdown of reproductive patterns at high population densities where many males mate with a female in rapid succession.

Within-Species Correlations Within-species correlations have been used less frequently in the study of copulatory behavior than has been the case for other methods. Several applications have been made in conjunction with the experimental approach just discussed. Naturally occurring variability is sometimes correlated with the probability of pregnancy. For example, Estep (1975) found that female roof rats mating for just one ejaculatory series had a higher probability of becoming pregnant if their partners delivered many preejaculatory intromissions rather than few. In this case, naturally occurring, rather than experimenter imposed, manipulations provided the relationship. Similar effects have been found in cactus mice (Dewsbury & Estep, 1975) and grasshopper mice (Lanier & Dewsbury, 1977).

Prey-Predator Interactions

Prey-predator interactions are focused on events that are critical to survival—acquisition of food for the predator and survival for the prey. Thus, they represent ideal loci for the study of adaptive significance because the consequences of success and failure are rather directly related to fitness.

Bird-Mouse Interactions Birds of prey have become highly adapted for predation. The visual acuity of many species exceeds that of most mammals (e.g., Fox et al., 1976). The remarkable capacities of owls to detect and locate sounds, and to correct their flight paths in midair, were stressed by Konishi (1973). Hunting birds appear to form a specific *searching image* in which they become selectively responsive to prey with which they have had recent experience (L. Tinbergen, 1960; Alcock, 1973). Thus, when prey of a particular kind are distributed at low densities, fewer of those prey are taken by predators than would be expected, presumably because the predators do not form an appropriate search image, as they do when higher densities prevail. It is as if you were hungry and, when searching for a banana, bypassed an orange that might have served equally well.

In responding to selective pressures from predators, prey such as mice have evolved numerous adaptations that are functional in predator avoidance. A number of investigators have studied differential predation by hawks or owls on mice of varying characteristics in large cages. Conspicuous white mice are more susceptible to predation by owls in this situation than are the naturally colored agouti brown animals (Kaufman, 1974a). Activity levels are important in exposing mice to predation. Morrison and Glickman found that mice that were more active in an open-field test were more likely to be taken by an owl than were mice with lower scores. Brain lesions known to increase activity in novel environments had similar effects (Glickman, 1971). Kaufman (1974b) carried the activity dimension to an extreme, contrasting predation on live and dead mice. Live mice were judged more conspicuous and were more likely to be taken. Metzgar (1967) varied time of residency in the cage before testing mice. He found that "resident" mice were less often eaten by an owl than "transients" were and concluded that residence in a limited home range probably confers a selective advantage related to protection from predators.

Surprisingly, laboratory rats are less susceptible to predation from owls in such settings than are wild Norway rats (Spiegel, Price, & Huck, 1974).

Mueller (1975) studied the effects of oddity on predation by hawks. Hawks were first offered mice of one color for 10 consecutive captures. They were then given a choice between mice of that and another color. The hawks generally selected the odd or unusual mice. Mueller argued that in nature, odd or unusual prey are likely to be unfit. It is to the advantage of the predator to prefer unfit animals. This allows animals of higher reproductive potential to live, breed, and produce greater food resources for the predator and its descendants.

Bat-Moth Interactions Bats have become highly proficient at locating small insects by using their echolocation system (Griffin, 1958; Griffin et al., 1960; also Chapter 10). In response to selective pressures from bat predation, insects have developed their own strategies (e.g., Roeder & Treat, 1961; Roeder, 1970). The auditory systems of moths are highly responsive to bat echolocation sounds, as demonstrated in the field by Roeder and Treat. In response to the sounds of bats at close range, moths display a rapid "power dive" in which they make erratic and unpredictable movements that appear to function in avoiding bats. Roeder and Treat used the method of within-species correlation to confirm that such behavior does indeed confer a selective advantage upon the moth. They noted that moths do not always display "evasive" behaviors in the presence of bats. Roeder and Treat observed 402 encounters and noted (1) whether or not evasive behavior was shown and (2) whether or not the moth was caught. They found a selective advantage of evasive action of 40 percent; for every 100 reacting moths that survived, just 60 nonreactors survived.

Breeding Behavior in Shore Birds The susceptibility of the eggs and young of breeding shore birds, together with the relative ease of locating and observing colonies, have led to a number of studies of the role of various aspects of the breeding behavior of such birds in providing protection from predators. The specific adaptations to cliff nesting were discussed above.

Patterson's (1965) use of the method of within-species correlation to study the role of spacing and breeding synchrony in gulls was discussed earlier. Tinbergen, Impekoven, and Franck (1967) conducted an experimental analysis of the role of spacing in protecting eggs from predation by wild carrion crows. Artificially camouflaged hens' eggs were laid out in plots of differing densities and crows were attracted to the plot with the addition of a conspicuous egg. Although crows spent more time searching the "scattered" plots, eggs in the "crowded" plots suffered a much higher mortality. Thus, it is apparently adaptive for birds to nest apart from each other.

A classical example of the use of experimentation in the field is provided by the studies of Tinbergen and his associates on removal of eggshells by gulls (Tinbergen, Broekhuysen, Feekes, Houghton, Kruuk, & Szulc, 1963; Tinbergen, Kruuk, Paillette, & Stamm, 1962; Tinbergen, 1963b). Like the birds of a number of other species, black-headed gulls remove eggshells from their nests shortly after young are hatched. The adaptive significance of this simple and brief behavioral pattern appears to be that it reduces the conspicuousness of the nest and thus reduces exposure to predation.

As one test of the extent to which eggshells attract predators, a group of camouflaged eggs was scattered about a dune valley. Half the eggs were single; the other half had empty eggshells placed 10 centimeters from them. Carrion crows and herring gulls preyed upon 65 percent of the eggs accompanied by a shell and on just 22 percent of the eggs without accompanying shells. In a subsequent experiment, it was found that the farther the shell is located from the camouflaged egg, the less is the risk from predation.

Tinbergen and his associates next used models to study the determinants of eggshell removal. They found that birds most readily removed objects colored white or khaki, the colors of normal gull eggs and shells. Shape of the object also was important. The presence of a thin edge appeared to provide the most salient cue in determining whether or not an eggshell would be removed.

Schooling in Fishes Among the functions proposed by different workers for the schooling tendency of fishes are reproductive facilitation, physiological advantages, and protection from predators (Shaw, 1970). There are several ways in which schools might function in reducing predation. According to one view, a school "mimics" a large fish and functions through such mimicry to repel predators. Alternatively, schooling might be understood as a response by individuals to "seek cover" in open habitat (e.g., Williams, 1966; Hamilton, 1971). Fish outside the school are often subject to predation, and schools typically become more compact in the presence of predators (Hobson, 1975). Alternatively, schooling fish may simply confuse predators with their great numbers.

Seghers (1974) used a within-species version of the method of adaptive correlation to study the relationship between schooling and predation in guppies. Seghers studied guppies from five different populations in Trinidad, each population being exposed to a different degree of predator pressure. He found that guppies from areas of little predator pressure were less likely to school than were those from regions of greater predation. By rearing animals in the laboratory, Seghers was able to demonstrate a genetic component to the differential schooling tendency of guppies from different populations.

Hobson (1975) provided an excellent overview of the varying patterns of interaction between predators and their prey among reef fishes. Various species are adapted for eating tiny crustaceans, cleaning other animals, ambushing styles of predation, schooling, and many other patterns of both predation and avoidance of predation.

Mimicry The subject of mimicry has been discussed in Chapters 4 and 13. Both Batesian and Mullerian mimicry are important sources of protection from predators (Brower, 1969; Kettlewell, 1965). An interesting form of mimicry was found in *Photuris* fireflies by Lloyd (1975). Females of several species of *Photuris* fireflies are predators on fireflies of several other genera. In the courtship pattern of fireflies, males fly about the environment, flashing in a species-characteristic pattern, until they observe a species-characteristic response from a female (Figure 5-2). The males then fly to the female, which remains on the substrate, and mate. The *Femmes fatales* of *Photuris* fireflies are able to mimic the species-characteristic flash patterns of the females of several different species. They thus attract the males in the same way as do females of the male's own species. However, *Photuris* females prey on their "suitors" rather than mating with them. Lloyd demonstrated the versatility of female *Photuris* using experimental methods in the field. After locating a female, he simulated the flash patterns of males of several different species in succession. The predatory mimics were able to make rapid adjustments and respond with the "correct" flash for the male pattern being simulated. This flexibility suggests that the capacities of the firefly brain are more complex than had previously been suspected.

Miscellany Three additional features of prey-predator interactions will be discussed. One aspect of predator avoidance is the control of diurnal activity patterns, so that animals are protected during periods of relatively high risk and active during periods of low risk. Lockard (1975) conducted a field experiment on the control of activity patterns in kangaroo rats. He used an "artificial moon" that was lighted with quartz iodide lamps and recorded the number of visits by wild kangaroo rats as a function of levels of illumination. The results suggested that the activity patterns of kangaroo rats are controlled by nocturnal illumination levels rather than an endogenous clock.

Predators learn to avoid certain kinds of prey. In a study of the role of stickleback spines as devices for avoiding predators, Hoogland, Morris, & Tinbergen (1957) found that pike and perch learned to avoid sticklebacks after a few encounters with their spines. The possession of an antipredator device capable of conditioning predators appears associated with sticklebacks' tendencies to nest in open habitat and develop relatively bold nuptial colors.

As mentioned above, it is important that predators not waste prey, as the prey are their major food source. One can almost conceptualize predators as adapted to "farm" prey. Thus, in "prudent predators" (Wilson, 1975), there are some adaptations that appear to ensure that unfit animals are culled (note that these also are the easiest to catch) and that excessive killings are minimized. While surplus kills have been reported, they are rare (Kruuk, 1972).

Exploratory Behavior

Glickman and Sroges (1966) used the method of adaptive correlation at the phyletic level to study object exploration in vertebrates at the Lincoln Park Zoo (Figure 4-3). Results of within-order comparisons suggested that object exploration in captivity is related to life-style in the natural habitat. Thus, species with complex food-getting patterns and little risk of predation often appeared to show more object investigation than species with strictly herbivorous habits and high risk of predation. Russell and Pearce (1971) tested these correlations with a group of Australian marsupials that have evolved varying food-getting habits independently of the placental mammals. Consistent with Glickman and Sroges's data and predictions, they found that the more carnivorous *Dasyuroides* displayed more object investigation than did marsupials that are primarily grazers or insectivorous feeders.

Locomotor-exploratory behavior has been a favored behavioral pattern for study by behavioral geneticists. Unfortunately, it is more difficult to relate such behavior to the natural situation than is the case with sexual and prey-predator behavior (see Bruell, 1967). It would seem most reasonable that intermediate levels of exploratory behavior would be maximally adaptive in the natural habitat. Animals that were too exploratory would be exposed to the kind of predation discussed above in relation to birds of prey. Animals displaying too little exploratory activity would fail to exploit environmental resources fully. Results of diallel-cross analyses of open-field activity are generally consistent with this expectation. A pattern of intermediate inheritance has been found in studies with inbred strains of mice (Henderson, 1967), selected strains of mice (Halcomb, Hegmann, & DeFries, 1975), and selected strains of rats (Broadhurst, 1960). The finding of substantial directional dominance for high levels of locomotor exploration in a maze (Bruell, 1964) is inconsistent with the considerations mentioned above.

The method of adaptive correlation at the species level was used in a study of open-field behavior in 12 species of muroid rodents (Wilson, Vacek, Lanier, & Dewsbury, 1976). Consistent with their general tendency to take less risk in locating food (Baker, 1971), field-dwelling species displayed lower activity levels than semiarboreal species or desert dwellers. Other relationships to ecological adaptations were seen with respect to the relative frequencies of different specific behavioral patterns. Desert dwellers were more likely to rear than were field dwellers, presumably because rearing improves the field of view in the desert but not in thick grass cover. Semiarboreal species showed the most jumping, locomotor-exploratory behavior, and climbing; field dwellers showed the most defecation, freezing, and grooming; desert dwellers showed the most gnawing and scratching of the apparatus. Such correlations between ecological adaptations and laboratory behavior should be tested with other species and groups of species.

Sleep

Allison and VanTwyver (1970) used the method of adaptive correlation at the phyletic level to study the adaptive significance of sleep. They found that they could divide species into "good sleepers," those that sleep for relatively long periods of time each day and show substantial periods of paradoxical sleep (see Figure 4-2), and "poor sleepers," those with less of both total sleep and paradoxical sleep. In general, Allison and VanTwyver found that good sleepers are either predatory species, with little risk from other predators, or prey species that have safe sleeping places. Prey species without safe sleeping places tend to be poor sleepers. Guinea pigs, rabbits, and most ungulates tend to display relatively little sleep compared with macaques, cats, and ground squirrels.

Much theorizing has been done with respect to the adaptive significance of sleep patterns. Traditionally, both in theory and in the application of everyday experience, it would seem that sleep is adaptive primarily in that it serves a restorative function. The organism is fatigued by its daily activity period and a period of quiescence is necessary for restoration of resources and function. While such a position seems reasonable, there is little evidence to support it. Thus, contemporary theorists generally seem to view sleep instead as "adaptive nonresponding" (Webb, 1974a, 1974b; Meddis, 1975). On this view, it is adaptive for animals to be inactive at certain times of day. Sleep may have evolved as a means of ensuring that such inactivity is enforced. According to this view, the diurnal placement, intermittency, and total amount of sleep displayed by a species is to be understood as an adaptation ensuring inactivity at times of minimal food availability and maximal danger from predation. The substantial amounts of sleep displayed by young altricial animals (e.g., Figure 8-1) may be adaptive in ensuring inactivity among helpless young animals. According to an alternative view, closer to traditional thinking, sleep functions to limit metabolic expenditure in a given day by enforcing rest (Zepelin & Rechtschaffen, 1974). Paradoxical sleep appears to serve a "sentinel" function by awakening the animal periodically and thus aiding in the detection of possible sources of danger (Snyder, 1966; Drew & Batt, 1972).

In a diallel-cross analysis of sleep in inbred strains of mice, Friedmann (1974) found overdominance for measures of the amount of paradoxical sleep, the number of long and short sleep episodes, and the diurnal ratio of total sleep. The results were interpreted as indicating that, in nature, it would be adaptive for mice to display appreciable total sleep, minimal fragmentation of sleep periods, a relatively even distribution of sleep, and little paradoxical sleep.

MAINTENANCE OF GENETIC VARIATION

It is clear that there is genetic variability among different animals of the same population. An interesting question is that of the significance of this variability. If there is one type of organism that is particularly well adapted to the environment, why do not all animals in the population come to possess identical genotypes correlated with that maximally adapted phenotype?

The issue of genetic variability received significant impetus in the 1960s and 1970s with the usage of starch-gel electrophoresis to estimate degrees of genetic heterozygosity. Essentially, electrophoresis is used to provide an estimate of the variation among the alleles that determine the structure of blood proteins. While considerable variation appears across species and groups of species, most estimates of heterozygosity indicate higher levels of heterozygosity within populations than would have been anticipated by classical theory. A sample of invertebrates—including land snails, *Drosophila*, field crickets, and horseshoe crabs—was found to have an average of 15 percent heterozygosity, whereas a sample of vertebrates—including fish, lizards, rodents, and humans—averaged about 6 percent (Selander & Kaufman, 1973). Different populations of the same species display different degrees of heterozygosity (e.g., Selander, Smith, Yang, Johnson, & Gentry, 1971). The gene pool of northern elephant seals shows very little variability, probably reflecting the decimation of populations by hunting in the last century. The uniform homozygosity of these elephant seals suggests that they lack the necessary pool of variability from which to adapt to future changing conditions (Bonnell & Selander, 1974).

One interpretation of the great genetic diversity of most populations is that it reflects the presence of transient genes in the process of being either eliminated or fixed in the population. It is more likely, however, that there are several active processes functioning to maintain genetic variability in populations. Indeed, the very process of sexual reproduction appears to function in this way (see Chapter 5). Wilson (1975) lists several possible mechanisms that would function to maintain genetic diversity, including the following:

1 Frequency-Dependent Selection

It often appears that selection does not favor or disfavor particular phenotypes but acts differentially on phenotypes as a function of their relative frequency in a population. If the less frequent of two alternatives is usually favored, the frequencies of alternative alleles in the population should stabilize at an intermediate level. The search images of predators should have such an effect. Where there is variability in the prey species, different forms would have an advantage when they were at relatively low frequencies, because search images of predators would be formed for the higher-probability form. Frequency-dependent sexual selection has been demonstrated in *Drosophilia* (e.g., Ehrman, 1970, 1972; Tardif & Murnik, 1975). Females prefer to mate with males of rare phenotype rather than of more common phenotype, thus creating a *rare-male advantage*.

2 Spatially Heterogeneous Environment with Migration

Where there is a spatially heterogeneous environment, different forms may be especially well adapted to different parts of the environment. Migration would spread the genes throughout the region.

3 Disruptive Selection

If selection acts against intermediate phenotypes, favoring either extreme (disruptive selection), genetic variability in the population would increase.

4 Cyclical Selection

Behavioral patterns adaptive in some conditions may be less adaptive under other conditions, and a regular cycle may develop. For example, correlated with variations in the population density cycles of voles, there appears to be cyclical selection for aggressiveness and the tendency to migrate at different phases of the cycle (Krebs, Gaines, Keller, Myers, & Tamarin, 1973; Myers & Krebs, 1974).

5 Heterozygote Superiority

Where the heterozygote has a higher fitness than the homozygote, there will be a tendency to maintain genetic variability in a population. Numerous examples of heterozygote superiority, as indicated by overdominance (hybrid vigor) in diallel-cross studies of inbred strains, have already been presented. However, the question remains as to whether such mechanisms might function in natural populations. An interesting approach to this question was undertaken by Garten (1976). He studied old-field mice from eight different naturally occurring populations and attempted to correlate behavior with the degree of genetic heterozygosity in the population as indicated by starch-gel electrophoresis. In general, high degrees of heterozygosity were associated with high social dominance, success in competing for food, and amount of approaching, attacking, and grooming (among mainland samples). Mice from the more polymorphic populations appear more likely to copulate under laboratory conditions than those from populations with less heterozygosity (Dewsbury & Lovecky, 1974). These data suggest that behavior may be important in maintaining heterozygosity and that heterozygosity may confer a competitive advantage. Behavioral genetics in natural populations has been given little study, but it is of great importance.

ORGANISMS AS ADAPTED WHOLES

An important point has been stated repeatedly throughout this book: while it is useful to consider in isolation the specific adaptations of organisms, it is important to remember that all the adaptations of an organism must interact to produce a complete, living, functioning, totally adapted organism. We shall close this chapter with some examples of the interaction of different adaptations to produce totally adapted life-styles.

Ungulates and Carnivores

An example of the pervasiveness of behavioral adaptations can be seen in the contrasting adaptations of large ungulates, which have considerable risk of predation, and of large carnivores, which have little risk of predation. Hediger (1965) made this comparison at the phyletic level and contrasted the patterns of copulation, sleep, drinking, parturition, and development of young in the two groups of animals (Table 14-1). It is apparent that in all these systems, the antelope pattern reflects the high predation pressure relative to the more leisurely patterns of behavior found among the large carnivores.

Breeding Patterns of Birds

Etkin's (1964) portrayal of the life-styles of three different types of birds—diurnal songbirds, marine birds, and flock-living species—was summarized in

Table 14-1 Comparison of Various Characteristics of Antelopes and Some Large Carnivores

Behavior	Bear/Lion	Antelope
Copulation	A matter of hours in bears	A matter of a few seconds or a fraction of a second
Sleep	Long and deep	Perhaps none at all or extremely short
Drinking	Especially slow in cats	Quick filling of the subdivided stomach
Parturition	Bears build nests	No nest whatever
Development of young	Extremely slow	Young ambulatory within 30 minutes

Source: After Hediger, 1965.

Chapter 6. Each pattern can be seen as a functional interconnection of various adaptations. Songbirds are small, diurnal birds that are subject to predation. They tend to be monomorphic and cryptically colored, thus gaining protection through camouflage. Songbirds typically rely more on auditory signals than visual displays in courtship, again a pattern than minimizes exposure to potential predators. Because there is just one pair, the young are confined to a territory; since the parents need not leave the territory in order to feed, individual recognition is minimal and responses to simple releasers predominate.

In marine species, by contrast, adults are relatively large and predation is concentrated upon the eggs and young. Because parents leave the territory to feed, individual recognition of both partners and young is well developed. Males share in parental duties, so that one parent always guards eggs and young. In contrast to songbirds, displays are highly visual and continue after pairing; there is little selection against such displays.

In flock-living, surface-feeding birds, there is usually considerable sexual dimorphism and males do not participate in care of the young. Females and young are protectively colored. Individual recognition is well developed and provides a basis for dominance relationships.

In summarizing such organizational patterns, Etkin concludes "that in all fundamentals, their behavioral factors—such as aggressiveness, hierarchy, territoriality, courtship, parental care, dimorphism, innate or learned communication—are all part of a co-ordinate complex which, in its entirety, is adaptive to the animal's ecology" (Etkin, 1964, pp. 269–270).

Brood Parasitism in Birds

Brood parasitism occurs when females of one species lay their eggs in the nests of another species and the parents of the second species care for the eggs. While reliable brood parasitism is found in about 80 species, the European cuckoos provide a particularly good example of the interrelatedness of adaptations, particularly adaptations for development. The female cuckoo locates a nest of the host species, removes one of the host's eggs, and replaces it with one that she lays in the nest. She then leaves the egg to compete on its own, a task for which it is well adapted. The egg resembles that of the host species, and the eggs of cuckoos living in different regions and parasitizing different hosts are differentially colored, functionally "mimicking" the eggs of the host (Southern, 1954). The incubation period of the cuckoo is shorter than that of the host; thus, the young cuckoo hatches early and has a competitive advantage for parental attention. The newly hatched cuckoo proceeds to roll the host's eggs from the nest, thus ensuring a monopoly on parental attention. Young cuckoos grow faster and are larger than the young of their hosts. Needless to say, the development of preferences for mating partners and of song patterns are largely programed within the individual, with little overt learning.

Thus, in cuckoos and other species of brood parasites, there is a complex of adaptations for rapid development, success in competition with young of the host for care from the parents, and minimal reliance on learning in social development.

r and K Strategists

In population biology, the symbol "r" represents the intrinsic rate of increase or growth of a population, whereas the symbol "K" represents the carrying capacity of the environment. Based on these parameters, population biologists differentiate *r strategists* from *K strategists* (MacArthur & Wilson, 1967; Wilson, 1975). The r strategists are opportunistic species that emphasize colonization of short-lived environments, rapid population increase, and full utilization of resources. By contrast, K strategists adapt to stable, predictable environments in which rate of population growth is unimportant but formation of stable social patterns is stressed. Each strategy entails a whole cluster of adaptations, some of which are presented in Table 14-2.

Examples of r and K strategies can be seen in different species of rodents common in North America. Prairie voles are r strategists, moving rapidly into beaver meadows or other appropriate habitats, reproducing rapidly, and experiencing dramatic fluctuations of populations size. By contrast, many *Peromyscus* mice live in more stable habitats and seldom experience the fluctuations in population characteristic of voles (Christian, 1970a).

Peromyscus Mice

With some 57 species and 235 subspecies, all native to the Western Hemisphere, mice of the genus *Peromyscus* have become a favored subject of scientists from a number of different disciplines (King, 1968, 1970). In nature, these animals have become adapted to a wide array of habitats and are an ecologically diverse but taxonomically closely related group. Many species adapt well to laboratory living, with respect to both breeding and behavior.

Layne (1969, 1970, 1971; Layne & Ehrhart, 1970) has conducted a series of

Table 14-2 Some Characteristics of r Strategists and K Strategists

Correlate	r strategists	K strategists
Climate	Variable or unpredictable	Relatively constant and predictable
Population size	Variable over time with wide fluctuations, usually below carrying capacity of environment	Relatively constant over time and often near carrying capacity of environment
Intraspecific & interspecific competition	Variable, often lax	Usually keen
Development	Rapid	Slower
Age for reproduction	Young	Delayed
Body size	Small	Relatively large
Frequency of reproduction	Once	Repeated
Emphasis in energy utilization	Productivity	Efficiency
Life-span	Short	Long
Colonizing ability	Substantial	Minimal
Social behavior	Weak, mostly schools	Frequently well developed

Source: Based on Pianka, 1970, and Wilson, 1975.

laboratory studies of different species of *Peromyscus* in an attempt to relate behavior in the laboratory to that in the natural habitat. The contrast between two species native to the southeastern United States—Florida mice (*P. floridanus*) and cotton mice (*P. gossypinus*)—is particularly instructive (Table 14-3). Florida mice are patchily distributed in relatively dry regions, where they use the abandoned burrows of other species. Cotton mice are much more broadly distributed and more variable in habit. In the laboratory, cotton mice prove to be better nest builders, diggers, and climbers than Florida mice. The activity patterns of cotton mice are less restricted to darkness than those of Florida mice, apparently in correlation with amount of available cover in nature. Together, these characteristics form a complex of adaptations equipping the mice for life in their particular habitats.

Lovebirds

The interrelatedness of different adaptations should be apparent from a consideration of the differences in the behavior of relatively primitive and more

Table 14-3 Some Characteristics of Two Species of *Peromyscus* Mice Native to the Southeastern United States

Characteristic	Florida mice	Cotton mice
Distribution of populations	Limited and patchy	Continuous and wide
Habitat	Dry, vegetated areas	Deciduous forests
Nest location	Use burrows of other species	Diverse (logs, stumps, rock crevices, human habitations, below ground)
Type of nest built	Simple platforms	Spherical and well constructed
Use of nest material in lab	Use little cotton and build poor nests	Use more cotton and build better nests
Digging behavior in lab	Slow and inefficient	Efficient
Climbing ability in lab	Poor	Good
Activity pattern in lab	Strictly nocturnal	More active and less rigidly tied to photoperiod

recently evolved forms of lovebirds (*Agapornis*) studied by Dilger (1960, 1962) and summarized in Table 13-1. The more primitive forms nest dispersed in pairs and form these pairs relatively late. By contrast, the colonial-breeding lovebirds are exposed to greater social interaction and form pair bonds earlier. The elaborate nest defense displays characteristic of the primitive forms are replaced with communal mobbing patterns in the colonial forms. Appropriately, the more social species display more ritualized aggression patterns than those that are dispersed, hence, presumably ensuring minimal injury due to fighting where social interactions are more common.

SUMMARY

In the study of adaptive significance, one attempts to determine the function served by behavioral patterns in the survival and reproduction of organisms. There is no presumption that the animal comprehends the significance of its behavior; rather, the behavioral pattern reflects the activity of a nervous sytem that is the product of the evolutionary history of the species.

Four primary methods are used in bringing the study of adaptive significance beyond mere armchair speculation: the experimental method, the method of within-species correlation, the method of adaptive correlation, and the behavior-genetic method.

All four major methods have been used in the study of the adaptive significance of patterns of sexual behavior.

Prey-predator interactions are rather directly related to survival and thus provide an ideal locus for the study of adaptive significance.

The possible adaptive significance of patterns of exploratory behavior and sleep are considered.

It appears to be advantageous to maintain certain levels of genetic variability in populations. Mechanisms functioning to maintain such variability are discussed.

The manner in which organisms are adapted to their environments as functional wholes is illustrated with considerations of ungulates and carnivores, breeding patterns of birds, brood parasitism in birds, r and K strategists, *Peromyscus* mice, and lovebirds.

Chapter 15

Sociobiology

Sociobiology is difficult to define. Its primary systematizer, Wilson (1975), states, "Sociobiology is defined as the systematic study of the biological basis of all social behavior" (p. 4). That is really too vague. The term has been applied to an approach, a theoretical orientation to the study of social behavior. The development of sociobiology in the 1960s and 1970s arose from a recognition and an examination of some of the implications of the theories of natural selection and population biology that had not previously been fully recognized and explored. This intense theoretical endeavor is forcing biologists and psychologists interested in social behavior to reexamine their premises and to consider social behavior in a new light. As with all important theoretical breakthroughs, sociobiology enabled animal behaviorists to see hitherto unrecognized relationships among seemingly diverse phenomena, and it suggested a veritable cornucopia of new hypotheses and of new data that must be collected. The sociobiological approach generated intellectual excitement of the sort that revitalizes a discipline only on very rare occasions.

The sociobiological approach will not appeal to all. It is highly theoretical and often far removed from data. Its theories can be very mathematical and highly abstract. Its apparent implications for ethical decision making are distasteful to many. Broad sociobiological theory does not permit the precise experimental testing that is possible with questions of mechanism and development. As important as sociobiology is, its role in the study of animal behavior has occasionally been exaggerated and it has been oversold (e.g., Wilson, 1975, Ch. 1 and 27). In spite of these difficulties, the sociobiological approach is important, and it behooves the well-rounded animal behaviorist to be familiar with at least its fundamentals.

After considering the cornerstones of sociobiology, we shall consider their applications in two main areas: reproductive strategies and the evolution of social behavior.

CORNERSTONES OF SOCIOBIOLOGY

The cornerstones of sociobiology are so important that they were introduced back in Chapter 3, so that the basic concepts would be available throughout the book. They bear restating at this point.

Genetic Fitness

Genetic fitness is measured by the contribution to the next generation of a particular genotype in a population in comparison to that of alternative genotypes. By definition, natural selection works to maximize fitness. Thus, the evolution of virtually all characteristics of the organism must be viewed ultimately in terms of the contribution they make to differential representation in the gene pool of the next generation. Fitness can be increased in many ways, as through both survival and reproduction, but in the long run it is the relative contribution to the gene pool of the next generation, and ultimately the generations that follow, that represents "the bottom line."

Natural Selection Works on Individuals

There is near unanimity among biologists that the actions of natural selection are to be understood at the level of the *individual.* Actions that appear on surface analysis to be performed solely for the good of the group or the species will be seen on further analysis to represent behavior that maximizes levels of individual fitness. The possibility that an individual might act through self-sacrifice for the good of the group and at its own expense was developed in its most refined and most testable form by Wynne-Edwards (1962). However, the notion that natural selection can favor characteristics that benefit the group at the expense of individual fitness has been generally discredited for most circumstances (Williams, 1966, 1971). Rather, natural selection works to maximize the fitness of individuals.

Inclusive Fitness

A critical early breakthrough in the development of sociobiology occurred when Hamilton (1964) proposed the notion of inclusive fitness. The basic principle underlying inclusive fitness is straightforward. Fitness is related to the relative representation of one's genes in the gene pool of the next generation.

However, one's genes are shared with close relatives as the result of common ancestry. Thus, a diploid vertebrate shares half its genes with its mother, father, sister, and brother. Reproduction by close relatives produces a greater-than-chance representation of one's own genes in the gene pool of the next generation. Thus, one can increase one's own fitness by facilitating the fitness of close relatives. This is termed *kin selection.* Following Hamilton, we can now see that fitness must be viewed more broadly than before. The term "inclusive fitness" reflects this increased breadth. With inclusive fitness, we recognize two major components of fitness: *personal (classical) fitness* and *kinship fitness.* Kinship fitness is represented by the relative contribution to the gene pool of the next generation by one's close relatives, devalued by a factor that accounts for the closeness of the relationship (one-half for a sister, one-quarter for an uncle, one-eighth for a first cousin, etc.). West Eberhard (1975) proposed the unit of *offspring equivalents* to specify the contribution of kinship fitness that would be measured by the number of offspring produced by kin devalued by the appropriate coefficient of relatedness.

Altruism

Altruistic behavior is any behavior that results in an increase in the personal (i.e., classical) fitness of another individual at the expense of one's own fitness. For example, one may jump into the water in an effort to save a drowning person. Self-sacrificing behavior toward one's own offspring typically would not be considered altruistic by most authors (e.g., Wilson, 1975) because one's own fitness is represented in one's offspring. However, altruistic behavior directed at a cousin would be considered altruistic because it would, by definition, increase the fitness of the cousin while decreasing one's own personal or classical fitness. One might expect altruistic behavior to evolve when the gain in kinship fitness resulting from an altruistic act exceeds the loss from personal fitness. Thus, inclusive fitness might be increased through altruistic behavior. One might use the term *ultimate altruism* to denote behavior that decreases one's inclusive fitness; ultimate altruism should never occur.

Reciprocal Altruism

Trivers (1971) detected an additional situation in which one would expect the evolution of altruistic behavior. *Reciprocal altruism* involves the trading of altruistic acts at different times by two possibly unrelated individuals. In effect, one person jumps into the water to save another from drowning in exchange for the "promise" that the one rescued will do the same for the rescuer under similar circumstances. Clearly, we would not expect so-called "reciprocal altruism" when there is a true decrease in inclusive fitness but only when the reciprocity results in no net loss of fitness for either individual. The three cases of reciprocal altruism discussed by Trivers were cleaning symbioses, warning cries in birds, and human reciprocal altruism. Evolution of apparently altruistic behavior through reciprocal altruism would require a rigid system of enforcement against "cheaters"—individuals who would benefit from the acts of others but not reciprocate under similar circumstances.

As always, this argument does not assume that animals consciously calculate relative benefits of alternative courses of behavior. Rather, individuals engaging in certain behavioral patterns have left more offspring, and thus natural selection has resulted in the spread of genes associated with the behavior in question.

Evolutionarily Stable Strategies

The sociobiological theorist must often deal with problems related to the determination of which of several "strategies" will be adopted by the individuals of a population, each of which is working to maximize its own fitness. Often there is a conflict of interest between different individuals. Often, the most effective behavioral pattern for an individual will vary with the situation and with the strategy of those with whom it is in conflict. To deal with such situations, John Maynard Smith (1976) has applied the mathematical theory of games to the study of evolution. The critical concept is that of the *evolutionarily stable strategy* (or ESS). An ESS is a strategy which, if adopted by most members of a population, cannot be bettered by an alternative strategy. Maynard Smith used mathematical models to consider the adoption of various strategies by individuals and the implications of such strategies for the population as a whole. The notion of the ESS can be applied to questions such as those of sex ratio, patterns of aggressive behavior, and patterns of social organization (Maynard Smith, 1976; Maynard Smith & Parker, 1976). The ESS strategy is presented in nonmathematical terms and expanded in several directions by Dawkins (1976). While the implications of the ESS approach have yet to be fully explored as of this writing, the approach appears to have great potential for the formulation of sociobiological theory.

REPRODUCTIVE STRATEGIES

If the essence of fitness is differential representation in the gene pool of the next generation, then much individual behavior is to be understood in relation to its facilitation of reproduction by the individual. The forces of natural selection, including sexual selection, should act to produce strategies of reproduction wherein each individual maximizes his or her own inclusive fitness. Natural selection, as related to reproduction, can act either on within-sex interactions, (intrasexual selection) or between-sex interactions (epigamic selection; see Chapter 3). For reasons to be discussed below, most such interactions involve male-male competition and female choice.

Male-Male Competition

If a male is to maximize his fitness, he must adopt a strategy that will maximize his production of viable, fertile offspring. Several methods of accomplishing this can be discussed.

Dominance Dominance entails a differential priority of access to resources, which in some cases may include mating partners (see Chapter 6). If dominance confers a differential ability to inseminate females, it would be important that males fight to achieve dominance and attain the increase in fitness to be gained through siring more offspring. This seems to occur under at least some conditions. Among elephant seals, for example, it has been estimated that less than one-third of the males in residence in a given season ever copulate (LeBoeuf, 1974). The higher the status of the male, as gained through male-male fighting, the more readily he approaches and copulates with females. In one study, 4 percent of the males inseminated 85 percent of the females (LeBoeuf & Peterson, 1969). Similar results have been obtained in a laboratory setting with house mice; dominant or more aggressive males have been shown to sire more offspring that subordinate or less aggressive males (Levine, 1958; DeFries & McClearn, 1970; Horn, 1974). The finding of heterotic inheritance for social dominance in genetic studies is consistent with its role in increasing reproductive fitness (Kuse & DeFries, 1976).

However, dominance does not ensure differential production of offspring in all situations. Using starch-gel electrophoresis (see Chapter 14), Duvall, Bernstein, and Gordon (1976) studied the offspring of a confined troop of rhesus monkeys. They found that the alpha male did not do all or even most of the successful mating. In studies of savanna baboons, Hausfater (1975) emphasized the degree to which dominance status changes over the lifetime of the individual. Thus, while dominance may confer differential reproductive ability to a male over the short run, it is unclear how it affects differential reproduction over the individual's complete lifetime. Bernstein (1976) proposed that the increased genetic fitness of the alpha males in some primate societies may lie not in their differential production of offspring but in their differential ability to ensure the survival of their offspring.

Territoriality Differential access to females may be gained in some species by males occupying choice territories. This is especially true of species with a *lek,* or system of reproduction involving a communal display ground. Sage grouse, for example, gather at communal display grouds (leks) at traditional sites on the sagebrush prairie. Males occupy small territories on the breeding ground. Females congregate at specific sites, or mating centers, within the breeding ground. These sites usually remain consistent from year to year. The result is that most copulation occurs within the territories of one or two males, although as many as 80 males may occupy adjacent territories. Thus, each year fewer than 10 percent of the males engage in more than 75 percent of all copulations (Wiley, 1973).

Sperm Competition Where it is likely that a female may mate with more than one male, it will be adaptive for each male to act to ensure that it is his sperm, rather than those of another male, that are effective in fertilizing the female's eggs. The situation in which the sperm from two or more males may compete to fertilize ova within a single female has been referred to as "sperm competition" (Parker, 1970). Where multiple inseminations can occur, one would predict the evolution of two kinds of mechanisms: (1) mechanisms that would enable a male somehow to cancel or neutralize the effects of sperm already in a female's reproductive tract as the result of matings with other males and (2) mechanisms that would prevent the male's own sperm from being similarly neutralized by a later male. Parker reviewed a considerable array of adaptations in insect reproduction that function in accordance with expectations derived from considerations of these two factors. For example, sperm precedence occurs in *Drosophila* and other insects when sperm from a second male directly displace from the sperm stores sperm from previous matings.

Adaptations that would reduce the probability or success of subsequent matings include: (1) mating plugs deposited after sperm ejaculation that block the female's reproductive tract against subsequent inseminations; (2) prolonged copulation, sometimes lasting hours or days; (3) passive phases, wherein the male remains mounted on or otherwise attached to the female without true genital contact; and (4) noncontact guarding phases, where the male guards the female although not in actual contact with her.

Similar adaptations have been observed in species other than insects, as, for example, the copulatory plugs of snakes (Devine, 1975) and coition-induced inhibition of sexual receptivity of female lizards (Crews, 1973). That the conditions for sperm competition occur in mammals is indicated by the finding of Birdsall and Nash (1973) that litters of wild-trapped *Peromyscus* mice had been sired by multiple males, as indicated by electrophoresis. Repeated copulations in rats and hamsters function through reducing reproductive success by subsequent males in situations of sperm competition (Dewsbury, Lanier, Estep, Oglesby, & Dudley, 1977). The copulatory plugs of guinea pigs also function to reduce the likelihood of success by subsequent males (Martan & Shepherd, 1976).

Strategies of the "Losers" As the gaining of high dominance rank and choice territory are often associated with age and size, it often appears adaptive for males with low status or poor territory to minimize risk so that they will survive to become effective breeders at a later time. However, alternative strategies sometimes occur. Among the bluehead wrasse (Atlantic reef fish), the females swim each day to a breeding ground wherein large, brightly colored males hold temporary breeding territories. Smaller males are excluded from these territories. The presence of external fertilization and the excessive congregation of females in the territories of a few males have fostered the development of three strategies by the smaller males. In "streaking," a small male rushes into a territory just as a pair are shedding gametes into the water; he then adds his gametes to the water. In "sneaking," the small male attempts to initiate early spawing by a female by providing tactile stimulation to a female as she rises to court the owner of the larger territory. In "group spawning," several males pursue the arriving females; it appears that the resulting tactile stimulation—in addition to the visual stimulation from other mating pairs—may result in spawning (Warner, Robertson, & Leigh, 1975).

Female Choice

In general, the problem for the female is not one of finding a mate; she may have many suitors. The female must choose from among males. Her behavior will be selected (1) to secure the "best" male genes for her offspring and (2) in some species, to choose a male that can provide the best resources for feeding and protecting her young. Thus, she acts to ensure both effective genes and a beneficial environment, so that her young benefit by the interaction of both in their development.

Many aspects of female choice appear functional in ensuring the selection of fit genes for her offspring. Essentially, she enhances the fitness of her offspring, and hence her own fitness, by mating with males that have demonstrated their fitness (Selander, 1972). Thus, the female sage grouse or bluehead wrasse that bypasses many vigorous and capable males in order to join the crowd in the region of males occupying choice territories appears to be selecting males that have "passed the test." Female dogs in heat are known to demonstrate clear mating preferences, either refusing to mate or mating with different individual males in rapid succession (Beach & LeBoeuf, 1967; LeBoeuf, 1967). Females of inbred strains of mice have been found to prefer to spend time and mate with males of strains other than their own, a pattern that may function to reduce inbreeding and maximize hybrid vigor (Yanai & McClearn, 1972).

Cox and LeBoeuf (1977) proposed that among elephant seals, females actively incite male-male competition. By protesting loudly when mounted, females attract the attention of males in the vicinity, especially the dominant male. This fosters male-male competition, decreases the likelihood of insemination by subordinate males, and functions to select genes for the female's offspring. When receptive, the females of many species—including sheep, roof rats, and bison—become quite active and run from males during solicitation (see Chapters 5 and 12). Such increased activity may function to attract numerous males and, again, incite male-male competition (Cox & LeBoeuf, 1977).

Females may select males not only because of the likelihood that they possess superior genes but also because they have privileged access to resources that are critical to the survival of the male, the female, and their offspring. This factor would be particularly important among species, such as territorial songbirds, in which there are prolonged pair bonds and a territorial site is occcupied for the entire breeding season (see Orians, 1969; Selander, 1972).

Bateman's Principle

Given all of the above, "Bateman's principle" should come as no surprise. It states that in most polygamous species, the variance of reproductive success among males will be greater than that among females. That is, there will be some males that make disproportionately large contributions to the gene pool of the next generation and others that make none at all. Most females will breed, and the number of offspring will vary very little among individual females. Bateman (1948) obtained empirical evidence to support this principle in a laboratory experiment in which *Drosophila melanogaster* with chromosome markers were permitted to mate in groups of five males and five females. In Bateman's experiment, 21 percent of the males failed to mate, whereas just 4 percent of the females failed to do so.

The Sociobiology of Parenthood

Many of the above phenomena can be understood through a consideration of parental investment, parental manipulation, and parent-offspring conflict.

Parental Investment The theory of parental investment was developed by Robert Trivers (1972) as an extension of some ideas proposed by Bateman. The basic phenomenon for which it provides an explanation is the disparity in certain characteristics of reproductive strategy between males and females. Why is it generally males that compete among themselves and females that exercise a choice, rather than females that compete among themselves with males making the choice? Why is there greater variability in reproductive success among males than among females? The answer may lie in the differential parental investment of the two sexes.

Trivers (1972) defined parental investment as "any investment by the parent in an individual offspring that increases the offspring's chance of surviving (and hence reproductive success) at the cost of the parent's ability to invest in other offspring" (p. 139). Investments may be of any of a variety of types, including the actual gametes, nourishment, care, protection, and the like. In all cases except breeding systems with complete monogamy, the relative investments of male and female will be unequal, with the investment of the female typically exceeding that of the male. When the investment by one sex exceeds that of the other, then the sex making the greater investment will become a limiting resource for the sex making the lesser investment. Individuals of the sex investing less will compete among themselves to breed with members of the sex investing more. This is because members of the sex investing less (typically males) can increase reproductive success by investing successively in the offspring of several members of the sex investing more (typically females). In essence, males can make their small investment readily and move from female to female, while females, having committed themselves with a greater investment, are stuck with the consequences. The disparity between male and female investment can probably be traced back to the initial differentiation of sex cells from gametes of equal size into females (producing larger gametes) and males (producing smaller ones). Because they have to compete for the limiting resources, members of the sex making the smaller investment, typically males, will evolve more extreme epigamic displays and will compete in intrasexual selection, as by becoming larger and more conspicuous.

What governs the operation of sexual selection is the differential parental investment of the sexes in their offspring. Where male and female investments are equal, there should be a symmetry of competition and choice. Where the male investment exceeds that of the female, female-female competition and male choice should predominate. While there has been no attempt at quantifying relative parental investment in species of birds, such as phalaropes, and fishes, such as seahorses and pipefish, in which males brood eggs and care for young, the presence of courtship by and brighter coloration of females is consistent with such expectations.

Desertion and Cuckoldry The intricacies of the theory of parental investment can be seen in considering the problem of desertion by the male or cuckoldry by the female of a monogamous pair. Each individual will act to maximize its own fitness. Each might increase its fitness by mating again with another partner if the first partner would continue to rear their young. If a partner leaves, the one remaining will be placed in a "cruel bind"; it will have to decide whether or not to attempt to raise the brood alone. The relative parental investment changes over time. In territorial songbirds, for example, the male's investment exceeds that of the female early in the season, when he has secured a territory. However, after an egg is laid, the female's investment exceeds that of the male. Whenever there is a disparity between the parental investments of the partners, one parent may be tempted to leave. As pointed out by Dawkins and Carlisle (1976), the primary factor is not past investment but the future investment that would be required to replace the lost offspring. Because the individual left with the offspring, typically the female, has greater investment in the offspring, in the sense that more would be required of her to replace them, she will be likely to continue to rear them. Thus she will increase the fitness of the deserter as well as her own, while the deserter is off attempting to find another partner. The monogamous pair, in the sociobiological view, look at each other not with stars in their eyes but with the question of whether their own fitness might be maximized by leaving and seeking additional partners or by staying to help rear the offspring already created. Once more, we must repeat our caveat. These processes do not require conscious deliberation by organisms. Rather, they represent behavioral patterns that have come about through evolutionary processes. Our shorthand discussions of "decisions" should always be translatable back to the operation of factors ultimately affecting differential reproduction in the past by individuals of different genotypes.

Selecting a Mate We can now return to the problem of selecting a mate. First, an animal must locate a partner of the appropriate sex and species who bears genes that are "fit." Further, in some situations, it will be to the advantage of the animal to select as a mate an individual that is likely to make a considerable parental investment and which is unlikely to depart before the task of rearing the offspring is completed. In some cases, this discrimination is easy. In insects of the family Bittacidae, for example, females select as mates those males that present large, palatable prey during courtship; the female consumes this prey during copulation (Thornhill, 1976). By feeding his mate, the male not only gets to inseminate her but probably increases her—and hence his—fitness. The female selects the male on the basis of appropriate parental investment.

In the parental behavior of common terns, the male provides most of the food for the young shortly after hatching. The female must select a male capable of supplying that food. Females appear to select males on the basis of male courtship patterns that entail the gathering and carrying of food and the feeding of the fish to the female. Nisbet (1973) found that there was a correlation between such variables as the total weight of the clutch (offspring), the amount of food fed to the young, and the amount of food fed to the female *during courtship*. Thus, the female appears to have a basis in courtship for predicting eventual male parental investment and reproductive success.

The courtship pattern of ring doves is complex (Chapter 5), and male courtship, nest building, and incubation are related to gonadal hormones (Chapter 12). Erickson and Martinez-Vargas (1976) argued that the mutual dependence of courtship and incubation behavior on the same hormones provides the female with a means of predicting the male's level of parental investment. As both courtship and parental behavior are controlled by the same mechanisms, a male showing appreciable courtship is likely to display appreciable parental behavior. Such would not be the case if courtship and parental behavior were under the control of different sets of hormones. The finding that male ring doves display less courtship and more aggression toward females that have recently associated with other males than to females that have been isolated may be interpretable as an adaptation reducing the chances of cuckoldry (Erickson & Zenone, 1976).

Although white ibis are pair-bonding shore birds, females sometimes mate outside the pair bond when males are away feeding (Rudegeair, 1975). They may thus gain insurance against the possibility that their mates may be sterile. However, females appear to "cheat" selectively, primarily with pair-bonded males nesting in the same region. Thus they may be selecting males with which to "cheat" as well as with which to pair on criteria related to nest location or tendency to form pair bonds.

Relationships to Young Other Than One's Own In many circumstances it is critical that one's resources be invested in caring for one's own offspring rather than the offspring of others. It stands to reason that if an organism has limited resources that can be apportioned to parental investment, it should act to avoid caring for the offspring of others in favor of producing and caring for its own offspring. One manifestation of such factors is infanticide, such as that which is quite prominent among langurs—leaf-eating monkeys in India (Hrdy,

1974). It appears that when a new male moves in to take over a troop, he first kills the infants of the troop. Presumably, this brings the females into estrus sooner than would have been the case had they been permitted to keep their young, and it enables the male to mate with the females and invest his resources in protecting his own offspring (Hrdy, 1974, 1976). Pregnancy blockage in mice (the Bruce effect discussed in Chapter 10) may function in a similar way. A male terminates a previous pregnancy, brings a female into estrus, mates with her, and expends his resources in the service of his own offspring rather than those of another male. In prairie voles, males are effective in blocking pregnancy up to the fifteenth day of gestation, well after the implantation of the fertilized ovum in the uterine wall (Stehn & Richmond, 1975).

In an attempt to determine whether foster parents would care for young, Power (1975) collected one member from each of 25 pairs of mountain bluebird parents and watched the behavior of the 10 new birds that naturally replaced the old mates. Whereas the collected fathers would feed offspring, clean nests, and vigorously defend the young, such behaviors were quite rare among the new consorts. Thus, the males did not exhibit any significant investment of effort or resources in the young of the collected males. (For a discussion of the controversy surrounding interpretation of this study, see Emlen, 1976, and Power, 1976.)

Power (1975) further suggested that "the stepwise hormonal preparation necessary for carrying out the successive stages of nesting in birds . . . is probably a proximate expression of ultimate selection for reproductive selfishness, making altruistic errors infrequent" (p. 143). It will be recalled from Chapter 12 that the hormonal conditions necessary to elicit short-latency maternal behavior in rodents are extremely complex and that such behavior is difficult to elicit. It would seem reasonable that this system too may be adapted, as much to guarantee that parental care is not directed toward young other than one's own as to ensure direction of parental care toward one's own offspring.

There are circumstances under which assistance in the care of young is provided by individuals other than the parents. This has been termed "alloparental care" (Wilson, 1975). Instances of alloparental care should be explicable as strategies for maximizing inclusive fitness under particular sets of circumstances. An example of alloparental care can be found among Florida scrub jays (Woolfenden, 1973, 1975). These birds form lifelong pair bonds and occupy permanent territories. About half the breeding pairs that Woolfenden studied were aided by "helpers" who assisted the parents in defending the territory against other jays, defending against predators, and feeding the young. By marking individual birds, Woolfenden found that the helpers were generally close kin of the territory holders. Thus, the "helping" phenomenon would appear to be an instance of kin selection. Helpers are effective in increasing successful reproduction by the territory holders. In one sample, the average production by unassisted pairs was 1.1 fledglings per pair; that among pairs with helpers was 2.1 fledglings per pair. The main reason for this difference appears to lie in defense against predators (see also Wilson, 1975).

Care of the young by individuals other than the mother is fairly common in troops of some primate species (Mitchell, 1969; Rowell, 1972; Wilson, 1975; Hrdy, 1976). There are a number of advantages that can be specified for alloparental care in primates. First, experience plays an important role in the development of maternal behavior in primates (e.g., Harlow, Harlow, & Suomi, 1971). By handling infants, the young females of the troop gain experience that may be beneficial when they become mothers. This, in turn, may benefit both the young helpers and the mothers themselves (through kin selection). Second, from the perspective of the mother, permitting other animals to handle the young may lead to care in times of danger and the mother's absence or to the formation of alliances that may be beneficial in intratroop conflict. Further, the presence of a "baby-sitter" gives the mother greater freedom in foraging (Hrdy, 1976).

Because paternity is uncertain in most primate social structures, it is difficult to separate allopaternal care from true paternal care. True allopaternal behavior directed at young females may lay the groundwork for eventual consortships (Ransom & Ransom, 1971) or function as an "agonistic buffer," as the presence of infants inhibits aggression (Deag & Crook, 1971).

Parent-Offspring Conflict and Parental Manipulation Classically, evolutionary theorists have considered parental behavior from the standpoint of the parent. However, Trivers (1974) pointed out that if one considers the offspring as *actors* in parent-offspring interactions, it becomes apparent that the interests of parent and offspring will come into conflict on some matters. A parent that is acting to maximize its fitness will generally distribute its attentions among all its offspring. An individual offspring, acting to maximize its fitness, may demand a greater proportion of the parent's attention than it is adaptive for the parent to give. Under extreme circumstances, it may be adaptive for a parent to kill one or more offspring if that increases the chances that at least some of them will live. Again, there may be parent-offspring conflict.

In Chapter 5 it was seen that the typical course of parent-offspring interaction among mammals progressed through a phase in which the parent initiated interactions in the early phase, the interactions were mutually initiated in the intermediate phase, and the offspring initiated most interactions in the terminal phase. This transition would appear to represent the progressive shift occurring when the young, as perceived by the mother, require less parental attention but nevertheless continue to demand such attention. Conflicts over weaning may be similarly understood. Trivers provided reasonable explanations for the effects of brief separations of mothers and their young in rhesus monkeys, studied by Hinde and Spencer-Booth (1971), in terms of parent-offspring conflict.

According to Alexander (1974), the parents are destined to win parent-offspring conflicts in the long run. This is because genes that reduce the mother's overall fitness as a result of competitive interactions among the young cannot spread regardless of their advantage to the offspring possessing them (Alexander, 1974). The same genes operative in the parent will become operative in the offspring when they become parents (Williams, 1966). If a juvenile mutates so as to possess genes that permit it to gain a disproportionate share of the parent's resources and thereby reduce its parent's fitness, the same gene will be transmitted to its own offspring, who in turn will reduce the fitness of the individual initially bearing the mutation. Thus, in the long run, no individual can benefit from possessing such an allele (Alexander, 1974, p. 340).

According to Alexander's view, the parent will manipulate the offspring in such a way as to maximize the parent's own fitness. According to this view, it is advantageous to both parent *and offspring* for the offspring to accede to such *parental manipulation.* Forms of parental manipulation of offspring that may be adaptive under particular circumstances include (1) limiting the amount of care to each offspring so that all survive and reproduce, (2) restricting or withholding parental care from some offspring when resources become insufficient for the entire brood, (3) killing of some offspring or feeding them to others, (4) causing some offspring to be temporarily or permanently sterile helpers at the nest, and (5) causing some offspring to adopt roles as permanent sterile workers or soldiers (Alexander, 1974). Parents should be selected to manipulate the sex ratio, producing a lower ratio of males to females as the condition of the mother declines (Trivers & Willard, 1973).

SOCIOBIOLOGY, BREEDING SYSTEMS, AND SOCIAL ORGANIZATION

An introduction to patterns of social organization was presented in Chapter 6. We can now consider some of the primary determinants of social organization as viewed from a sociobiological perspective. Elucidation of the fundamental determinants of patterns of social organization has been one of the primary goals of the sociobiological approach. In sociobiological perspective, social organizations can be viewed as the product of the actions of a group of individuals each adopting a strategy for survival and reproduction that will maximize their individual fitnesses. Where living in an organized society will increase levels of individual fitness, organized societies would be expected to develop; where they would not increase individual fitness levels, no such development would be expected.

Advantages and Disadvantages of Sociality

Although there are many advantages to sociality, there are disadvantages as well (Wilson, 1972, 1975). Among the "antisocial" aspects of phylogenetic inertia are the following:

1 *Limitations of food:* A large concentration of food is necessary to support a large group.

2 *"Reproductivity effect":* Animals living in large groups tend to produce fewer offspring per animal than those living in smaller groups.

3 *Sexual selection:* When sexual selection has favored development of

large, aggressive males, they are less likely to be integrated into organized societies.

4 *Danger of inbreeding:* Where a tightly knit social structure limits breeding choices, inbreeding can occur and thus reduce individual fitness levels.

5 *Exposure to disease and predators:* A grouping of animals increases the chances of spread of disease and, under some conditions, may increase the risk of detection by predators.

Since social organization has evolved repeatedly, there are obviously many factors that counteract the antisocial forces and act to produce the evolution of organized societies (Wilson, 1975; Klopfer, 1974). Among the factors working to increase sociality are:

1 *Defense against predators:* While groups of animals may be more likely than isolates to be detected by predators, they also have improved defenses through group action.

2 *Increased competitive ability:* Just as groups are able to thwart predators, they can overwhelm animals of another species with which they compete for identical resources.

3 *Buffers against environmental pressures:* Animals in groups buffer each other from the environment, as in the protection against wind and the dilution of environmental toxins.

4 *Increased feeding efficiency:* Feeding efficiency may be increased through either imitation of foraging habits or cooperation in the procurement of food.

5 *Penetration of new adaptive zones:* Grouping may on occasion permit animals to enter novel habitats or to assume a new life-style.

6 *Increased reproductive efficiency:* Animals in groups can more easily locate mates and synchronize breeding.

7 *Increased survival at birth:* Groups of animals are often more able to ensure survival of offspring than are isolates.

8 *Improved population stability:* Social living may increase individual fitness levels by ensuring that population size remains within tolerable levels appropriate for the habitat.

9 *Modification of the environment:* Large groups can alter the environment—as in building structures, regulating environmental temperature, or altering plant growth—as the result of their presence and behavior.

Evolution of Social Behavior in Insects

Perhaps the most remarkable example of the evolution of a pattern of social organization and breeding can be seen in the 24,000 or so species of eusocial insects (see Chapter 6). The critical defining attribute of eusociality is the existence of a sterile worker caste. In each eusocial species there exist individuals that have forsaken their ability to breed in favor of the opportunity to help their mother, the queen, in her reproductive endeavors. If social organizational systems have evolved as means of maximizing individual fitness, how do we explain such a complete loss of reproduction—a trait that would appear counter to fitness at the most fundamental level? There is disagreement regarding the major influences that have shaped the evolution of eusociality in insects. Three major hypotheses have been proposed (West Eberhard, 1975).

Haplodiploid Hypothesis The haplodiploid hypothesis was a major contribution of the theories of kin selection as developed by Hamilton (1964). According to this hypothesis, each individual can increase its inclusive fitness by increasing the fitness of its close relatives. The increase in fitness is devalued as a function of the remoteness of the family relationship. The unique attribute of the Hymenoptera (ants, bees, and wasps), the group of species in which eusociality is most characteristic, is that of *haplodiploidy*. In vertebrates, all cells except the gametes are diploid (i.e., possess the diploid chromosome number). In the Hymenoptera, by contrast, only females develop from fertilized eggs and are thus diploid; males develop from unfertilized eggs and possess the haploid chromosome number. Queens appear often to mate with just one male.

As a result of these unusual circumstances, the degree of relatedness among individuals belonging to the Hymenoptera is shifted in some important ways. Thus, while a female shares the usual half of her genes with her father, mother, and offspring, she shares three-quarters with her sisters and just one-quarter with her brothers. Hamilton's deduction was straightforward. Because of these kinship relations, a female can make a greater contribution to her fitness by helping her mother rear her sisters, with whom she shares three-quarters of her genes, than in producing offspring, with whom she would share only half her genes. This simple derivation is of critical importance and has far-reaching implications. A considerable number of predictions can be derived from this set of genetic relationships. Thus by combining kin-selection theory with Fisher's sex-ratio theory, Trivers and Hare (1976) were able to predict sex ratios in a variety of insect species living under different conditions. The observed data were consistent with predictions from the theories. On the other hand, several objections have been raised against kin-selection theory as the only explanation of the evolution of eusociality (West Eberhard, 1975; Alexander, 1974).

Mutualistic Hypothesis According to Michener (e.g., Lin & Michener, 1972), the haplodiploid hypothesis is not sufficient, in and of itself, to account for the evolution of eusociality. He proposed a different evolutionary sequence than that proposed by Hamilton (termed a "semisocial" rather than a "subsocial" route) and emphasized the value of group living, with and without kinship, in cooperative defense against both parasites and predators. Because such groups exist among the Hymenoptera, it is possible that they represent a preadaptation from which eusociality arose. According to this view, cooperation, or *mutualism*, was a critical aspect of the evolution of eusociality.

Maternal-Control Hypothesis Alexander (1974) argued that the evolution of eusociality represented an instance of parental manipulation of the sort discussed above. Thus, for Alexander, the queen "uses" some of her offspring to help rear others in much the same way that a mammal might resorb some fetuses in order to nourish others. The mother has adopted a strategy in which she produces some of her offspring only in order that they might be sacrificed for the sake of her other offspring, who will therefore enjoy maximal fitness and, in the long run, maximize the fitness of their mother.

Synthesis These hypotheses are all relatively recent and the debate is current. Thus, it is difficult to decide among them, if indeed such a decision is necessary. In an attempt at synthesis, West Eberhard (1975) has put the situation well:

> There is no reason, as I have argued, why Hamiltonian kin selection, Michnerian mutualism, and Alexanderian parental manipulation cannot all have contributed simultaneously or sequentially to produce the evolution of sterile castes; and it seems fruitless, *in making generalizations*, to argue in favor of one idea to the exclusion of the others (although this may be an important question in discussing a particularly well-studied case) (p. 29).

The Breeding Systems of Birds

Birds adopt a variety of organizational strategies during the breeding season. The conditions under which different organizational strategies are adopted ought to be derivable from sociobiological theory. In *monogamy*, a single male and female pair for at least a breeding season and sometimes for life. *Polygamy* is a general term used to include instances where an individual mates with more than one individual of the opposite sex. In *polygyny*, one male mates with more than one female; in *polyandry*, one female mates with more than one male.

The vast majority of species of birds are monogamous. Monogamy appears to represent an adaptation to situations in which a pair of birds can more efficiently maintain a territory and provide food for the young than could a lone mother. On occasion, birds develop polygamous organizational patterns, most typically with a single male and more than one female occupying a given territory. Verner and Willson (1966) noted polygyny or promiscuity in 14 of 291 species of North American passerine birds. Orians (1969) developed a model to explain the evolution of different organizational patterns adopted by breeding birds. Fundamental to the model is the notion, discussed above, that the female will maximize her fitness by mating with a male that is both genetically fit and in possession of a territory containing optimal resources for the development of the young.

Essentially, reproductive success is related to the quality of the environment in which the animals live. Thus, a female should mate with the male possessing the best territory. However, as more females pair with that male, the availability of the resources in the territory will be reduced. There is thus a trade-off between the quality of the habitat and number of females already bonded to the territory holder. The female should elect to mate with a male already possessing one mate only when the gain from living in a superior territory outweighs the loss incurred by being the second female on that territory. Thus, there should be a restricted set of conditions under which polygyny should evolve. These relationships are portrayed graphically in Figure 15-1. From this basic model and a few additional factors, Orians (1969) derived a number of predictions regarding conditions under which particular breeding patterns should exist. A few representative predictions follow:

1 *Polyandry should be rare:* This prediction follows from considerations of parental investment, as discussed above.

2 *Monogamy should be rare among mammals, but common among birds:* Because of the presence of mammary glands in the female mammal, the male mammal is destined to play a relatively minor role in providing care of the offspring. By contrast, the male bird is as adept as the female in everything except egg laying and thus can more easily take a more active role in the care of the young—an essential factor in monogamy.

3 *Polygyny should evolve more readily in precocial than in altricial birds:* Because the young of precocial species can find food at an earlier age, they are less dependent than are altricial young on a continuing parental investment by the male of the sort that occurs in monogamy.

4 *Polygyny is likely to evolve among species having altricial young and*

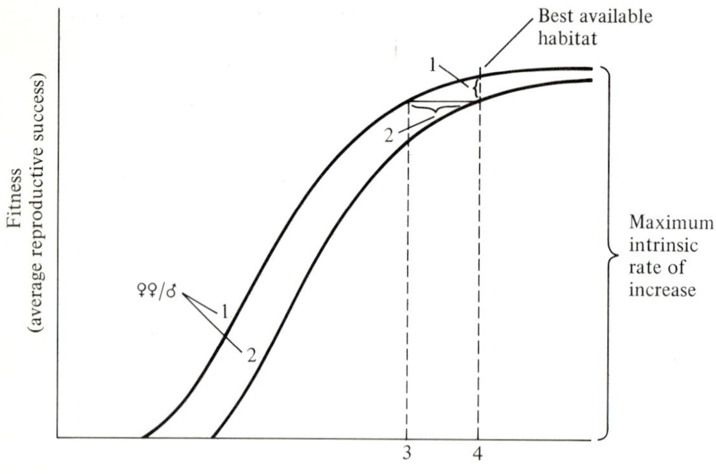

Figure 15-1 Graphic model of the conditions necessary for the evolution of polygynous mating patterns in birds. Average reproductive success is assumed to be correlated with environmental quality. Females select from among males on the basis of whether or not another female already is paired with the male and on an estimate of quality of the territory. The distance "1" is the difference in fitness between females mated monogamously and bigamously in the same environment; the distance "2" is referred to as the "polygynous threshold," the minimal distance between habitat qualities that will result in a favoring of bigamous matings by females. *(From Orians, 1969.)*

nesting in marshes: This prediction follows from the facts that in marsh environments there is a great variability in the resources existing in different areas and that depletion of resources, such as insects, is less of a problem for newly arriving individuals.

Like all good models and theories, this model is testable. For example, Carey and Nolan (1975) used the model of Verner and Willson and Orians to predict the existence of polygyny in indigo buntings. The prediction was confirmed and females mated to polygynists were found to produce as many young as those mated to monogamists.

Correlates of Mammalian Social Organization

The patterns of social organization are extremely diverse. They range from monogamy to promiscuity and from solitary living to life in large groups (see Eisenberg, 1965). Some of these forms of mammalian social organization were discussed in Chapter 6. The correlates of mammalian social organization are extremely complicated. There is no simple way to summarize the array of factors that affect group size and the patterns of social relationships that occur within groups. Patterns of social organization are often not even consistent within a species but vary in relation to local variations in habitat, season, etc. The task of determining the correlates and prime movers of mammalian social organization represents a difficult challenge for sociobiological theory.

Numerous authors have proposed models of varying levels of complexity and sophistication for the explanation of diversity in mammalian organizational systems. Most models focus on two factors: *resource distribution* and *predator pressure*. The distribution of resources is critical in permitting the formation of large groups. In most cases, a plains-dwelling grazer can form large groups with relatively little danger of depleting resources. By contrast, a species at the top of a food chain (that may lead from plants to herbivores to carnivores and top carnivores) cannot live so densely. Top carnivores, such as wolves and big cats, must often travel considerable distances in order to locate prey.

Danger from predation is a second major determinant of group size. As a general rule, large groups are better equipped to deal with predators than are small ones. Predation pressure, then, will work in conjunction with the distribution of food and other resources to determine the most effective group size. As an example, the number of males one would expect in a group can be seen to vary according to resource distribution and predation pressure. Insemination of a group of females requires neither a large number of males nor the continued association of males with the group. Thus, where resource distribution dictates, the number of males in a group can be reduced. However, as males are often larger and stronger than females, they are particularly well equipped to defend the troop against predators. Thus, in areas of great predation pressure, one would expect more males in groups. To oversimplify severely, males have to earn their share of the resources through predator protection (or other means).

These general principles can be illustrated and their strengths and limitations better understood through a consideration of the correlates of social organization in a few selected mammalian species.

Bovidae The Bovidae are a group of artiodactyls (even-toed ungulates, including cattle, antelopes, duikers, sheep, and goats). Estes (1974) analyzed the patterns of social organization of some 70 species of African Bovidae, 29 of which were found to be solitary and 41 gregarious. He found that species that live in open country generally are gregarious and of large stature, with prominent horns. They are mobile and tend to run from danger. Such species are generally grazers and have distinctive markings. By contrast, species living in closed habitats are generally solitary, small, sedentary, cryptically-colored browsers. They tend to have simple horns and to hide from danger. Open habitat in both Bovidae and other mammalian groups tends to be associated with increased predation pressure and increased group size. Estes noted that when attracted to open ground, solitary bovids tended to group more closely together. Such behavior may have provided a precursor of organized herds.

Prairie Dogs Black-tailed prairie dogs are a species of Sciuridae (squirrel-like rodents) that inhabit the northern plains of the United States (Figure 15-2). Prairie-dog towns may contain up to 1,000 individuals but are

Figure 15-2 Prairie dog. *(Photograph by the author.)*

subdivided into successively smaller groups termed "wards" and "coteries" (King, 1955, 1959; Smith, Smith, Oppenheimer, deVilla, & Ulmer, 1973). Social organization within and between coteries is of a degree that is well beyond that of most of the Sciuridae. The evolution of large groups and complex social organization appears to represent adaptations to the open habitat characteristic of prairie dogs.

Wallabies Whiptail wallabies are a species of macropodid marsupial, resembling kangaroos, and are probably the most social of the marsupial species. They live in open habitats and form loosely knit groups, termed "mobs," of about 30 to 50 members (Kaufmann, 1974a, 1974b). The mob is organized into a dominance hierarchy among the adults and subadults. The formation of groups in the marsupials appears to represent adaptation convergent with that among plains-dwelling ungulates. As the groups move toward grazing habits in open habitats, they increase in size and become more social. The grazing pattern, diurnal activity pattern, presence of predators, and other factors can be seen to interact in favoring complex social organization (Figure 15-3).

Vicuña Vicuña are members of the camel family that inhabit the region of the Andes mountains in South America. Their pattern of social organization is interesting and somewhat atypical, as males defend territories and keep harems the year round rather than just during a brief breeding season (Koford, 1957; Franklin, 1974). Two separate territories are maintained: a sleeping territory and a feeding territory. Each morning, the male, the harem of females, and the associated young move from the sleeping territory through a "neutral corridor" to the feeding territory. All feeding, sleeping, and reproduction take place within the group's territories. The role of the territorial male is critical, as he determines the location, limits, and size of the territory and regulates the composition of his group. Nonterritorial males form all-male bands.

The social pattern of vicuña appears adapted to the limited food supply in the high mountain environment. By defending a feeding territory, the male and his group ensure the existence of the resources necessary for their survival and reproduction. As one would expect, territory size tends to be larger where plant growth is less.

Marmots Marmots are a genus of sciurid rodents that occupy a variety of different habitats in the United States. By using the method of adaptive correlation at the species level, it is possible to correlate differences in the ecology and social behavior of different species (Armitage, 1962; Barash, 1973a, 1973b, 1974). The pattern of social organization of the Olympic marmot of the state of Washington is more closely integrated and less aggressively organized than that of the yellow-bellied marmot of the Rocky Mountains, which, in turn, is more social than the eastern woodchuck. Barash believes that the progressive shortening of the growing season in the habitats of woodchucks, yellow-bellied marmots, and Olympic marmots is of great importance

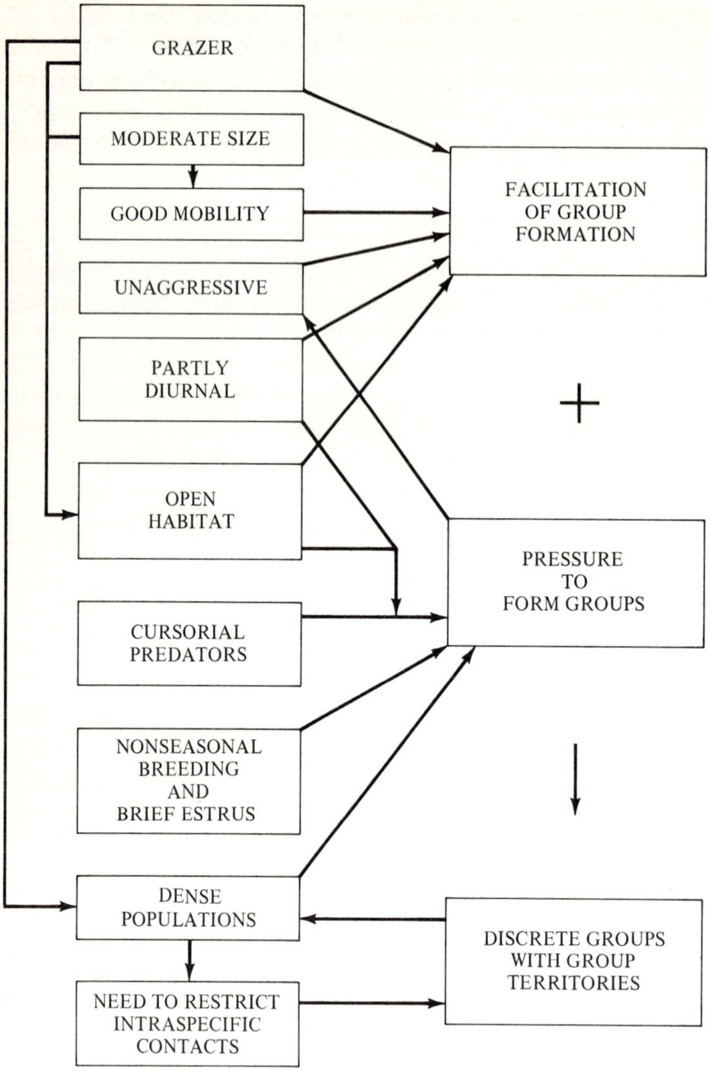

Figure 15-3 Diagram of some behavioral and ecological determinants of social organization in whiptail wallabies. *(From Kaufmann, 1974a.)*

in determining the pattern of social organization. Associated with the decrease in length of the growing season are (1) decreases in proportional growth rates, (2) decreases in breeding frequency, (3) increases in age at dispersal, (4) increases in age at sexual maturity, and (5) increased degrees of sociality. Andersen, Armitage, and Hoffmann (1976) believe that the selective pressures acting on the evolution of social systems in marmots are more complex than was suggested by Barash.

Primates Primates of different species are organized into different social systems. Six different grades of grouping patterns—from a solitary pattern through monogamous pairs, single-male groups, aggregate single-male groups, multiple-male groups, and diffuse social parties—were distinguished by Southwick and Siddiqi (1974) and in Chapter 6. Numerous attempts have been made to relate these variations in social organization to ecological variables (e.g., Crook & Gartlan, 1966; Jolly, 1972; Eisenberg, Muckenhirn, & Rudran, 1972; Wilson, 1975).

Predation pressure and the pattern of food distribution appear to be among the most critical determinants of primate social organization, as they are for the social organization in other groups of mammals. The multimale system appears to be a specialized form of social organization that is a particular adaptation to terrestrial foraging by primates of intermediate body size. The unimale system or the age-graded male system is favored in arboreal species that eat primarily fruits or leaves. In particular, an arboreal, diurnal, leaf-eating pattern appears correlated with a social structure that tends toward a one-male system with a small home range and the utilization of vocal choruses to maintain spacing (Eisenberg et al., 1972).

Particular attention has been devoted to the study of the correlates of patterns of social organization among baboons (e.g., Kummer, 1971; Altmann, 1974). Yellow baboons typically live in relatively open country, where there may be trees but the predominant cover is grass. They live in relatively large groups. In the Amboseli Reserve in Kenya, the mean group composition included 14.8 adult males, 16.5 adult females, 12.7 juveniles, and 10.5 infants (Altmann, 1974). These groups are relatively permanent and the breeding pattern is characterized by promiscuity.

Hamadryas baboons live in the Ethiopian desert, an extremely harsh environment for nonhuman primates. The most elemental organizational level among hamadryas baboons is the one-male unit, the members of which typically forage together during the day. These one-male units group together at a particular sleeping rock at night to form a band. The composition of bands is relatively constant. Bands may temporarily join together to form a larger troop.

Gelada baboons of the Ethiopian mountainous grassland also form herds from one-male groups. However, the geladas forage en masse and break up into harems in the late afternoon.

Kummer (1971) provided an excellent summary of the method of adaptive correlation as applied to the study of social organization among baboons. Altmann (1974) refined the analysis and developed a relatively complex set of 11 principles underlying baboon social organization. The fundamentals underlying these 11 principles are based in resource distribution and predation pressure. The yellow baboons find food relatively easy to obtain; predation pressure on them, as from large carnivores, is considerable. Multimale groups can be supported by the resources and are adaptive in defense against predators. By contrast, in regions of poor resources, predation pressure is less and food limitations are more restricting. The unimale pattern is typically observed under such circumstances.

Labrid Fish

Perhaps the ultimate in the regulation of sex ratio and the control of the number of adult males can be found in the labrid fish, *Labroides dimidiatus* (Robertson, 1972). The basic social unit in these fish is a male with a harem of about three to six mature females and several immature fish, all living within a male's territory. Larger individuals dominate smaller ones. On the death of the male, the dominant female of the harem undergoes a sex reversal to become a functional, dominant male. Thus, the male in each harem actively inhibits a sex reversal on the part of the subordinate females by actively dominating them.

In this system, the genotype of the male is that which is maximally recombined, because the male spawns regularly with all females in the group whereas the female spawns primarily with the dominant male. The male genotype may be that which is best adapted to local conditions, because the male was derived from the oldest and most dominant female in the group.

SUMMARY

Sociobiology is based on an attempt to bring information regarding evolutionary and population biology to bear on the problems of reproductive strategies and patterns of social organization. The cornerstones of sociobiology are the concept of genetic fitness; the notion that selection works on individuals; and the concepts of inclusive fitness, kin selection, altruism, and reciprocal altruism.

Diverse patterns of reproductive strategy can be seen as attempts by individuals to maximize their fitness under diverse ecological conditions. These strategies can be seen to be related to factors such as parental investment, female choice, parent-offspring conflict, and parental manipulation.

Patterns of social organization are the result of the banding together of groups of individuals in ways that maximize their levels of individual fitness. Different patterns of social organization develop in response to different ecological pressures, as can be seen in the evolution of social behavior of insects, the breeding systems of birds, the correlates of mammalian social organization, and the sex-reversal pattern of labrid fish.

Part Seven

Learning: A Synthesis

The study of learning presents a problem for one attempting to provide a rational organization of material in the study of animal behavior. Learning can be viewed descriptively, as a critical aspect of development, as an important mechanism in the control of behavior, as a product of evolutionary history, or as an adaptation. Depending on one's emphasis, material on learning could thus be placed in any of the major sections of a book such as this. Alternatively, it could be scattered throughout all sections. Yet a third approach is taken in this book: an entire major section is devoted to the study of learning. An attempt will be made to introduce the student to all of these aspects of the study of learning. The organizational structure of this book (i.e., observation and description followed by the consideration of questions of development, mechanism, evolution, and function) provides a reasonable framework for considering research on learning. Further, by using this framework for problems of learning, its utility can be further demonstrated.

In Chapter 16, a description is provided of various forms of learning, the ontogeny of learning and learning ability, and the mechanisms related to learning. Chapter 17 is concerned with evolutionary changes in learning and learning ability and with the adaptive significance of learning and learned behavior.

Chapter 16

Learning: Description, Development, and Mechanisms

We begin Chapter 16 with a consideration of the definition of learning. This will be followed by an introduction to the various forms of learning that have been studied by animal behaviorists. The latter introduction will include representative experiments and should familiarize the reader with the basic phenomena to be considered in the study of learning. With preliminary matters out of the way, we shall then proceed to consider the ontogeny of learning and the nature of the mechanisms associated with learning. The goal of this chapter is to provide a firm foundation for the material in Chapter 17.

This presentation of basic procedures and paradigms used in the study of learning will, of necessity, be brief. For a more complete treatment, the reader may consult a textbook on the psychology of learning (e.g., Mackintosh, 1974; Hulse, Deese, & Egeth, 1975; Tarpy, 1975).

DEFINITION OF LEARNING

We shall use the term "learning" to indicate a rather broad range of behavioral modifications. As was the case with the term "aggression," the term "learning" is one that has been borrowed from everyday language. It is useful in designating, in a very general way, the nature of the material to be discussed.

However, there are many instances of behavioral modification, especially among invertebrates, that may be classified either as learning or as something else, depending on the definitions one accepts. We shall briefly consider a few of the problems encountered in defining learning. For more detailed discussions of these problems, the reader is referred to Kimble (1961), Dyal and Corning (1973), Miller (1967), and Jensen (1970).

The major difficulty with definitions of learning is that they usually include too much (i.e., phenomena that would not normally be regarded as learned) or too little (i.e., they exclude phenomena that would normally be regarded as learned). We prefer to err in the former direction. We shall define learning in a very broad sense as a change in behavior that occurs as the result of practice. The inclusiveness of the definition should be immediately apparent. Let us consider some problems in defining learning.

A Change in Behavior

Learning entails a change in behavior. Changes in behavior can entail (1) the appearance of a behavioral pattern that is new to the organism, (2) a change in stimulus-response relationships so that a response is given in the presence of a stimulus not previously associated with it, or (3) a change in the probability or form of a response that has previously occurred in a given situation.

A distinction must be made between learning and performance. It is quite possible that the potential for a change in behavior occurs as a change in the organism, but that the potential is never manifested in overt behavior. Therefore, Kimble (1961) would specify that learning represents a change in *behavior potential* rather than in behavior. As we can detect learning only when there is a change in behavior, we shall continue to discuss learning as a change in behavior. However, it must be remembered that changes in behavior potential may occur that would not be perceived as instances of learning under a given set of testing conditions but that might be so perceived under a different set.

According to many ethologists (e.g., Thorpe, 1963), learning represents an *adaptive* change in behavior. Adaptiveness is not included in the definition adopted here. There are many instances wherein a specific behavioral modification occurs even though it will not be adaptive to the organism. Not all instances of learning lead to an "improvement" in behavior. Although the ability to modify behavior and the processes underlying such modification certainly are adaptive in the long run, adaptiveness should not be a prerequisite to the identification of learned behavior in particular instances.

Many would define learning as a *relatively permanent* modification of behavior. Miller (1967) would demand that the behavioral modification persist for days or weeks rather than seconds or minutes in order to qualify as "grade-A certified learning." The advantage of including permanence as a criterion for learning is that it easily excludes such phenomena as changes in behavior due to motivational changes, sensory adaptation, and fatigue. Permanence certainly is an important characteristic of many vertebrate learning systems. However, in many invertebrate systems, behavioral modifications may persist for only a matter of minutes. It would not seem profitable to exclude such modifications from the realm of the learned solely on the basis of our mammalian time scale.

Nature of Practice

By specifying that the modification of behavior must result from practice rather than simply experience, we attempt to exclude changes occurring as the result of maturation, senescence, injury, and various physiological manipulations.

Kimble (1961) would specify that the practice must be *reinforced practice*. However, because of the difficulties encountered in defining reinforcement, such a specification would seem to complicate rather than clarify the issues.

While recognizing that our definition of learning is broad, it is one that is useful when we regard "learning" simply as a general term with which we can bring together a group of diverse phenomena having something in common.

FORMS OF LEARNING

Animal behaviorists have used many different procedures in studying learning. We shall next consider some of these. A description of the various forms of learning is analogous to the observation and description that provides the starting point for the study of all animal behavior (Tinbergen, 1963a; also Chapter 1). Many of the distinctions to be made among different "forms" of learning are based on differences in the procedures used in the study of learning. The fact that the procedures used differ need not necessarily imply that the processes underlying these different forms of learning are either similar or different. There may be one, two, or many kinds of processes underlying different forms of learning.

Habituation

Habituation may be defined as the decrease in probability or amplitude of a response, occurring when a stimulus which elicits that response is presented repeatedly. In an excellent discussion of the phenomenon of habituation, Thompson and Spencer (1966) listed nine characteristics of habituation that are manifested in most systems. We shall consider a few of these. If, after repeated stimulus presentation and habituation of the response, the stimulus is no longer presented, the response tends once again to be elicited by the stimulus. Within limits, the longer one withholds the stimulus, the greater is the likelihood that the response originally elicited by the stimulus will again be elicited by it. This phenomenon is called *spontaneous recovery*. It may require a few minutes to several days, depending on the conditions. If a repeated series of habituation training trials and spontaneous recovery trials is given, the habituation occurs progressively more rapidly (*potentiation of habituation*). Habituation of a response to a given stimulus shows generalization to other similar stimuli (*stimulus generalization*). Thus, if an animal habituates a response to a tone of a given frequency, it is likely that responses to tones of frequencies close to the training stimulus will also show habituation. The more dissimilar the test tone from the habituation tone, the less would stimulus generalization be apparent. Finally, presentation of another (usually very intense) stimulus results in a sudden recovery of the habituated response (*dishabituation*). In order to be considered true habituation, the response decrement must clearly be the result of changes in the central nervous system rather than of *adaptation* in the sensory system or fatigue in the effectors.

When a puff of air is applied to the gills of a horseshoe crab, it responds with a movement of its telson (tail). Lahue, Kokkinidis, and Corning (1975) studied the characteristics of habituation of the telson reflex, monitoring the muscle movements electrophysiologically. They found a clear habituation of the telson reflex to repeated presentations of the air-puff stimulus. A brief squirt of saline from a syringe aimed at the site of habituation resulted in dishabituation. Initial levels of responsivity were normal after an interval of 1 hour without stimulus presentation (spontaneous recovery). Potentiation of habituation was observed after about 2 hours. No generalization of the habituation was observed when stimuli were applied to other regions (no stimulus generalization). Habituation of contraction responses of sea anemones to water-stream stimulation was observed by Logan (1975). Ratner and Gilpin (1974) studied the habituation and retention of habituation of the response to air puffs in both normal and decerebrate earthworms.

Classical Conditioning

There are two major classes of procedures used in the study of associative learning: classical conditioning and instrumental conditioning. The difference between the two procedures is simple. In the classical conditioning paradigm, the sequence of events on a trial is in no way affected by the behavior of the animal. No matter what the animal does, the consequences established either by the experimenter or the apparatus that has been programed are unaltered. In instrumental learning, by contrast, the behavior of the animal is an important determinant of the sequence of events. Typically, in the instrumental paradigm, a reward or punishment is given or taken away upon the occurrence of a response designated as either correct or incorrect (Kimble, 1961).

In a typical experiment on classical conditioning, one begins with a stimulus (the unconditioned stimulus or UCS) that elicits a particular response (the unconditioned response or UCR). One then presents a second stimulus that does not reliably elicit the UCR (the potential conditioned stimulus or CS) at about the same time as the UCS for a number of trials. As a result of the repeated pairing of the CS and UCS, a conditioned response (CR) comes to be elicited by the CS. The CR is similar but not necessarily identical to the UCR. Establishment of a CR is most effective if the CS precedes the UCS by a brief interval on each trial. The classical example of classical conditioning is that of the salivary reflex of dogs (e.g., Pavlov, 1927). At the beginning of the experiment, presentation of meat powder (the UCS) to the dog elicits salivation (the UCR), whereas presentation of a tone (the potential CS) does not. After repeated pairings in which the tone occurs just before the meat powder, the tone comes to elicit salivation (a CR). If, after establishing a CR, one repeatedly presents the CS without the UCS, the CS will lose its tendency to elicit the CR, a phenomenon known as *extinction.*

It should be noted that the conditioned and unconditioned responses represent only a small fraction of the total pattern of behavior of an animal in a conditioning experiment. A dog in a Pavlovian experiment may look toward the tone or food pan and display head movements, bodily orientation movements, chewing, and licking (Zener, 1937).

Considerable controversy has surrounded the problem of the proper control procedures for the demonstration of classical conditioning. The primary problem of control is that of separating instances of true classical conditioning from pseudoconditioning and sensitization. *Pseudoconditioning* is the strengthening of a response to a previously ineffective stimulus as the result of repeated elicitation of the response by another stimulus with which it is not paired. Thus, if pseudoconditioning occurred, the tone in the Pavlovian experiment would come to elicit salivation after the salivation reflex had been repeatedly elicited by meat powder, even though the tone and meat powder were never paired. *Sensitization* refers to an increase in the strength of a response originally evoked by the CS as a result of its pairing with the UCS and UR. It differs from classical conditioning in that the strengthening is of a response appropriate to the CS rather than the UCS. In the salivary-conditioning situation, sensitization would be demonstrated if some response appropriate to the tone were strengthened as the result of the experimental procedure. While pseudoconditioning and sensitization are often treated as nuisances only to be controlled for, Dyal and Corning (1973) have emphasized that they are interesting and important phenomena, worthy of study in their own right.

As controls in studies of classical conditioning, experimenters often use presentations of (1) the CS alone, (2) the UCS alone, (3) the UCS and CS explicitly unpaired, (4) the CS and UCS presented simultaneously (*simultaneous conditioning*), or (5) the CS and UCS paired, but with the UCS first (*backward conditioning*). Alternatively, a *discrimination conditioning* procedure may be used. Two stimuli are presented repeatedly, only one of them being paired with the UCS. Typically, only the stimulus paired with the UCS comes to elicit the CR. Rescorla (1967) argued that none of these procedures provided adequate controls and advocated a *truly random group* in which there was no predictability of the occurrence of the two stimuli (e.g., the UCS might occasionally follow the CS). Seligman (1969) found problems even with the truly random procedure.

An example of classical conditioning in the comparative literature can be found in studies of flatworms (planarians) (e.g., Jacobson, Horowitz, & Fried, 1967). The CS in the studies reported by Jacobson et al. was either a vibration or a light. The UCS was shock. The UCRs and CRs were turns or contractions made to the UCS or CS. In this procedure, the CS (light) is turned on for 3 seconds. The UCS (shock) is turned on so as to overlap with the last 1 second of the CS presentation. Turns or contractions occurring in the first 2 seconds of CS presentation (i.e., before the UCS comes on) are counted as CRs. Jacobson et al. demonstrated significant classical conditioning in relation to backward conditioning, simultaneous conditioning, and CS-only control groups. Further, they successfully used a differential conditioning procedure wherein half their subjects came to show the CR to vibration and not light when shock was paired with vibration, while the other half of their subjects experienced shock following light but not vibration and came to show the CR only in response to light. After this was accomplished, both groups were successfully reversed by reversing the contingencies (i.e., worms formerly responding to light were conditioned to respond to vibration and vice versa).

Instrumental Conditioning

Studies of instrumental conditioning differ from those of classical conditioning in that in studies of instrumental conditioning the sequence of events is contingent (i.e., dependent) upon the behavior of the organism. Typically, some kind of reward or punishment is utilized. A bewildering array of different procedures, all falling within the general realm of instrumental conditioning, has been used in different experiments. An interesting attempt at formulating a taxonomy of instrumental conditioning procedures has been presented by Dyal and Corning (1973). We shall consider some of the paradigms that have been of particular importance in comparative studies.

Appetitive Instrumental Learning In the appetitive paradigm for studying instrumental learning, some kind of reward, often termed a "positive reinforcement," is given whenever an animal makes a response designated by the experimenter as correct. Two major classes of procedures may be distinguished—the discrete trials and free-operant procedures.

Discrete-trials procedures In the discrete-trials procedure, the experimenter controls the pacing of the experiment by giving the animal a set of clearly defined test trials. A common procedure is to have an animal traverse a straight runway in order to obtain a reward. For example, Flood and Overmier (1971) trained goldfish to swim down an alley from a start chamber to a goal chamber using food as the reinforcement. They found that the time taken to leave the start chamber and to traverse the alley decreased progressively over trials. Ablation of the entire telencephalon of the fish produced no significant difference in the speed of leaving the start chamber.

Free-operant procedure In the free-operant procedure, the animal is not given discrete trials but rather is free to respond at any time. Typically, a device (called a manipulandum) is placed in the apparatus. Operation of the manipulandum is followed by the delivery of reinforcement. In the best-known example, a laboratory rat depresses a lever and obtains food reinforcement (Skinner, 1938; Rachlin, 1976). As a result of the contingency between the response (i.e., lever pressing) and the reinforcement (i.e., food), the probability of emission of the response is increased.

In some situations, the animal displays the operant response at a low rate at the beginning of the experiment; reinforcement serves to increase the probability or rate of lever pressing from this low *operant rate.* In other situations, the operant response does not occur. For such situations, Skinner pioneered use of a shaping procedure, wherein the animal is progressively reinforced for *successive approximations*—i.e., responses coming closer and closer—to the desired response. Thus, the criterion for reinforcement is progressively upgraded as the rat must move near the bar, orient toward it, lift its paws, and so on until the complete response occurs.

Reinforcement need not be delivered for every response. Sometimes reinforcement may be intermittent. Thus, for example, one may reward only every *n*th response (a ratio schedule) or only the first response after a given time interval (an interval schedule). Psychologists studying operant conditioning have used a wide variety of *schedules of reinforcement* and found them to be important determinants of behavior in these situations (Ferster & Skinner, 1957). If reinforcement is discontinued altogether, a decrease in the probability and rate of response (*extinction*) is observed.

Operant conditioning experiments have been conducted with a considerable variety of different species. The nature of the response and reinforcement are adjusted to suit the species under study. For example, Powell (1973) shaped common crows to peck a key by using live mealworms or dog food as reinforcers. Shumake and Caudill (1974) conditioned vampire bats to lick a tube with cattle blood as the reinforcer. Male Siamese fighting fish have been conditioned to emit an operant response with presentation of a model of another fish in its aggressive display as a reinforcer (e.g., Thompson, 1966; Johnson & Johnson, 1973).

An interesting variation on the operant procedure was used by Pryor and her associates (Pryor, 1969; Pryor, Haag, & O'Reilly, 1969) in studies of rough-toothed porpoises. Operant techniques were used, but a different response was rewarded in each session. After the porpoise had been reinforced for a particular behavioral pattern in a given session, that response would never again be effective in securing reinforcement in later sessions. After exhausting their usual repertory of responses, Pryor's porpoises exhibited a wide range of novel and "creative" responses.

Active Avoidance Learning In the active avoidance paradigm, the animal must make an appropriate response if it is not to be exposed to an aversive stimulus. The animal thus comes to learn to respond in the manner designated by the experimenter in order to *avoid* a noxious stimulus.

In the most common procedure in avoidance learning studies, a *shuttle box* is used (see Figure 16-1). Typically, a shuttle box is made up of two chambers, one painted black and the other white. The two chambers are connected by a door, which may be either a simple archway or a guillotine-type door that can

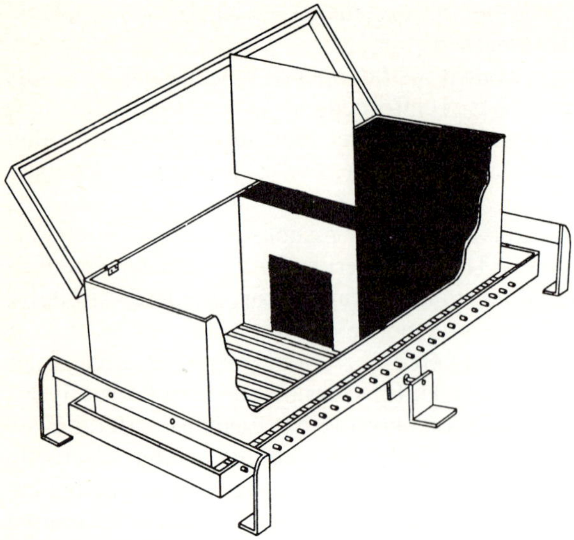

Figure 16-1 A shuttle box. (*From Campbell, 1967.*)

be opened and closed. In the discrete-trials procedure—the most common method—a trial is started with the onset of a signal, such as a buzzer, and the opening of the door. The animal then has several seconds in which to leave the start chamber and move to the other chamber of the apparatus. If the animal fails to leave the chamber within the allotted period, it receives electric shock (delivered through metal electrodes). With practice, there is typically an increase in the number of avoidance responses (i.e., departures from the start chamber before the onset of the shock) per block of trials. In some experiments, the animal is moved back to the original box between trials, so that it must always move in the same direction in order to make an avoidance response. This is termed "one-way avoidance learning." In other experiments, the animal starts from whatever chamber it was in at the end of the previous trial and must move to the opposite chamber. This procedure is termed "two-way avoidance learning." Because the animal must learn to enter a chamber in which it may recently have been shocked, two-way avoidance learning often requires more trials for learning than the one-way method.

Boice (1970a) studied active avoidance learning in four species of frogs and toads. A one-way, discrete-trials procedure was used, so that the frogs and toads always had to go from the same (shock) side to the other (safe) side. The animals had 4 seconds from the opening of the door in which to make an avoidance response. A species of toad that is noted as an active hunter showed good active avoidance learning in this situation. Other species of frogs and toads, that are typically more passive, showed little or no learning.

Active avoidance learning can also be studied using a free-response paradigm. This procedure is called "Sidman avoidance" after Sidman (1953). In a Sidman avoidance experiment, a response simply postpones shock for a predetermined interval of time (e.g., 30 seconds). Thus, if an animal never permits a 30-second interval to pass without making a response, it will never receive a shock. If it fails to respond within an interval, shock will be delivered. There is no signal and there are no discrete trials. Behrend and Bitterman (1963) used the Sidman avoidance paradigm with goldfish in a shuttle box. The fish had to change compartments every 20 seconds in order to avoid shock. Appreciable learning was demonstrated.

Escape Learning Some experiments are conducted in such a manner as to permit only *escape learning.* The task of the animal is to escape from a noxious stimulus or situation in a short time; it has no opportunity to avoid the noxious stimulus. A shuttle box could be used in such an experiment by turning on shock at the same time that the door is opened. In some studies, rats must learn to escape from a tub of water by way of a ramp (e.g., Festing, 1974).

Passive Avoidance Learning In studying passive avoidance learning, one begins with an animal that already performs a response reliably. It may be either a species-characteristic response or one that was trained by the experimenter. The experimenter then delivers a noxious stimulus every time the response occurs. Passive avoidance learning is observed when the animal inhibits the response and no longer displays the response that was previously a frequent occurrence. This procedure is essentially a punishment paradigm.

Passive avoidance learning was demonstrated in land snails by Stephens and McGaugh (1972) and Siegel and Jarvik (1974). Snails display a negative geotaxis (tendency to climb in a direction against gravity). In these studies, the snails were allowed to climb up a pole. When the snails encountered loops of thread containing quinine or associated with shock at the top of the pole, they displayed increased latencies to climb and avoidance of the pole. They thus displayed passive avoidance in response to the aversive situation.

Brower and Brower (1962) used a passive avoidance paradigm in studying mimicry. Toads were found to eat mealworms and drone flies in the experimental situation. Drone flies are harmless mimics of honeybees. After several trials in which the toads ate mealworms or drone flies, they were presented with and stung by honeybees. After a few such experiences, toads typically continued to eat mealworms but avoided both honeybees and the mimic drone flies. The toads thus displayed passive avoidance of the responses of eating honeybees and drone flies.

Discrimination Learning In the study of discrimination learning, the animal is presented with two or more stimuli and the opportunity to make a response. Responses occurring in the presence of one of the several stimuli are reinforced, while those occurring in the presence of the other stimuli are not. The stimuli do not elicit a response but rather set the occasion for the response. Learning is manifested as an increase in the percent of responses to the "correct" stimulus.

In simultaneous discrimination learning, the two (or more) stimuli are presented at the same time. The task of the animal is to choose and respond to the correct stimulus. In successive discrimination learning, the various stimuli are presented one at a time. The animal must learn to respond to the correct stimulus but inhibit responding in the presence of the incorrect stimuli (a "go–no go" problem).

A frequently used apparatus for the study of simultaneous discrimination learning in primates is the Wisconsin General Test Apparatus (WGTA; Figure 16-2). The apparatus permits the experimenter to make controlled presentations of pairs of stimuli to the monkey. The animal must learn to select the correct stimulus in order to obtain food, which is typically hidden in a food well located in the stimulus tray beneath the objects. The location of the two stimuli and that of the food may be changed from trial to trial. Either the specific stimulus or the position of the stimulus (left or right) can be indicative of the correct response.

One of several alternative procedures may be used when an animal makes an incorrect response. With the *noncorrection procedure*, a trial ends with the occurrence of the incorrect response. With the *correction* procedure, an incorrect choice is followed by another trial with a stimulus configuration identical to that of the previous trial (i.e., the position of the stimuli remains unchanged). With the *guidance* procedure, the animal is forced to make the correct choice after it makes an incorrect one, as it is given no alternative.

Discrimination learning studies can also be conducted in the free-operant situation. Simultaneous discrimination experiments can be performed by

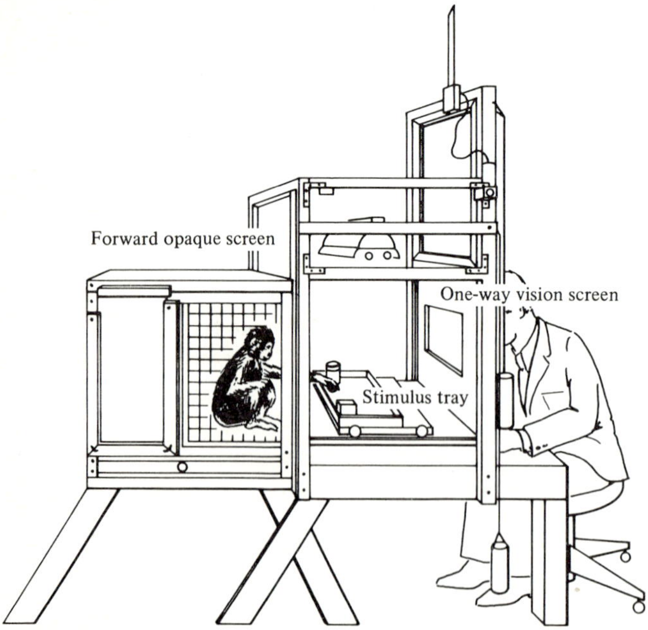

Figure 16-2 Wisconsin General Test Apparatus (WGTA). (*From Harlow, 1949.*)

introducing two manipulanda and permitting the animal to choose between them. In simple successive discrimination studies, the animal learns to make the operant response (e.g., a lever press) in the presence of one stimulus (e.g., a light) but not another (e.g., a tone).

An example of a discrimination learning study is that of Warren (1959), who used the WGTA with rhesus monkeys. Experimentally sophisticated rhesus monkeys, which had learned many previous problems, were found to learn discriminations quite rapidly. The fewest errors were made when both the position of the object (left or right) and its cues (appearance) were relevant to the discrimination. The presence of relevant object cues facilitated positional discrimination learning, whereas irrelevant object cues interfered with such learning.

Many of the more complex paradigms to be discussed are simply variants on the basic discrimination learning paradigm.

Reversal learning A reversal-learning paradigm is simply a series of discrimination learning problems in which the identity of the correct stimulus is changed from problem to problem. Suppose that an animal has to choose between a red and a green stimulus, with the red being correct. The discrimination problem is continued until the animal either reaches a given criterion for learning (e.g., it is correct in 9 out of 10 consecutive trials) or receives a predetermined number of trials. At this point, the experimenter reverses the correct stimulus (with no indication of the reversal to the animal). The animal must now learn to go to the green stimulus rather than the red one. On reaching the criterion on that problem, the identity of the correct stimulus is again reversed, so that the animal must once again select the red stimulus. Many species show a tendency to improve in problem-solving ability in successive problems as they require fewer and fewer trials to reach the criterion.

Rumbaugh and Steinmetz (1971) studied reversal learning in lowland gorillas. The animals were tested in a modified WGTA and were rewarded with grapes for selecting correctly between a red square and a red circle. Although there were considerable differences among different individuals, the animals generally improved with successive reversals.

Interproblem learning sets Like reversal learning problems, interproblem learning-set tasks consist of a series of discrimination learning problems. However, whereas the reversal-learning paradigm uses the same pair of stimuli throughout the experiment, in the interproblem learning-set paradigm a different pair of stimuli is used in each discrimination problem. Thus, if an animal has to choose between a red square and a red circle on the first problem, it may have to choose between a green triangle and a blue cross on the second problem. Nevertheless, performance improves on successive problems as the animal "learns to learn" or demonstrates "learning sets" (Harlow, 1949; Miles, 1965).

Kamil and Hunter (1970) studied object-discrimination learning-set performance in Greater Hill myna birds. Four mynas each received 1,000 discrimination problems in a modified WGTA. One hundred multidimensional stimulus objects were used. Considerable improvement over successive problems was observed.

Delayed response The delayed-response paradigm, like most learning procedures, can be run with many variations. The simplest form of the delayed-response problem is just a variation on the old "shell game." A piece of food is placed under one of two objects in full view of the animal. The animal need only select the object under which it has seen the experimenter place the food. The trick is that delays of varying lengths are imposed between the time the food is covered and the time at which a response can be made. Many factors—such as the animal's orientation during the delay period, the physical arrangement of the apparatus, and the distribution of trials—can affect an animal's performance in this situation (Fletcher, 1965).

In a classical study of delayed-response learning, Harlow, Uehling, and Maslow (1932) compared 24 primates of 10 different species. Both individual and species differences were evident.

Probability learning In all the problems we have discussed thus far, the patterning of reinforcement has been consistent within a particular problem (i.e., it is always the left stimulus, the red circle, etc. that is reinforced). Thus, the pattern of reinforcement may be described as 100:0. One alternative is rewarded 100 percent of the time and the other 0 percent. In probability learning experiments, different ratios of reinforcement are associated with the two stimuli in such a manner that the animal has no way to determine which of the stimuli is correct on a given trial. For example, one may reward only responses to the red stimulus on 70 percent of the trials and only responses to the green stimulus on 30 percent of the trials (a 70:30 problem). While it is clear that red will be rewarded 70 percent of the time in the long run, there is no way to predict which will be correct on a given trial. A maximal number of reinforcements will be obtained if the animal always selects the more frequently reinforced side. Thus, in the hypothetical problem above, it will receive reward on 70 percent of the trials if it always selects red. This strategy is termed "maximizing," because the animal maximizes the number of reinforcements it receives. Any other distribution of responses will, in the long run, produce a lesser percentage of reinforcement. Some animals display *probability matching* (see Figure 17-3). In a matching strategy, the animal distributes its responses between the two stimuli in the same proportion as the reinforcements are distributed (e.g., 70 percent red and 30 percent green). In the long run, an animal adopting such a strategy will receive reinforcement on just 58 percent of the trials ($0.7 \times 0.7 + 0.3 \times 0.3$).

Woodard and Bitterman (1973) studied probability learning in goldfish with 100:0, 70:30, and 50:50 ratios of reinforcement. In visual probability learning problems (red versus green), probability matching was observed. However, in spatial probability problems (right versus left rewarded), the fish tended more toward maximization.

Maze learning Maze learning is a form of discrimination learning that has been studied by psychologists since the turn of the century. The simplest mazes are shaped in the form of a "T" or a "Y" and are called, appropriately enough, T-mazes or Y-mazes. The animal must run down the stem of the maze until it reaches the choice point. It is then differentially rewarded, depending on whether it goes right or left or responds appropriately to any of a variety of cues that can be provided. *Complex mazes* are made up of many choice points; the animal must find its way from the start to the goal. The elapsed time and number of errors in the complex maze are typical dependent variables. Schneirla (1946) studied learning as displayed by ants in a variety of mazes.

Other Complex Problems Numerous other complex learning tasks have been used by animal behaviorists, primarily with primates. In the *oddity problem*, the animal is typically presented with three stimuli, two of which are identical. It must learn to choose the odd stimulus in order to gain reward. As with all complex learning procedures, there are many variants on this procedure. Administering a series of oddity problems in which different sets of stimuli are used results in an *oddity learning set*.

In a *matching task*, the animal must select from among several stimuli the one that matches a sample stimulus. In *alternation learning*, an animal must learn to alternate responses on successive trials (e.g., left, right, left, right, etc.). A more complex variant is the *double-alternation* problem, wherein the animal must alternate in couplets (e.g., left, left, right, right, left, left, etc.). The procedures of alternation learning and delayed-response learning can be combined to make a *delayed-alternation* task. Masterton and Skeen (1972) studied hedgehogs, tree shrews, and bush babies in such a task by requiring the animals to alternate responses and imposing delays of varying lengths between successive responses.

DEVELOPMENT OF LEARNING

Having described the basic learning paradigms, we can now consider their use in studies of the development of learning. In the study of development, we assess the role of the genes and environment in the ontogeny of behavior. In the case of the development of learning, we can consider genetic factors, effects of domestication, effects of early experience, and ontogenetic changes in learning ability and learning processes.

Behavior-Genetic Analyses of Learning

Different species differ markedly in their performance in learning tasks. Although many of these differences are of genetic origin, they are not traditionally included in the area of behavioral genetics and will be discussed in Chapter 17. Most behavioral-genetic analyses of learning have concentrated on within-species genetic comparisons.

The classical example of the use of the selection method in the study of learning is that of Tryon (1940), who studied learning in a 17-unit, complex, multiple T-maze. "Maze-bright" and "maze-dull" lines of rats were selectively bred from base stock. At the end of 18 generations of selection, the two lines diverged markedly. Subsequent analyses (e.g., Searle, 1949) revealed that while the two strains did differ with respect to the number of errors made in Tryon's maze, they differed in many other ways as well. The maze-bright rats learned better than maze-dull rats in hunger-motivated situations but more poorly in a task in which they had to learn to escape from water. The two strains also appeared to differ in "emotionality." Thus, as in any selection study, the sources of the observed differences in phenotype are complex. While Tryon's rat strains differ, they appear not to differ in intelligence per se. Other selection studies of learning have been conducted on complex-maze learning (Tolman, 1924; Heron, 1941; Thompson, 1954), avoidance learning (Bignami, 1965), and water-escape learning (Festing, 1973b).

Different strains of rats and mice have been found to differ markedly in their learning performance (e.g., Wahlsten, 1972; Caul and Barrell, 1973; Festing, 1973a), and the standard behavior-genetic methods have been used in a study of reversal learning in two inbred strains of house mice (Carran, 1975).

The diallel-cross method has been especially popular in studies of the determinants and implications of strain differences in learning. Wahlsten (1972) reviewed diallel-cross studies of learning and observed that directional dominance or heterotic inheritance generally was found in studies using inbred strains, although not in studies in which noninbred parents were used. A pattern of inheritance with substantial directional dominance has been reported for avoidance conditioning (e.g., Collins, 1964; Messeri, Oliverio, & Bovet, 1972; Holmes, Aksel, & Royce, 1974), T-maze learning with shock motivation (Stasik, 1970), acquisition of conditioned fear (Henderson, 1968b), acquisition of conditioned suppression (Connor & Winston, 1972), and water-escape learning (Festing, 1974; Hyde, 1974). Given the theory presented in Chapters 7 and 14, such data would suggest that these forms of learning are adaptive in the natural habitat. Reversal learning in rats has been found to display a pattern of intermediate inheritance (Carran, 1972).

Wahlsten (1972) reviewed estimates of the heritability of learning from eight studies with mice, rats, and swine. Values generally were intermediate, ranging from 0.21 to 0.56.

Various other methods have been used in behavior-genetic analyses of learning. Tyler and McClearn (1970) studied parent-offspring correlations among mice in a simple runway learning situation. Oliverio and Messeri (1973) studied the effects of 31 different single-gene mutations on active avoidance and maze learning in mice. Winston, Lindzey, and Connor (1967) found an effect of the albino gene on avoidance learning.

The importance of strain differences in interacting with experimental treatments was stressed in Chapter 7. Manipulations of intertrial interval produce very different effects on avoidance learning in different inbred strains of mice (Bovet et al., 1969; Figure 7-4). Strain differences may have been the cause of many of the disagreements and failures to replicate experiments in the study of animal learning.

Effects of Domestication

Counter to the dictates of traditional beliefs in animal behavior, domestication appears to have no detrimental effect on the performance of rats in learning situations (Boice, 1973). Indeed, laboratory rats often appear to perform better than wild Norway rats in such situations. Such results have been found in studies of two-way shuttle box active avoidance learning (Boice, 1972), discrimination-reversal learning in a free-operant situation (Millar, 1975), and complex-maze learning (Bamber & Boice, 1972). These differences do not necessarily mean that laboratory rats are more "intelligent" than wild rats. More likely, the difference in performance reflects differences in activity levels, reactions to the test situation, and the like.

Effects of Early Experience

Studies of the effects of early experience on performance in learning situations have produced mixed results, with effects of early experience observed in some studies but not in others (see Daly, 1973). The most consistent effects of early experience in rats have been observed in studies of avoidance learning. Both Levine (1956) and Levine and Wetzel (1963) found early handling of rats to result in better avoidance learning when the rats matured. Even with avoidance learning, however, some negative results have been reported (see Daly, 1973). Whatever the result, it appears that effects of early handling on performance in avoidance learning tasks result more from differences in "emotionality" than in learning ability per se (see Denenberg, 1964). Generally, manipulation of neonatal rats has failed to alter their adult performance levels in learning situations based on appetitive rather than aversive motivation. Thus, early handling appears not to affect performance of rats in the Hebb-Williams test, a series of maze-learning problems designed to assess the "intelligence" of rats (Denenberg & Morton, 1962). Environmental manipulations occurring at slightly later developmental stages, at or after weaning, appear to have more pervasive effects on performance in learning tasks (e.g., Forgays & Read, 1962).

Thompson and Heron (1954) found that dogs reared under conditions of environmental impoverishment (i.e., restricted perceptual experience) were inferior on a battery of tests—including orientation and barrier problem solving, delayed-response learning, and an open-field maze test—to dogs reared as pets. Fuller (1967) found that dogs reared as pets made fewer errors in a spatial-reversal learning task than did isolates. However, other differences in performance attributable to environmental impoverishment were marginal.

Effects of variations in the early experience of rhesus monkeys have produced minimal effects on performance in learning situations (Harlow, Harlow, & Suomi, 1971). Harlow, on comparing the performance of rhesus monkeys reared in his nuclear family apparatus with those reared in total isolation, found no differences on a variety of tasks including discrimination learning, delayed-response learning, and complex learning-set performance. The socially enriched animals did perform better than isolates on an oddity learning-set problem. Consistent with the general lack of effects of early experience, Singh (1969: Harlow, Harlow, & Suomi, 1971) found no difference between feral rhesus monkeys in India reared in cities versus forests and Harlow's deprived monkeys when the animals were tested on a battery including discrimination learning, delayed-response tests, learning sets, and oddity learning sets.

Some substantial effects of early experience have been reported in chimpanzees. Davenport, Rogers, and Menzel (1969) found that chimpanzees reared in restricted laboratory environments were inferior to wild-born animals when tested in a discrimination-reversal task. Davenport and Rogers (1968) found similar effects on delayed-response learning. In general, animals reared under conditions of environmental restriction appeared to experience difficulty in inhibiting old responses and were easily distracted by irrelevant stimuli.

Ontogeny of Learning and Memory

The performance of animals in demonstrating learning and retention has been repeatedly shown to vary as a function of age. However, the extent of variation with age varies greatly across different species and tasks.

Sensitive Periods In some species, there exist periods—called "sensitive periods"—in which the organism is particularly susceptible to effects of environmental input. Thus, the developing organism may show good learning and retention if specific input occurs at the appropriate developmental stage but little or no such learning and retention if the input occurs too early or too late. Some examples have already been discussed. The classical example is that of imprinting (Hess, 1959; also Chapter 8). If the young chick or duckling is exposed to a moving stimulus too early or too late, imprinting is much poorer than that occurring at the peak of the sensitive period.

A second clear example of a sensitive period can be found in the ontogeny of birdsong (Marler, 1970; also Chapter 9). The developing white-crowned sparrow must hear the male species-characteristic song during the period when it is from 10 to about 50 days of age if it is to develop the ability to sing a normal song.

Learning in Young Animals The learning abilities of young animals have been well established. However, not all abilities appear to mature at the same rate. Whereas chicks are able to learn an active avoidance response on the day of hatching, the ability to learn a passive avoidance response does not appear until about the third day (Peters & Isaacson, 1963). Delayed appearance of passive avoidance in chicks was found in a different situation by Fischer and Campbell (1964).

Neonatal dogs have been found capable of learning a variety of tasks (Stanley, 1970). Stanley, Bacon, and Fehr (1970) found that young beagles were able to master discrimination learning tasks within the first week after birth. Stanley et al. used a simultaneous discrimination procedure in which the puppies had to discriminate between approaching cloth surfaces and approaching wire surfaces in order to obtain milk reinforcement. Bacon and Stanley (1970) found 2-week-old beagles able to show progressive improvement in reversal learning in the same apparatus. Dogs less than 2 weeks of age were also found to respond appropriately in classical conditioning experiments and in instrumental escape and avoidance learning (Stanley, 1970).

Misanin, Nagy, Keiser, and Bowen (1971) studied the emergence of memory in neonatal rats required to learn an escape response with shock motivation. Retention improved over the first 9 days of life.

File and Scott (1974) studied habituation in rats to an air puff over the first 19 days of life. They found habituation to grow progressively slower from day 1 to day 8 but then to improve and reach a peak at about day 16.

Ontogenetic Changes in Learning and Memory Studies of the ontogeny of learning and retention in rats have been reviewed by Campbell (1967) and Campbell and Spear (1972). Campbell (1967) stressed the many difficulties in ensuring that learning situations designed for young animals are comparable to those for older animals. The difficulties of equating motivational levels and levels of reinforcement and of taking factors such as patterns of unlearned behavior and previous experience into consideration constitute important problems in the design of such experiments.

The general pattern of the more recent studies in rats has been that there are few developmental changes in the rate of acquisition in various tasks but considerable developmental changes in retention. Campbell and Campbell (1962) studied rats in a conditioned-fear task. The animals were shocked in one side of a shuttle box and were tested for the retention of learned fear, as manifested in avoidance of the side in which they had been shocked. Eighteen-day-old rats showed nearly 100 percent avoidance when tested immediately after training. However, their performance levels dropped to chance levels with 21- and 42-day retention intervals. In rats that were 54 or 100 days of age, there was little forgetting at any retention interval.

Results similar to those on the conditioned-fear problem have been obtained with other tasks. Kirby (1963) compared active avoidance learning in rats at 25, 50, and 100 days of age. The rats did not differ in either acquisition or extinction as a function of age. However, retention was poorer in the younger animals. The pattern of no difference in acquisition but an age difference in retention was found by N. Smith (1968) in a study of an escape response in a T-maze. Retention of a discrimination learned in conjunction with a lever-press task was found to be poorer in young rats than in older animals (Campbell, Jaynes, & Misanin, 1968).

The patterns of developmental changes in learning and retention in other species differ from those of rats. In guinea pigs, for example, which are born at a more advanced developmental stage than rats, no developmental changes were found in acquisition or retention of either an active or passive avoidance learning task (Campbell, Misanin, White, & Lytle, 1974).

Developmental patterns in rhesus monkeys vary greatly for different learning problems (Zimmerman & Torrey, 1965). Classical conditioning has been demonstrated in very young monkeys; a decision as to whether or not the rate of classical conditioning changes with age depends on the index one selects for consideration. The ability to learn single-problem discrimination tasks appears very early in the life of the young monkey and changes little as a function of age (Zimmerman & Torrey, 1965).

Considerable age differences are apparent when monkeys are required to learn more complicated tasks. Learning-set formation and performance in delayed-response tasks, for example, show considerable improvement as a function of age (Harlow, Harlow, Rueping, & Mason, 1960). According to Harlow, Harlow, and Suomi (1971), efficient solution of certain simple discrimination learning problems occurs in rhesus monkeys at about 6 months of age, of delayed-response tasks at about 10 months of age, and of oddity learning-set tasks at 36 months of age.

Reinstatement There is a major paradox in the literature on retention in young animals. It has been repeatedly demonstrated that the early experiences of nonhumans and humans alike are of pervasive importance to the life of the organism (see Chapter 8). However, the retention of experiences learned early in life appears quite poor in comparison to that of experiences occurring in adulthood. How can one resolve the paradox of poor retention of experience occurring at a time when experience has pervasive influence? One consideration is the fact that normal development involves the acquisition of many sequentially dependent responses. Thus, if an animal fails to learn a particular segment, there may be interference with segments which are to be learned at later developmental stages and are dependent on the earlier segments (Campbell, 1967).

A second factor in resolving the paradox of poor retention and pervasive effects of experience in the young animal has been termed "reinstatement" (Campbell & Jaynes, 1966). "By reinstatement we denote a small amount of partial practice or repetition of an experience over the developmental period which is enough to maintain an early learned response at a high level, but is not enough to produce any effect in animals which have not had the early experience" (Campbell & Jaynes, 1966, p. 478). As applied to humans, reinstatement would occur when a parent repeatedly reminds a growing youngster of some traumatic event that occurred earlier in the child's life. Although such reminders may be insufficient by themselves to produce effects of trauma, they potentiate effects of the earlier experience.

In order to illustrate the operation of reinstatement, Campbell and Jaynes studied the retention of conditioned fear using the same paradigm as that of the developmental study of Campbell and Campbell (1962) discussed above. Rats that had been shocked when 25 days of age showed very poor retention of learned fear when tested at 53 days of age. However, if the developing rats received just three shocks—one each at 7, 14, and 21 days after initial learning—retention was greatly improved. Control animals that received only the reinstatement shocks displayed little evidence of conditioned fear. It thus appears that, through the process of reinstatement, experiences occurring after a learning experience can function to improve retention of that learning.

MECHANISMS AND LEARNING

A detailed consideration of the mechanisms subserving learning would take us far afield of the traditional material included in courses in comparative animal behavior. Such material is generally covered in courses in learning or physiological psychology and is not appropriate for detailed coverage here. Therefore, we shall present a very brief review of the kinds of research that can be done in this area and refer the reader to a course in learning or physiological psychology or to the references cited herein (e.g., Kupfermann, 1975; Rosenzweig & Leiman, 1968; Thompson, Patterson, & Teyler, 1972; Thompson, 1976).

"Black-Box" Approaches

While studies aimed at determining the mechanisms subserving learning often utilize physiological or anatomical intervention, such procedures do not represent the only possible approach. An alternative strategy is to treat the organism as a "black box" and to search for orderly relationships between input to the system (i.e., stimuli) and output from the system (i.e., responses). Thus, Skinner (1950, 1953) contends that the essential task confronting the student of learning is not that of constructing a theory or physiological model of learning but rather of defining orderly stimulus-response relationships that will make it possible to predict and control behavior. Skinner contends that the analysis of processes occurring at levels other than that of the observable stimulus and response may direct the researcher away from systematic exploration of the nature of the variables that control behavior. Whatever one may think of this view, it is clear that it has represented a powerful strategy for the elucidation of an appreciable and important body of experimental data (e.g., Honig, 1966). The record of operant or Skinnerian psychologists in developing a systematic experimental analysis of behavior of considerable utility in applied situations has been impressive.

An alternative black-box strategy stems from mathematical theory and computer science and is concerned with the development of mathematical and computer models that simulate the processes mediating learning and problem-solving behavior (e.g., Hunt, 1968; Hilgard & Bower, 1966; Greeno & Bjork, 1973). As with the operant strategy, emphasis is placed upon an effort to delineate orderly relationships between input and output, with minimal concern for the actual nature of the intervening processes.

The Search for the Engram

A very large number of neuroscientists from a variety of traditional disciplines is actively engaged in a search for neural, anatomical, and chemical correlates of learning and memory. As behavior is controlled in part by the central nervous system and is modified as the result of learning, it is assumed that there must exist some physical manifestation of learning within the central nervous system. Such a hypothetical entity is termed a "memory trace" or "engram." The difficulties of research in this area were graphically portrayed by the distinguished neuropsychologist Karl Lashley in summarizing a career devoted largely to the search for the engram. Lashley wrote, "I sometimes feel, in reviewing the evidence on the localization of the memory trace, that the necessary conclusion is that learning just is not possible" (Lashley, 1950, pp. 477–478).

Clearly, as Lashley was well aware, learning is possible, and some progress has been made in the search for the engram using a variety of approaches. The fundamental methods used in the physiological search for the engram are identical to those used in the physiological studies of other behavioral patterns and discussed in Chapter 11 (i.e., lesion or ablation, electrical recording, electrical stimulation, and chemical techniques). These techniques have been applied in a number of ways.

Tracing Pathways Much effort has been devoted to attempts to trace the pathways of processes mediating learning. For example, the locus of habituation in a variety of organisms appears to lie in a depression of activity across the synapse between two adjacent neurons. The fundamental mechanism appears to be a decrease in the release of transmitter substance as the result of repeated stimulation (Groves & Thompson, 1970; Thompson, 1976). It appears as though habituation and dishabituation may be mediated by two very different processes.

Thompson and his associates (Thompson, 1976) have attempted to trace the pathway for the mechanism underlying classical conditioning of the nictitating membrane of the eyes of rabbits. The response is controlled by the sixth cranial nerve; modifications of motoneurons in that nerve and in the hippocampus have been found to be associated with classical conditioning. Olds and his associates (e.g., Olds, Disterhoft, Segal, Kornblith, & Hirsh, 1972) attempted to trace the pathway of short-latency neural responses throughout the rat brain in conditioning situations. Woody and his associates (e.g., Woody & Black-Cleworth, 1973) have attempted to analyze the cellular mechanisms associated with classical conditioning in cats.

The basic strategy of studies in this area is to probe the various regions of the brain, often using electrophysiological techniques, in an attempt to find specific alterations consequent upon experience. The difficulties of controlling for various changes resulting from processes other than learning are considerable. "With few notable exceptions, studies of neurophysiological correlates of learning fall into two categories, those with no controls and those with inadequate controls" (Thompson et al., 1972, p. 73).

Chemical Approaches Consistent with the general research emphasis on the chemical control of behavior, many neuroscientists have used chemical techniques in the study of learning and memory.

Chemical changes following training If memory is stored by specific chemicals, it ought to be possible to detect chemical changes in the brain consequent upon learning experiences. Such changes have often been reported for a variety of chemicals including RNA (e.g., Hydén & Lange, 1965), acetylcholine and related cholinergic chemicals (e.g., Deutsch, 1969, 1971), and brain peptides (e.g., Ungar, Desiderio, & Parr, 1972). Interpretation of such changes is difficult and no definitive chemical basis for the engram has yet been established.

Trace consolidation According to the theory of trace consolidation, the memory trace is labile and susceptible to destruction for a short period after learning. After this period, the trace "consolidates" into a permanent physical alteration of the brain that is resistant to many experimental manipulations. Numerous studies have been directed at attempts to interfere with the consolidation of memory traces using electroconvulsive shock and various chemicals injected into the organism (e.g., McGaugh, 1966; McGaugh & Herz, 1972; Davis & Agranoff, 1966).

Memory transfer The most controversial studies in the area of substrates of learning and memory have centered on attempts to transfer memory from one organism to another. Typically, there are four groups of organisms, one trained and three untrained. Chemicals are extracted from the brains of the trained group and one untrained group and injected into the other untrained groups. Recipients of extracts from the brains of organisms that experienced training have often been reported to perform at levels superior to those achieved by recipients of extracts from untrained animals. The inference is that memory has been coded in specific chemicals and transferred by injection from one animal to another (see review by Smith, 1974). The basic phenomenon of memory transfer has been observed so frequently and in so many different laboratories that it cannot easily be written off as completely artifactual. However, many problems have been encountered in replicating results and in determining the specificity of these effects. The nature of the chemicals underlying these phenomena has yet to be definitively established.

Hormones As discussed in Chapter 12, hormones are of pervasive influence in the control of a wide variety of behavioral patterns. Thus, it should come as no surprise that hormones have been found important in learning. Their most consistent and dramatic effects are exemplified by the pituitary-adrenal system's hormones (i.e., adrenaline, noradrenaline, ACTH, and glucocorticoids) on the acquisition of tasks based on fear motivation (see Levine, 1968; DiGiusto, Cairncross, & King, 1971; Bohus & deWied, 1976).

Simple Systems A popular strategy in the search for correlates of learning and memory has been the utilization of invertebrate preparations as "models" for the study of behavioral modification (e.g., Eisenstein, 1967; Horridge, 1965; Kupfermann, 1975). Various species of marine mollusks, insects, and octopuses have been popular subjects in these studies. Although the forms of behavioral modification that can be demonstrated in many of these preparations are somewhat limited, they possess the advantage of being mediated by a limited number of neurons that are relatively accessible to the scientist. The hope is that mechanisms governing the control of behavioral modification in relatively simple systems will provide significant insight into the nature of such systems in organisms with nervous systems of greater complexity. We shall consider learning in invertebrates in greater detail in Chapter 17.

SUMMARY

The topic of learning can be studied from the perspectives of description, development, mechanisms, evolution, and function.

While there are difficulties with most definitions of learning, it is convenient to define learning as a change in behavior that occurs as the result of practice.

Many forms of learning have been studied in comparative animal behavior, including habituation, classical conditioning, and instrumental learning (including appetitive instrumental learning, active-avoidance learning, escape learning, passive-avoidance learning, and discrimination learning). Variations in the study of discrimination learning include reversal learning, interproblem learning sets, delayed-response learning, probability learning, and other complex problems.

In the study of the development of learning, the topic of interest is the manner in which the genotype and environment interact to produce complex ontogenetic patterns of learning. Included in such an approach are studies of behavior-genetic bases of learning, effects of domestication, effects of early experience, and the ontogeny of learning and memory.

Studies of the mechanisms of learning can be divided into black-box approaches and those directed at understanding the physiological bases of learning and memory. Included in the latter category are attempts to trace neural pathways, to understand the biochemical bases of learning, and to use simple systems as analogs to learning processes in complicated nervous systems.

Chapter 17

Learning: Evolution and Adaptive Significance

Having described the basic phenomena of learning and considered questions of development and mechanism, we now turn to a consideration of the final two classes of questions to be asked in the study of learning—questions of evolution and adaptive significance.

EVOLUTION OF LEARNING ABILITY

Studies of the evolution of learning have been focused upon two general sets of questions. The first is that of the phylogenetic generality of learning ability. Research has been focused on the question of whether or not individuals of particular species, especially invertebrate species, are able to learn. The second general set of questions is concerned with the nature of differences in learning ability and in the processes underlying learning. Given that several diverse species are able to learn, one can ask whether they differ either in their learning ability or in the processes subserving learning. While some process-oriented research has been conducted on invertebrates, emphasis has been placed on studies of vertebrates.

In the study of the evolution of learning, virtually all comparative analyses have been conducted at the phyletic level. There have been few if any systematic studies of the comparative learning performance of a group of closely related species. Further, evolution has generally been treated as history. There has been little attempt to understand the process of the evolution of learning ability. At the same time, it is generally recognized that species are differentially adapted to different habitats and would thus be expected to show specific adaptations appropriate to them. Such adaptations will be considered in the section on adaptive significance.

Learning in Invertebrates

Research on learning in invertebrates has focused on the question of whether organisms of different phyla are able to display behavioral modifications of the sort that we would label as "learned" and, to a lesser degree, on the nature of the processes underlying behavioral modification. Some research has been focused on the nature of physiological mechanisms of behavioral modification in "simple systems" (see Chapter 16).

Animal behaviorists' apparent relative neglect of research on invertebrate learning and behavior has been repeatedly noted (e.g., McConnell, 1966; Corning & Lahue, 1972). Despite this "neglect," the literature on invertebrate learning is so large that we can merely scratch its surface in a brief review. This considerable literature has been reviewed in an excellent three-volume set on learning in invertebrates (Corning, Dyal, & Willows, 1973a, 1973b, 1975), and in numerous other review papers (e.g., Thorpe, 1963; McConnell, 1966; Eisenstein, 1967). We shall briefly consider the evidence for the occurrence of learning in several phyla.

Protozoa While there is good evidence for habituation among protozoans, their capacity for associative learning is yet to be clearly established.

Habituation Habituation of responses to repeated presentation of stimuli has been convincingly demonstrated in several species of protozoans. Further, the resemblance to habituation in metazoans is indicated in the fact that virtually all the defining attributes of habituation proposed by Thompson and Spencer (1966) have been demonstrated in protozoans (Corning & Lahue, 1972; Wood, 1973). In addition, stimulus-specific habituation has been demonstrated, as, for example, in *Stentor coeruleus* (Wood, 1973). *Stentor* contracts in response to both mechanical and photic stimuli. Repeated presentation of either the mechanical or photic stimulus produces a decrement in the probability of contraction in response to the repeated stimulus. The response to the nonrepeated stimulus is unaffected, thus showing that the decrement in responding is not a function of fatigue or exhaustion of the response system.

Stimulus-specific habituation of the contraction response of *Vorticella convallaria* to electrical and mechanical stimuli was demonstrated by Patterson (1973). A popular subject for studies of protozoan habituation has been the cigar-shaped ciliary *Spirostomum ambiguum* (e.g., Applewhite, 1971; Hamilton, Thompson, & Eisenstein, 1974).

Research on habituation in acellular organisms has progressed well beyond the question of the existence of habituation into an exploration of the mechanisms underlying the phenomenon. Techniques employed have included the recording of receptor potentials and action potentials and manipulation of ionic concentrations in the surrounding medium (Applewhite & Gardner, 1971; Patterson, 1973; Wood, 1973).

Associative learning (general) Although much attention has been directed at the problem of associative learning in protozoans, there is as yet no convincing demonstration of the phenomenon. As the older work in this area is not definitive and has been reviewed elsewhere (e.g., Thorpe, 1963; Jensen, 1965; Corning & vonBurg, 1973), we shall concentrate on two more recent controversies.

Associative learning: The Gelber-Jensen controversy A behavioral modification resembling approach learning was demonstrated in the ciliate protozoan *Paramecium aurelia* by Gelber (1952). A bare wire was first introduced into a culture of paramecia. Few animals approached or clung to the wire. Next, the animals in the culture received a group of 40 "training" trials. Prior to every third trial, the wire was "baited" by dipping it into a suspension of bacteria, a food for the paramecia. As training progressed, there was an increase in the number of paramecia that approached and clung to the wire. At the end of training, introduction of a bare wire resulted in approach and clinging in cultures that had been trained but not in either naive (untrained) cultures or cultures that had received an equal number of trials with insertion of the wire but without the use of bacteria.

Jensen (1957a, 1957b, 1965) offered an explanation alternative to that based on learning for Gelber's results in an effort to explain the behavioral modification observed by Gelber. According to Jensen, Gelber's training procedure creates a "food-rich zone" in the center of the culture. Jensen argued that paramecia are attracted to food-rich zones and that when they are in such zones, one of their characteristic (unlearned) responses is a positive thigmotropism, or tendency to cling to things. Thus, Jensen concluded that the modification of behavior in Gelber's experiment was the result of a modification of the environment rather than of the organisms.

The debate regarding the correct interpretation of these experiments was intense (Jensen, 1957a, 1957b, 1965; Gelber, 1957, 1965; Katz & Deterline, 1958). It was shown that an increase in the tendency to approach a wire could be established merely by creating a food-rich zone with no training. But Gelber claimed that "many thousands of times as many bacteria" were used in these demonstrations as compared with her studies. Stirring of the culture after training was found to interfere with performance on the test trial. However, such an effect could be the result either of the breakup of the food-rich zone or of some other deleterious consequence of the mechanical stimulation created by stirring per se. Gelber (1957) claimed to have observed conditioning when the bacteria were presented at the side of the culture at the same time as the wire was introduced into its center. She (1965) claimed that flooding the culture with amounts of food sufficient to induce reproductive fission did not appreciably interfere with retention of the wire-clinging response.

After reading both the primary literature and summaries thereof (e.g.,

McConnell, 1966; Corning & vonBurg, 1973), the present author is unable to conclude that learning either did or did not occur in Gelber's situation. We must await better evidence before it can be concluded that protozoans are capable of associative learning.

Associative learning: Tube-escape behavior French (1940) reported an experiment in which he repeatedly sucked paramecia up into a glass tube. The response of the paramecia was to swim out of the tube back to the culture medium. Over the course of 30 trials, some of the animals displayed decreases in the amount of time required to escape from the tube. The phenomenon bears a considerable resemblance to that of escape learning in vertebrates. The tube-escape phenomenon was replicated and further studied by Hanzel and Rucker (1972) and Huber, Rucker, and McDiarmid (1974).

More recent research makes it appear that the decreases in escape response time do not reflect escape learning. Applewhite and Gardner (1973) were able to replicate the earlier results on tube-escape behavior when conditions similar to those in the earlier experiments were used. However, when the tubes were cleaned in a prescribed manner before use and the organisms were adapted to the process of being sucked up before experiments began, there was no change in escape time over trials. Thus, it is possible that some characteristic of the tubes, rather than of the animals, changed as a result of training.

Bennett and Francis (1972) found decreases in tube-escape latencies in *Stentor* when a vertical tube was used and the animals had to swim downward to escape. However, with wider tubes, horizontal tubes, or vertical tubes, in which the animals had to swim up in order to escape, no improvement over trials was observed. To explain these results, the authors suggested that the response observed is not one of learned escape but a delayed geotaxis (tendency to swim toward gravity) that is released only after a period of mechanical stimulation, such as that resulting from bumping into the walls of the tube.

The behavior of aquatic protozoans is so closely linked to minute chemical changes in the surrounding medium that it may never be possible to dissociate true learning from changes in the medium, even if true associative learning occurs in protozoans. It has even been suggested that we not require that true learning in protozoans be independent of the external environment (McConnell, 1965; Applewhite & Gardner, 1973). Can the engram be stored outside the organism?

Platyhelminthes Considerable publicity and controversy have surrounded the research conducted on learning in planarians. These flatworms are interesting because it is in the Platyhelminthes that a nervous system with bilateral symmetry and a primitive brain appears. As would be expected, there is evidence for the occurrence of habituation in planarians (e.g., Westerman, 1963).

Classical conditioning The modern era of research on learning in flatworms was initiated with the publication of a paper by Thompson and McConnell (1955) reporting classical conditioning. This original experiment lacked some important controls (Jensen, 1965). Further, several workers either failed to observe classical conditioning in flatworms or obtained results that suggested the possibility of sensitization or pseudoconditioning (e.g., Bennett & Calvin, 1964; Brown, Dustman, & Beck, 1966; Halas, James, & Knutson, 1962; Hullett & Homzie, 1966). The fires of controversy over the existence of classical conditioning were fueled by two events apparently irrelevant to such research. First, McConnell and his associates reported a series of controversial results on the retention of memory after regeneration, cannibalism, and interanimal transfer of chemicals (McConnell, 1966). Second, McConnell edits a journal, called the *Worm Runner's Digest*, that represents a delightful blend of comparative psychology and humor but which appears to have led some very skeptical critics to question the credibility of the scientific endeavor.

The more recent studies on classical conditioning in flatworms have incorporated better controls than the earlier studies, and convincing demonstrations of conditioning are now available (e.g., Jacobson et al., 1967; Block & McConnell, 1967). In the study of Jacobson et al. discussed in Chapter 16, numerous control groups were employed and differential conditioning to visual and vibratory stimuli was accomplished. Like that of protozoans, the behavior of flatworms is influenced by a wide array of subtle factors, and it is these influences that appear to explain many of the failures to obtain classical conditioning (McConnell, 1967). For example, planarians secrete a kind of slime. Conditioning is much more rapid if a layer of slime is allowed to accumulate in the apparatus (McConnell & Mpitsos, 1965). Frequent cleaning of the apparatus can interfere with planarian behavior.

Instrumental learning Numerous devices and procedures have been used to demonstrate instrumental learning in flatworms, including mazes of various sorts (e.g., Best, 1965) and a free-operant response technique (Lee, 1963; see review by Corning & Kelly, 1973).

It now seems clear that the phenomena of habituation, classical conditioning, and instrumental learning can be demonstrated in flatworms. However, the processes underlying such learning may be quite different from those of vertebrate learning. Some phenomena of learning in flatworms, such as extinction, appear different from those of vertebrates, suggesting the possibility of different underlying mechanisms.

Annelida Convincing demonstrations of habituation in earthworms, the most popular annelid species in behavioral research, have been provided for the backward-movement response to air puffs (Ratner & Gilpin, 1974) and the responses of contracting and hooking elicited by vibratory stimuli (Gardner, 1968).

Well-controlled studies of classical conditioning in earthworms have also been reported (Ratner & Miller, 1959; Peeke, Herz, & Wyers, 1967). The response of rearing and withdrawing to a light (UCS) was studied with a mild vibration as the CS. The probability of response to the CS increased over trials, whereas such changes did not occur in groups tested to control for effects of sensitization and pseudoconditioning.

A long tradition of studies of instrumental learning by earthworms in T-mazes dates back to the study by Yerkes (1912). While it is clear that earthworms do indeed come to make the correct of two choices in a T-maze with shock as a punishment for incorrect choices, the basis of this behavioral modification is not clear. Rosenkoetter and Boice (1975) observed improvement over trials only when chemical cues were permitted to accumulate in the maze. They suggested that T-maze "learning" in earthworms may be an artifact of the accumulation of pheromones that guide the worm to the correct choice.

Arthropoda Habituation has been demonstrated in several arthropod species including horseshoe crabs (Lahue et al., 1975), crayfish (Krasne, 1969), and the pupae of grain beetles (Hollis, 1963). A well-controlled study of classical conditioning of proboscis extension in blowflies was reported by Nelson (1971). The UCS was an application of a sugar solution to the mouthparts of the fly, with the CS being the application of water or a saline solution to the feet.

Much attention has been devoted to the study of instrumental learning in arthropods. The studies of maze-learning in ants conducted by Schneirla (e.g., 1946) are classics in comparative psychology. Effects of maze-learning experiences in larvae have been shown to survive through metamorphosis in adult grain beetles (Borsellino, Pierantoni, & Schieti-Cavazza, 1970). Conditioning of responses related to food getting in honeybees has been demonstrated by Bermant and Gary (1966) and Wenner and Johnson (1966).

Horridge (1962, 1965) studied the acquisition of avoidance responses in insects, with and without their heads, by using a rather clever procedure (Figure 17-1). In the training situation, the legs of two cockroaches are connected in series to an electrical stimulator. The electrical circuit is completed and both animals receive shock whenever animal P lowers its leg into a saline solution, thus completing the electrical circuit. With training, animal P comes to hold up its leg and thus avoid shock. Animal R is a "yoked control" that receives the identical number and temporal patterning of shocks, but they are delivered in such a way as to not be contingent on its behavior. Learning is demonstrated as a difference between animals P and R in a test situation when the animals are shocked independently for lowering their legs. Animal P experiences fewer shocks in the test situation than animal R. Both intact and headless locusts and cockroaches have displayed avoidance learning in this test situation (Horridge, 1962, 1965; Disterhoft, 1972).

Considerable difficulty has been encountered in demonstrating learning in fruit flies, a species that, in other respects, provides ideal subjects for behavioral studies. Although some learning studies have been quite controversial (e.g., Murphey, 1967; Yeatman & Hirsch, 1971; Hay, 1971; Bicker & Spatz,

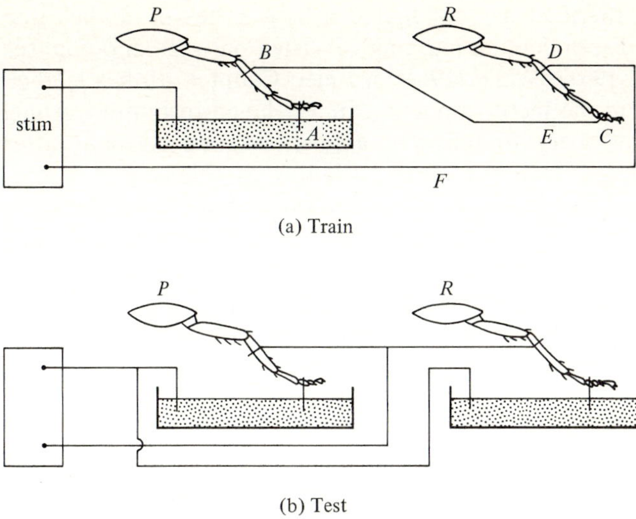

Figure 17-1 Experimental arrangement for the conditioning and testing of insect avoidance learning. *(a)* During training the two animals are arranged so that both animal P and animal R receive shocks when P lowers its leg beyond a critical point. *(b)* In the retest situation, the animals are connected in parallel so that either receives electric shock separately if its leg is lowered. When trained and tested in this situation, animal P typically demonstrates passive avoidance learning, as indicated by fewer instances of the lowering of its leg than of animal R, the "yoked control." (*From Horridge, 1962.*)

1976), it now appears that fruit flies can learn a discriminated avoidance task based on odor cues (Quinn, Harris, & Benzer, 1974).

Mollusca Many scientists have become interested in the gastropod mollusks because of the relative accessibility of their nervous sytems. They are ideal "simple systems" for physiological analyses. Kandel and his associates (see Kupfermann, 1975) have conducted an extensive series of studies on the habituation of the gill-withdrawal reflex in the marine mollusk *Aplysia*. Both classical conditioning and avoidance learning have been found in the marine gastropod *Pleurobranchaea californica* (Mpitsos and Davis, 1973; Mpitsos and Collins, 1975). Food-aversion learning has been demonstrated in terrestrial slugs (Gelperin, 1975).

The ability to learn appears well developed in the cephalopod mollusks and has been widely studied in the octopus (Boycott and Young, 1950; Wells, 1965). As it is apparent that habituation and classical conditioning occur in octopuses (see Sanders, 1975), attention has been focused on performance in more complex learning situations. The procedures used in many of these experiments have been criticized for imprecision and lack of automation by Bitterman (1975a).

Conditioning of a free-operant response in octopuses was demonstrated using lever pressing by Dews (1959) and using a response in which the octopus had to extend its arm up a vertical tube by Crancher, King, Bennett, and

Montgomery (1972). Sutherland has conducted a long series of studies that demonstrate excellent discrimination learning of visual patterns in octopuses (e.g., Sutherland, 1957, 1961). Wells (1961; see also Chapter 10) has studied discrimination learning in the tactile modality. In a delayed-response learning situation, octopuses have been found to be able to respond correctly after delays of up to 2 minutes, with performance falling to chance levels after 4-minute delays (Dilly, 1963; Sanders, 1975). Octopuses have been shown to display improvement with successive reversals in a reversal-learning task using a simultaneous-discrimination learning procedure, stationary shapes as stimuli, and sardines as reinforcement (Mackintosh & Mackintosh, 1964).

As it is apparent that octopuses are capable of relatively complex forms of learning and possess a nervous system quite different from that of vertebrates, much attention has been devoted to the study of effects of physiological intervention on learning performance in octopuses (see Sanders, 1975).

Plants Although there have been a few attempts to study learning in plants, a "botanical psychology" has yet to come of age. Efforts have been concentrated on the sensitive plant *Mimosa.* In response to a variety of stimuli—including mechanical stimulation, electrical shock, cutting, burning, or darkness—the leaflets of this plant close and the stem droops rapidly. *Mimosa* show a loss of responsivity to electrical and mechanical stimuli after repeated presentations that resembles the phenomenon of habituation in animals in several of its characteristics (Holmes & Gruenberg, 1965; Holmes & Yost, 1966; Applewhite, 1972). There is evidence for stimulus specificity, and effects of variations in stimulus intensity and intertrial interval resemble those observed in animals.

Although there have been reports of successful conditioning experiments in plants (Haney, 1969; Armus, 1970), the necessary controls have been lacking and the phenomenon has not yet proved replicable (Levy, Allen, Caton, & Holmes, 1970; Applewhite, 1975). A primary difficulty in conducting a definitive experiment with *Mimosa* appears to lie in the problem of finding a stimulus to which the plant is responsive but which does not already produce the response to be conditioned.

Learning in Vertebrates

Comparative research on learning in vertebrates has generally been focused on questions of the evolution of the ability to learn and of the processes underlying that ability. Some investigators have concentrated on *quantitative differences* in learning by comparing the relative abilities and rates of acquisition by different species in certain standard learning situations. Other investigators have searched for *qualitative differences* in the ways in which different species perform in standard learning situations.

A number of assumptions underlie these approaches. Implicit in many quantitative comparisons is the belief that there exists a unidimensional trait of "intelligence" that varies among different species and can be assessed using the standardized techniques of the learning laboratory. Most investigators are not content merely with demonstrations of differences in learning ability but seek differences in certain directions. It is anticipated that in a "valid" comparison, humans "ought" to do better than great apes, while the apes ought to do better, in turn, than monkeys, carnivores, rodents, birds, reptiles, amphibians, and fish. For some, this expectation has been based on a belief in a *scala natura*, or phylogenetic scale. The lack of congruence between belief in a unitary phylogenetic scale and the contemporary conceptualization of the phylogenetic tree was discussed by Hodos and Campbell (1969) and in Chapter 3. Alternative reasons for expecting such orderings stem from a consideration either of relative brain complexity (Hodos, 1970; Warren, 1972; Masterton & Skeen, 1972) or of recency of common ancestry with humans (Bitterman, 1975b). An alternative expectation for the ordering of species was expressed by Rensch (1956, 1957), who proposed that learning ability varies with absolute brain size.

For a recent review of information on learning, brain function, and evolution, the reader is referred to the volumes edited by Masterton, Hodos, & Jerison (1976) and Masterton, Bitterman, Campbell, and Hotton (1976).

Quantitative Comparisons A large number of different tasks have been used in making quantitative comparisons among species.

Simple tasks Species differences in the rate of acquisition of relatively simple learning tasks have been reviewed by Warren (1965a, 1965b) and Brookshire (1970). There is no substantial evidence to suggest that there exist orderly species differences in the rate of acquisition of simple tasks. Brookshire (1970) reviewed a considerable number of studies of classical conditioning and found considerable differences in the ease of conditioning of different stimulus-response relationships within individuals, within species, and between species. In rabbits, for example, acquisition of a conditioned eyelid response requires about twice the number of trials as acquisition of conditioned extention of the nictitating membrane of the eye (Schneiderman, Fuentes, & Gormezano, 1962; Schneiderman & Gormezano, 1964). Given these and other results, Brookshire concluded that "These considerations make it extremely difficult to support the contention that the quest for quantitative differences in the rate of acquisition of classically conditioned responses will permit a systematic ordering of species in terms of learning ability" (Brookshire, 1970, p. 300).

Similarly, there is little basis for optimism regarding the emergence of consistent cross-species orderings on other simple tasks. Sidman avoidance learning occurs at about the same rate in goldfish, pigeons, rats, cats, dogs, squirrel monkeys, rhesus monkeys, and chimpanzees (Warren, 1965a). Similar results appear when one considers simple reward training and discrimination learning (Warren, 1965b). Rats have been shown to learn a discrimination in a Y-maze more rapidly than either squirrel monkeys or rhesus monkeys (Rumbaugh, 1968).

Complex problems Given the general lack of discriminability of the learning performance of different species in simple tasks, experimenters have turned to a variety of more complex problems. Several complex tasks can be learned by some primates and no other species. For example, it is difficult to imagine a nonprimate acquiring the American Sign Language in the manner of chimpanzees (Gardner & Gardner, 1969; Fouts, 1973; also Chapter 6). No nonprimate has solved the Weigl oddity problem. This is a conditional oddity problem in which, for example, a subject must select the object of the odd *shape* when objects are on a white background but of the odd *color* when they are on a black background (Warren, 1965a). However, these problems have not been useful in making broad comparisons, as the nonprimate species fail to solve them. Tasks such as the delayed-response problem (Maier & Schneirla, 1935) and the double-alternation problem (Warren, 1965a) have failed to produce meaningful orderings of species. Faced with these failures, students of vertebrate learning have turned to a number of other tasks.

Reversal learning The utility for cross-species comparisons of reversal learning in a spatial discrimination situation has been stressed by Gossette (1970). Thus, Gossette (1969) found differences in performance that appear to separate "more primitive birds" from those that are "more advanced." Gossette argued that the spatial discrimination learning task could be used to scale quantitative differences in learning ability across a wide range of vertebrates. However, many of the data from different species of mammals are difficult to interpret both with respect to data from other mammals and in relation to those from birds (Gossette, Kraus, & Speiss, 1968; Gossette & Gossette, 1967). Thus, it is difficult to understand why skunks should outperform squirrel monkeys and why red-billed blue magpies should outperform both.

Learning sets The learning-set task has been among the most widely used in the study of comparative vertebrate learning. Some comparative data are presented in Figure 17-2. The reasons for interest in this problem should be readily apparent on gross examination. The performance of apes is outstanding, whereas tree shrews, rats, and squirrels display little or no evidence of the development of true learning sets. Even with learning sets, however, there are some complications in interpretation. Why should squirrel monkeys and marmosets score so low in comparison to carnivores? Why should pigeons score so well? Whereas rapid learning-set acquisition—comparable to that of cats, marmosets, and squirrel monkeys—was demonstrated in myna birds and blue jays (Kamil & Hunter, 1970; Hunter & Kamil, 1971), marginal performance was found in crows (Hunter & Kamil, 1975).

Delayed alternation Masterton and Skeen (1972) studied delayed alternation of three species of mammals. Rather than rely on a broad sample of vertebrate species with no consideration of phylogenetic history, they selected three species that may lie along the evolutionary stem leading to the Old World primates and apes—hedgehogs, tree shrews, and bush babies. As one progresses through these three species, one finds (1) closer probable recency of

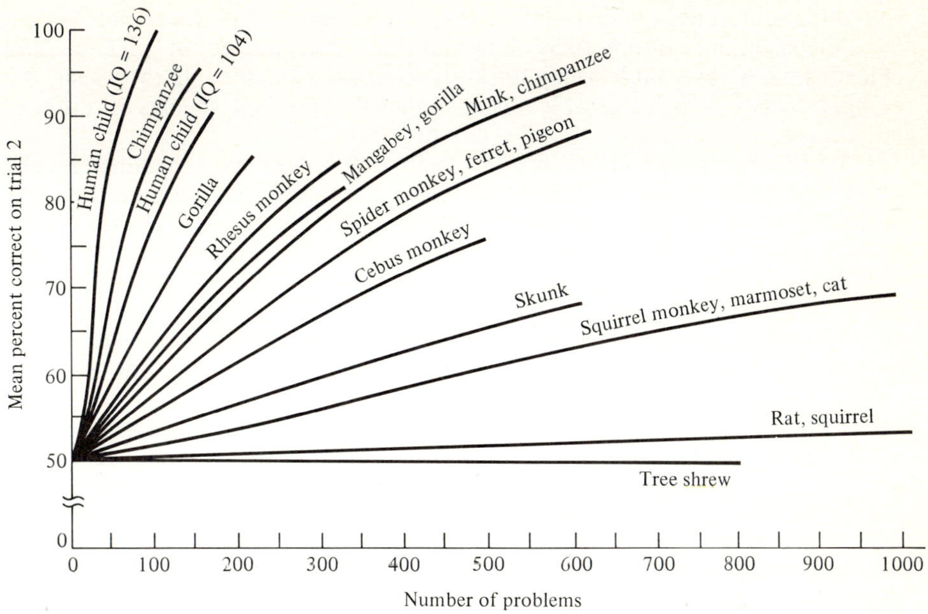

Figure 17-2 Performance of representatives of a number of vertebrate species on learning set problems. The dependent variable is the mean percent correct choices on the second trial of each problem. (*From Hodos, 1970.*)

ancestry to anthropoids, (2) increases in the size of the prefrontal system of the forebrain, and (3) increased ability to perform on delayed-alternation tasks at long delays.

Criticisms of quantitative comparisons Attempts to find orderly patterns of species differences with respect to quantitative differences in learning have received much criticism (e.g., Warren, 1973; Bitterman, 1975b). At least six bases for criticism of these approaches are apparent:

1 *Individual differences:* One problem with most quantitative interspecies comparisons on learning tasks lies in the use of small samples and lack of consideration of individual differences (Warren, 1973). There are sizable individual differences in performance in tasks such as learning sets. Rumbaugh (1968) summarized individual differences in squirrel monkeys (*Saimiri*): "*Saimiri* specimens vary widely in their basic learning-set skills, to the extent that if one used trial 2 performance as the criterion, equivalent training conditions will net essentially perfect and efficient performance on the part of a few (better than 90% correct—as good as the best macaques and apes) and equally impressive and inefficient performance on the part of others (to the extent that they remain "learning-setless") (p. 309). With individual differences so pervasive, it is difficult to interpret differences in the mean performance levels of different species as indicative of species-typical levels of "intelligence" or as a reflection of differential development of the nervous system.

2 *Differences in motivation:* It has often been observed that motivational differences affect performance in learning tasks. For example, Levine (1972) found that differences in drive level, as measured in percent of weight loss with food deprivation, differentially affected the performance of pigeons and chickens in a reversal-learning task. It is extremely difficult to assess the comparability of motivational levels in studies that produce species differences in learning.

3 *Differences in reinforcement:* As with differences in motivational levels, differences in magnitude of reward have repeatedly been found to affect performance in learning situations. Although techniques for the scaling of reward magnitude, such as the progressive ratio method of Hodos (1961), are available, they have yet to be systematically employed in the comparative study of species differences in learning.

4 *Differences in sensory function:* Different species differ with respect to the relative development of different sensory modalities and the ease of conditionability of stimuli presented in those modalities. Devine (1970) found the ranking of cebus and rhesus monkeys to vary according not only to the training procedures used but also to the attributes of the stimuli that were to be discriminated. Whereas primates are a highly visual species, rats respond more readily to olfactory stimuli (Nigrosh, Slotnick, & Nevin, 1975). When tested for the development of learning sets with the use of the usual visual stimuli, rats do poorly (Figure 17-2). However, "Learning-set performance of rats trained with odor stimuli is comparable to that reported for primates trained on visual cues" (Slotnick & Katz, 1974, p. 796). Given data such as these, one must wonder about the significance of comparisons such as those in Figure 17-2.

5 *Differences in motor capacities:* The performance of animals in learning tasks is in part a function of their possession of the requisite motor skills. For example, gibbons were thought unable to solve certain patterned-string problems used to demonstrate "insight" learning. However, the testing situation in which strings are placed on flat surfaces is a difficult one for the gibbon hand. When strings were elevated rather than placed on flat surfaces, gibbons were found capable of solving such problems (Beck, 1967).

6 *Species-typical preparedness:* Different species are adapted in such ways that they appear "prepared" to learn certain things and "contraprepared" to learn others (Seligman, 1970). Species-typical preparedness can have a great influence on performance in learning situations and will be discussed in greater detail below.

Conclusions regarding quantitative comparisons There is no evidence that a scale of intelligence along which different species can be ordered does not exist. However, if such a scale should exist, the problems to be solved in describing it appear for the present to be of such a magnitude as to be almost insurmountable. In principle, there are techniques available to solve most of these problems. It is true that all motivational, sensory-perceptual, motor, and other factors other than "intelligence" that affect performance will act to underestimate an animal's learning capacity and errors should be self-correcting with additional research (Hodos, 1970). However, the researcher looking for orderly quantitative differences in learning capacity should be prepared for a long struggle in attempting to control and equate a variety of factors that can influence performance.

Qualitative Differences The history of the comparative psychology of vertebrate learning has been dominated by a *continuity theory*, according to which the processes underlying learning are assumed to be essentially identical across all vertebrate species (Thorndike, 1911; Skinner, 1956; Voronin, 1962). An attempt to counter continuity theory with demonstrations of systematic qualitative differences in performance in learning situations has been undertaken in a comprehensive series of well-controlled experiments by M. E. Bitterman (1960, 1965a, 1975b).

The basic approach The fundamentals of Bitterman's approach can be expressed in two phases, *control by systematic variation* and *comparison of functional relations*. Bitterman (1960) noted the difficulty of equating situational variables in learning tasks for any two different species (control by equation). He therefore proposed the notion of control by systematic variation. While we may be unable to equate situational variables, we can vary them systematically in order to assess their effects on learning performance. Thus, for example, if length of food deprivation is believed to be contributing to a difference between two species, one can systematically vary amount of food deprivation through the total available range in both species and thus assess the effect of this variable. If the difference between species lies in deprivation level, some levels should be found at which differences would disappear. If they do not, one could have confidence that the difference is not a function of the factor being systematically varied.

Rather than search for quantitative differences in learning, Bitterman would look for qualitative differences that might reflect the operation of qualitatively different underlying processes. Thus, Bitterman advocates comparison of functional relationships between variables rather than comparison of absolute scores. These principles should become more clear as Bitterman's research is reviewed.

Reversal learning Bitterman (1965a, 1965b) compared the performance of various species in reversal-learning situations. He noted, as discussed above, that a wide range of vertebrate species—including monkeys, rats, and pigeons—show improvement as they learn successive reversal problems. By contrast, teleost fish, including both goldfish and mouthbreeders, were found to show no such improvement (e.g., Bitterman, Wodinsky, & Candland, 1958). Bitterman (1965b) noted, "Progressive improvement in habit reversal has been sought without success in fish under a wide variety of conditions, whereas the rat and pigeon do progress under an equally wide range of conditions." Painted turtles and decorticate rats (rats with the cerebral cortex removed, as discussed in Chapter 11) showed improvement in reversal learning based on spatial cues but not visual cues. Bitterman noted, "We conclude simply that experiments on habit reversal tap an intellectual capacity of higher animals that is not at all developed in the fish and is manifested by the turtle only in a restricted class of problems" (Bitterman, 1965b).

In the time since 1965, it has become clear that fish can show improvement in habit reversal under some conditions (e.g., Engelhardt, Woodard, & Bitterman, 1973). Bitterman (1975b) noted that "improvement *occurs* only rarely, although it does occur." Bitterman (1975b) was more guarded in his conclusions than in earlier papers but nevertheless maintained that species differences in reversal learning reflect qualitative differences among species. After discussing the procedures necessary for the demonstration of improvement in fish, he noted: "The fact that such stratagems are quite unnecessary to produce substantial improvement in choice experiments with other vertebrates leads to the suspicion of differences in process, a suspicion which is strengthened by analysis of the course of improvement."

Probability learning A second problem that Bitterman regarded as effectively separating fish from rats in a qualitative manner is performance in the probability learning situation. Whereas fish show random probability matching (see Chapter 16), rats tend toward maximizing reinforcements, approaching 100 percent choice of the more frequently reinforced alternative (Figure 17-3). When they do not maximize, rats appear to display systematic strategies; they never show random matching (Bitterman, 1965a). Painted turtles, decorticate rats, and pigeons tend to probability match on spatial problems although not on visual problems (Bitterman, 1965a). Once again, however, the differences do not appear absolute. "Random matching is not always found in African mouthbreeders, goldfish, painted turtles, or pigeons—the conditions required to produce it have not yet been fully defined—but it never has been demonstrated in mammals" (Bitterman, 1975b).

Other problems Bitterman (1975b) cited a number of other tasks in which the performance of rats and fish appears to differ qualitatively. One example is the *depression effect*, an extremely important phenomenon in the development of learning theory. If two groups of animals are trained on different amounts of

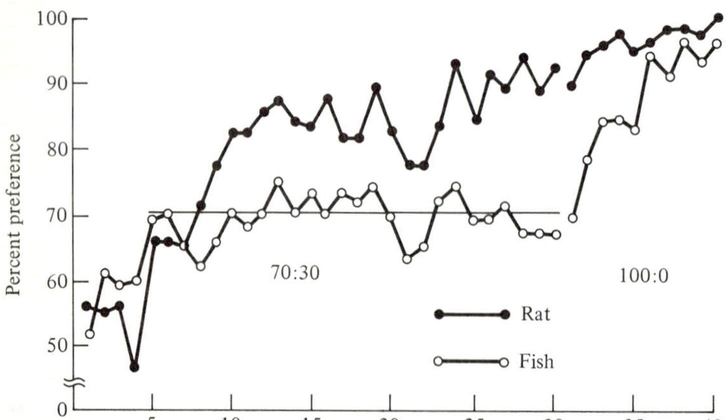

Figure 17-3 Performance of laboratory rats and fish (mouthbreeders) in a probability learning task. Whereas the rats approach maximization of the number of reinforcements, the fish show random probability matching. (*From Bitterman, Wodinsky, & Candland, 1968.*)

reward, there will develop a difference in the performance levels between the groups. Animals receiving the larger rewards will run faster. One can then shift the amount of reward for half of the animals in each group while keeping it the same for control animals for each group. If this experiment is performed with rats, the animals that had received the large reward and were shifted to the small reward slow down to the point that their running speeds are slower than those of animals trained throughout on the smaller reward; this is the depression effect. If the same experiment is performed on fish, there is no such shift in performance; the fish that are shifted to the small reward continue to swim as fast as the controls that continue on the large reward (Lowes & Bitterman, 1967).

Another difference in a functional relationship between rats and fish can be seen in the effect of differences in amount of reward on resistance to extinction—the number of responses given under extinction conditions (Chapter 16)—before some criterion of nonresponding is attained. In rats, animals receiving larger rewards show less resistance to extinction (Gonzalez & Bitterman, 1969), whereas in fish, there is greater resistance to extinction after training with large rewards (Gonzalez, Potts, Pitcoff, & Bitterman, 1972).

Several additional problems were cited by Bitterman (1975b) as producing different functional relationships between rats and fish (e.g., "the dimensional transfer effect" and "the partial reinforcement effect").

While all these problems appear to discriminate between processes underlying learning in rats and fish, it must be noted that much less research has been conducted on these problems than on probability learning and reversal learning. Whether these differences are truly absolute or a matter of degree will be determined only after additional research.

Conclusions regarding qualitative comparisons In the search for different functional relations rather than quantitative differences in learning scores, one minimizes the effects of the difficulties discussed in the section on quantitative comparisons. Nevertheless, such problems can never be completely eliminated. While control by systematic variation would appear to be a sound procedure, it is obviously impossible to apply for every potentially relevant variable in every species (unless one finds the key to immortality). While the earlier results of qualitative comparison suggested the existence of true quantum differences, the magnitude of these differences has shrunk as more research has been conducted on problems such as reversal learning and probability learning. Reference to the differences as "qualitative" appears strained when they occur under some conditions but not others. Although it appears quite reasonable that there may indeed be substantial differences in the processes underlying learning in species as diverse as laboratory rats and teleost fishes, the qualitative nature of such differences has not yet been convincingly demonstrated. However, the student of comparative learning has much to learn from the controlled procedures and parametric variations used by Bitterman in his research program.

ADAPTIVE SIGNIFICANCE OF LEARNING

The study of learning from the viewpoint of evolutionary history has been focused on general learning processes and abilities as revealed by performance in a set of standardized laboratory tasks. Although the nature of the apparatus is typically modified somewhat in order to allow for differences in the gross sensory and motor capacities of different species, the intent is to permit an equivalent assessment of general processes in standardized situations.

An alternative approach to the study of the evolution of the capacities for and processes underlying learning developed from the continued interaction of ethologists and comparative psychologists in the 1960s and 1970s. This approach is based on a consideration of learning in relation to *specific* adaptations appropriate to survival and reproduction in the particular habitat in which each species evolved. Rather than assess communalities of process using standardized laboratory tasks, researchers influenced by this approach emphasize specific adaptations for the learning of biologically relevant relationships. It was apparent in Chapters 14 and 15 that a great number of behavioral patterns—including patterns of sexual behavior, sleep, prey-predator interactions, and social behavior—can be understood as specific adaptations that are functional for a particular life-style in the natural habitat. As stated by Rozin & Kalat (1971), "We propose to treat learning and memory as any other biological characteristic, subject to natural selection and therefore adapted to handle specific types of problems" (p. 459).

As a result of the interaction of ethology and comparative psychology, there has occurred a dramatic surge of interest in considering the phenomena of the learning laboratory in relation to species-characteristic behavioral patterns that are adapted for functioning in the natural habitat. This interest has led to the development of a variety of theories designed to interpret a variety of phenomena of the learning laboratory, not as instances of the operation of general laws but rather as the result of the occurrence of patterns of species-characteristic behavior adapted for the natural habitat (e.g., Breland & Breland, 1961; Glickman & Schiff, 1967; Glickman, 1973; Bolles, 1970; Staddon & Simmelhag, 1971; Shettleworth, 1975).

As it is impossible to cover in a few pages the full range of implications of the adoption of this perspective, the interested reader may wish to pursue any of several good reviews and sources of related information (e.g., Seligman, 1970; Seligman & Hager, 1972; Shettleworth, 1972; Hinde & Stevenson-Hinde, 1973). We shall attempt to consider a few of the implications of the adoption of an ecological-ethological strategy in the study of learning.

Capacities and Processes in Relation to Species-Characteristic Behavior

Adoption of an ecological-ethological strategy in the study of learning permits one to view two important problems in a new light. First, it has frequently been found that animals are able to perform in biologically relevant situations in a manner that greatly exceeds estimates of their capacities as determined in standard laboratory situations. Thus, estimates of the learning ability of a given species may be far too low if based only on well-controlled but arbitrary laboratory tasks.

Second, it is quite possible that different laws may apply in different situations. "Biologically speaking, there is no reason to assume that there should exist an extensive set of generally applicable laws of learning, independent of the situation in which they are manifested. . . . The 'laws' of learning in the feeding system need not be the same as those in other systems; manifestation of a learning capacity in one area does not imply that it will be accessible in other areas" (Rozin & Kalat, 1971, pp. 460, 459). Although devotees of the ecological-ethological strategy differ among themselves with regard to the probable degree of generality of the laws of learning (see Seligman, 1973), many believe that there exist some differences in relation to the particular function of the behavioral patterns under study.

Learned Taste Aversions The findings that, more than any other, have had impact on the development of an ecological-ethological strategy in the study of learning are those that have dealt with learned taste aversions. The results of experiments on the taste aversions appear to indicate that rats are adapted or "prepared" to make certain associations rather than others. Garcia and Koelling (1966) found that when illness, induced by x-irradition or injections of lithium chloride, was paired with the ingestion of substances having a salient taste, the taste was associated with illness and the rats decreased their intake of substances having that taste. By contrast, when an auditory-visual stimulus was paired with illness, there was no such reduction in intake. When electric shock was substituted for the illness, associations were formed more readily with the auditory-visual stimulus than with the taste. Thus, it would appear that rats are prepared to associate illness with taste cues and electric shock with auditory-visual cues. The probable adaptive significance of this preparedness is readily apparent. If a wild rat should ingest small quantities of a poison or other toxic food, it would be beneficial for the animal to learn to avoid ingestion of foods of similar taste. Consistent with this inference is the finding that wild rats that have survived a poisoning attempt become "bait-shy" and are very difficult to poison (Barnett, 1963).

Perhaps the most remarkable characteristic of learned food aversions is their time course. In most laboratory conditioning situations, the reinforcement must follow the stimulus within a matter of a few seconds. By contrast, food aversions can be learned when the occurrence of illness follows ingestion of unique substances by more than an hour (Garcia, Ervin, & Koelling, 1966). The advantage of such a system in the natural habitat is obvious. The rat that has been poisoned is unlikely to become ill for some time after ingestion. As the rat may have visited many places in the interim, it would be advantageous for illness to be associated with the novel taste experiences occurring some time previously. The finding that exceptionally long response-reinforcement intervals can be tolerated in the learning of biologically relevant associations provides an instance of the altered perspective on animal capacities and processes that can be gained when learning is viewed in relation to adaptation in the natural habitat.

Subsequent research on food aversions has proceeded in various directions (see Garcia, Hankins, & Rusiniak, 1974). In contrast to rats, bobwhite quail—a more visual species than rats—have been found able to learn illness-induced aversions with 1/2-hour delays using either visual or gustatory cues. Indeed, the visual cue is the more salient for quail (Wilcoxon, Dragoin, & Kral, 1971). Taste aversions have been demonstrated in garter snakes, using earthworms as a novel food with lithium chloride injected 30 minutes later (Burghardt, Wilcoxon, & Czaplicki, 1973), and in infant guinea pigs, using sucrose ingestion followed by lithium chloride injections 30 minutes later (Kalat, 1975). Indeed, the vaginal secretions of female hamsters, that normally serve as a sexual attractant, have been conditioned using these techniques (Johnston & Zahorik, 1975). Prey-lithium associations have been studied in coyotes and wolves, with preliminary evidence gathered to support the possibility that these techniques might be used to condition coyotes to avoid killing and eating sheep (Gustavson, Kelly, Sweeney, & Garcia, 1976).

It should be noted that the interpretations of these experiments by Garcia and his associates have been challenged and subjected to some controversy (e.g., Bitterman, 1975b, 1976; Garcia, Hankins, & Rusiniak, 1976; Krane & Wagner, 1975).

Delayed-Response Performance in Digger Wasps Delayed-response learning in the laboratory generally produces estimated values of seconds or minutes as the maximal delays tolerated by some mammalian species in laboratory situations (Maier & Schneirla, 1935). A behavioral pattern in digger wasps that is roughly analogous to that studied in delayed-response problems in the laboratory was studied by Baerends (Tinbergen, 1951). A female digger wasp supplies food (caterpillars) for larvae in several different holes in a day. She determines the number of caterpillars to be provided for a given hole on the basis of a visit made each morning. Manipulation of the contents of the hole prior to that visit affects the number of caterpillars brought that day; manipulations made after the female's visit are without effect. Thus, the female retains information about the amount of material to be provided for each nest hole throughout the day. In this biologically relevant situation, the ability of the wasp to deal with relatively complex information thus appears well beyond that which would be likely to appear in a standardized laboratory test.

Other Biologically Relevant Performance Performance by animals in learning a number of other biologically relevant tasks appears to exceed that typically observed in laboratory learning situations. For example, avoidance learning by frogs and toads can be quite slow in a number of situations (Boice,

1970a; Yaremko, Jette, & Utter, 1974). By contrast, toads learn a passive avoidance of ingestion of bees and their mimics after just a few encounters (Brower & Brower, 1962).

In his studies of celestial orientation in indigo buntings, Emlen (1975) found evidence that young birds learn about star patterns by observing the rotation of the stars. One wonders about the difficulty of creating such performance were the animals not adapted for it.

Species Differences in What Is Learned

There are considerable differences among different species with respect to the role that learning plays in the development of the species-characteristic behavioral repertoire. Animals of different species learn different things.

Birdsong Perhaps the most clear-cut illustration of species differences can be found in birdsong (see Chapter 9). Whereas some species are able to display the full species-characteristic song without ever hearing the songs of other individuals, birds of other species must hear the song (Nottebohm, 1972). The stimuli that are effective in altering the development of birdsong differ across species. In white-crowned sparrows, for example, the songs of other species have little effect on song development (Marler, 1970). It appears as though the birds possess a "template" that selectively rejects some songs but that becomes modified by experience with the species-characteristic song. Vocal imitation by myna birds in the field is characterized by a similar specificity (Bertram, 1970). Mynas not only do not imitate birds of other species but appear selectively responsive to calls made by individuals of their own sex.

Learning What to Eat A critical aspect of the development of behavior in many species is the development of food preferences. In many species, the nature of the food appropriate for ingestion must be learned. Young rats appear to learn feeding patterns through two mechanisms (Galef, 1976). Gustatory cues in the milk of the lactating female are sufficient to permit the young animal to recognize her diet and cause the young to ingest that type of food preferentially (Galef & Sherry, 1973). In addition, young rats tend to take their first meals of solid food in the vicinity of an adult and thus to ingest the same food as the adults (Galef & Clark, 1971).

The development of food recognition by young chicks has been analyzed in a series of experiments by Hogan (1973a, 1973b, 1973c). Hogan noted that "many of the mechanisms that lead the chick to its goal appear to be specially adapted to the conditions encountered by a newly-hatched chick, and these mechanisms frequently seem to defy some of the laws of traditional learning theory. Thus, critical or sensitive periods, associations over long delays, and reinforcement of the pecking response with no concomitant discrimination of the object being pecked, all play a role in the development of food recognition" (Hogan, 1973c, p. 136).

Individual Recognition Different species of birds behave differently with respect to individual recognition. In general, there appears to be little evidence of individual recognition in diurnal, territorial songbirds (Etkin, 1964). The young of flock-living birds, such as chickens and ducks, appear well able to recognize their mothers and to follow them about, as discussed in the section in filial imprinting in Chapter 8. Individual recognition in adult flock-living birds is critical to the formation of dominance hierarchies. In chickens, the comb—in configuration with the beak or wattle—appears to provide the most likely basis for individual recognition (Candland, 1969).

Individual recognition in colonial-nesting shore birds is essential both between partners and between parents and young. Without such recognition, parents' resources might be expended on young that are not their own (see Chapter 15). Remarkably, recognition of individual vocal patterns provides a basis for many of these discriminations (Beer, 1970).

Species-Specific Defense Reactions Bolles (1970), in a very influential paper, criticized traditional learning-theory conceptualizations of avoidance learning. He noted that animals learn some avoidance tasks in the laboratory much more easily than others. Bolles proposed that the differences in ease of acquisition of different avoidance responses can be understood in relation to species-specific defense reactions (SSDRs). According to Bolles, animals in nature do not learn to avoid danger gradually, as would be suggested by laboratory experiments. If they had to do so, they would perish before learning was complete. Rather, new or sudden stimuli release a set of *innate* SSDRs. When the task to be learned by an animal represents either an instance of its SSDRs in that situation or a response compatible with them, "learning" will be rapid. When the response to be learned is incompatible with the SSDRs, learning will be quite slow. Thus, it is much more difficult to get rats to turn a wheel or press a bar in order to avoid shock than to run from the area in which they receive shock. Bolles's proposals have provided a catalyst for much research on SSDRs and avoidance learning, and the results obtained have been generally compatible with his proposals (e.g., Petersen & Lyon, 1975; Osborne, Caul, & Vanstrum, 1976).

Constraints on Learning

In analyzing the mechanisms through which selectivity is imparted to the learning performance of different species, different authors write of "preparedness" (Seligman, 1970), "biological boundaries of learning" (Seligman & Hager, 1972), or "constraints on learning" (Shettleworth, 1972; Hinde & Stevenson-Hinde, 1973). The point to be made with all these terms is that there are certain predispositions and limitations to what animals of different species will learn. These predispositions and limitations stem from the characteristics of the relevant stimuli, responses, and reinforcers—as well as the interaction of all three—in addition to the time of occurrence of the learning experience (Shettleworth, 1972; Hinde, 1973).

Constraints on the Stimuli to Be Learned Some constraints on learning stem from the nature of the stimuli to be learned. Thus, rats are generally less responsive to visual stimuli than to stimuli in the olfactory modality (Nigrosh, Slotnick, & Nevin, 1975). Whereas octopuses appear responsive to differences in the weights of different objects, they appear unable to use this information to learn a discrimination based on differences in weights (Wells, 1961; also Chapter 10).

Constraints on the Responses to Be Learned Constraints on learning may be imposed by the nature of the response to be learned. For example, the requirement that normally passive frogs and toads make an active avoidance response may be incompatible with the predominant tendencies of these animals in the test situation (Van Bergeijk, 1967; Boice, 1970a).

Constraints as a Function of the Nature of Reinforcers Some constraints on learning are imposed as a result of the nature of the particular reinforcer used in a particular situation. Hogan (1967) compared the performance of Siamese fighting fish (bettas) when learning to swim for food reinforcement versus the opportunity to display to a mirror. While the rate of acquisition was similar for both tasks, the asymptotic rate of swimming to display was lower and more variable than that for food. Swimming for display showed much more rapid extinction than swimming for food. When schedules of reinforcement were imposed, the number of responses increased as a function of increases in the ratio of responses required to produce a food reinforcement. No such increase was seen with display reinforcement (Hogan, Kleist, & Hutchings, 1970).

Constraints Resulting from Stimulus-Response Interactions Some constraints are imposed by the interaction of particular stimuli and responses. Thus, in song learning by white-crowned sparrows, only the species-characteristic song is effective in the development of singing (Marler, 1970). While it is likely that the birds could be conditioned to respond to the songs of other species as a cue in a discriminated operant task, that stimulus is ineffective in the learning of the particular response of singing.

Constraints Resulting from Stimulus-Reinforcer Interactions An excellent example of constraints on learning imposed by the nature of stimulus-reinforcer interactions is provided by research on the Garcia effect in rats that was cited above (e.g., Garcia & Koelling, 1966). Thus, an auditory-visual stimulus is readily associated with shock reinforcement, whereas a taste stimulus is not. By contrast, a taste stimulus is more readily associated with illness than is an auditory-visual stimulus.

Constraints Resulting from Response-Reinforcer Interactions Constraints on learning can be imposed by the particular interaction between the response to be learned and the reinforcer used in a particular learning task. For example, sticklebacks can be readily trained to either swim through a ring or bite a rod when the opportunity to display to a ripe female is used as a reinforcer (Sevenster, 1968). However, whereas swimming through the ring produced a high rate of response, biting the rod produced a low rate. This difference was interpreted as the result of the incompatibility of the biting response, common in aggressive contexts, and the sexual nature of the reinforcement.

A second example of response-reinforcer interactions can be seen in the phenomenon of *autoshaping*. A hungry pigeon is placed in an operant chamber and presented with food on an irregular schedule regardless of its behavior. If a key is briefly lighted before each food presentation, the bird will come to peck the key even though it is not reinforced for doing so (Brown & Jenkins, 1968). Indeed, birds will continue to peck the key even when pecking specifically prevents the presentation of food (Williams & Williams, 1969).

Shettleworth (1975) studied a set of 20 to 24 mutually exclusive behavioral patterns in hamsters. Some behavioral patterns, particularly those related to locomotion and active contact with the environment, increased as feeding time approached for food-deprived hamsters. The occurrence of other behavioral patterns, such as grooming and scent-marking behavior, was depressed near feeding time. In a later experiment, Shettleworth rewarded hamsters for spending appreciable amounts of time engaged in particular behavioral patterns. Whereas those behavioral patterns that were common prior to feeding showed substantial increases in response to the reward contingency, other responses did not. Thus, when rewarded for digging, rearing in the open, or scrabbling, hamsters showed rapid increases in time spent in these behavioral patterns. No such increases were seen with reinforcement for face washing, scratching, or scent marking. In interpreting these data, Shettleworth suggested that conditioning mechanisms facilitate "only behaviors appropriate to anticipation of the reinforcer rather than any reinforced response, as is usually assumed" (p. 56).

Constraints Resulting from Sensitive Periods A final kind of constraint on learning results from the different developmental stages. Such sensitive periods for learning have been clearly demonstrated for filial imprinting (Hess, 1959) and song learning (Marler, 1970).

Conclusions Regarding Adaptive Significance and Learning

The surge of interest that has been expressed in the study of learning in an adaptive context has produced some intriguing new results and has spurred a resurgence of the comparative approach to studies of learning. As different workers consider the implications of specific learning abilities and the constraints on learning, a whole new and refreshing perspective is gained on the study of learning. However, it would be a mistake to discard completely all research conducted in the process-oriented tradition. Research conducted in that tradition has produced a body of methodologically sound research together with some important principles. The comprehensive understanding of learning must come from an integration of different approaches to different questions. The challenge for the 1980s may lie in integrating diverse information on the development, mechanisms, evolution, and adaptive significance of learning. Although these aspects of behavior are often studied by different individuals who have different goals, the importance of these differences can easily be exaggerated. We return to the general theme of this book. A comprehensive understanding of behavior requires observation and description, followed by a consideration of all four classes of questions to be asked of behavior—questions of development, mechanism, evolution, and adaptive significance. It is to be hoped that information on learning gathered from different perspectives can be integrated to produce a balanced understanding of this important phenomenon.

SUMMARY

The study of learning in invertebrates has focused on the question of whether or not invertebrates of various taxa are able to learn. Whereas there is considerable evidence for habituation in protozoans, a definitive demonstration of associative learning is not available. Both classical and instrumental learning have been demonstrated in planarians. Associative learning has been demonstrated in segmented worms, insects, and mollusks.

Studies of quantitative differences in learning among different species are fraught with difficulties of equivalency and interpretation. Some species differences have been demonstrated for performance in problems such as reversal learning, learning sets, and delayed alternation. Bases for skepticism regarding the interpretation of these performance differences stem from considerations of individual differences as well as differences in motivation, reinforcement, sensory function, motor capacities, and species-typical preparedness.

The search for qualitative differences in the processes subserving learning is based on comparisons of functional relations and control by systematic variation. While some differences have been found, their interpretation as "qualitative" remains somewhat controversial.

As an outgrowth of the interaction of ethologists and comparative psychologists, increased attention has been paid to specific adaptations related to learning. This research has been referred to under such rubrics as "preparedness" and "constraints on learning." An ecological-ethological analysis of learning reveals new perspectives on the capacities of different species for learning and on the nature of the processes subsuming learning. There are considerable differences among different species in the nature of what is learned. Constraints on learning can arise from the nature of the stimulus, response, or reinforcer; the interaction of these three; and from the developmental stage of the animal.

A comprehensive understanding of behavior can be obtained when observation and description are followed by the consideration of questions of development, mechanism, evolution, and adaptive significance.

References

Adkins, E. K. Functional castration of the female Japanese quail. *Physiology and Behavior*, 1973, **10**, 619–621.

Adkins, E. K. Hormonal basis of sexual differentiation in the Japanese quail. *Journal of Comparative and Physiological Psychology*, 1975, **89**, 61–71.

Adkins, E. K., & Adler, N. T. Hormonal control of behavior in the Japanese quail. *Journal of Comparative and Physiological Psychology*, 1972, **81**, 27–36.

Adler, N. T. Effects of the male's copulatory behavior on successful pregnancy of the female rat. *Journal of Comparative and Physiological Psychology*, 1969, **69**, 613–622.

Adler, N. T. The behavioral control of reproductive physiology. In W. Montagna & W. A. Sadler (Eds.), *Reproductive behavior*. New York: Plenum, 1974. Pp. 259–286.

Adler, N., & Bermant, G. Sexual behavior of male rats: Effects of reduced sensory feedback. *Journal of Comparative and Physiological Psychology*, 1966, **61**, 240–243.

Adler, N. T., Resko, J. A., & Goy, R. W. The effect of copulatory behavior on hormonal change in the female rat prior to implantation. *Physiology and Behavior*, 1970, **5**, 1003–1007.

Adler, N. T., & Zoloth, S. R. Copulatory behavior can inhibit pregnancy in female rats. *Science*, 1970, **168**, 1480–1482.

Ahlskog, J. E., Randall, P. K., & Hoebel, B. G. Hypothalamic hyperphagia: Dissociation from hyperphagia following destruction of noradrenergic neurons. *Science*, 1975, **190**, 399–401.

Alcock, J. The evolution of the use of tools by feeding animals. *Evolution*, 1972, **26**, 464–473.

Alcock, J. Cues used in searching for food by red-wing blackbirds (*Agelaius phoeniceus*). *Behaviour*, 1973, **46**, 174–187.

Alcock, J. *Animal behavior: An evolutionary approach.* Sunderland, Mass.: Sinauer, 1975.

Alexander, R. D. The evolution of mating behaviour in arthropods. In K. C. Highnam (Ed.), *Insect reproduction.* London: Royal Entomological Society, 1964. Pp. 78–94.

Alexander, R. D. Arthropods. In T. A. Sebeok (Ed.), *Animal communication: Techniques of study and results of research.* Bloomington, Ind.: Indiana University Press, 1968. Pp. 167–216.

Alexander, R. D. The evolution of social behavior. *Annual Review of Ecology and Systematics*, 1974, **5**, 325–383.

Alexander, R. D., & Brown, W. L., Jr. Mating behavior and the origin of insect wings. *Occasional Papers of the Museum of Zoology, University of Michigan*, 1963, **628**, 1–19.

Allison, T., & VanTwyver, H. The evolution of sleep. *Natural History*, 1970, **79**(2), 56–65.

Altman, J., & Sudarshan, K. Postnatal development of locomotion in the laboratory rat. *Animal Behaviour*, 1975, **23**, 896–920.

Altmann, M. Naturalistic studies of maternal care in moose and elk. In H. L. Rheingold (Ed.), *Maternal behavior in mammals.* New York: Wiley, 1963. Pp. 233–253.

Altmann, S. A. Baboons, space, time, and energy. *American Zoologist*, 1974, **14**, 221–248.

Andersen, D. C., Armitage, K. B., & Hoffmann, R. S. Socioecology of marmots: Female reproductive strategies. *Ecology*, 1976, **57**, 552–560.

Andrew, R. J. The origin and evolution of the calls and facial expressions of the primates. *Behaviour*, 1963, **20**, 1–109.

Andrew, R. J. Recognition processes and behavior, with special reference to effects of testosterone on persistence. *Advances in the Study of Behavior*, 1972, **4**, 175–208.

Anonymous. Effects of sexual activity on beard growth in man. *Nature*, 1970, **226**, 869–870.

Applewhite, P. B. Similarities in protozoan and flatworm habituation behavior. *Nature New Biology*, 1971, **230**, 284–285.

Applewhite, P. B. Behavioral plasticity in the sensitive plant, *Mimosa. Behavioral Biology*, 1972, **7**, 47–53.

Applewhite, P. B. Learning in bacteria, fungi, and plants. In W. C. Corning, J. A. Dyal, & A. O. D. Willows (Eds.), *Invertebrate learning: Vol. 3. Cephalopods and echinoderms.* New York: Plenum, 1975. Pp. 179–186.

Applewhite, P. B., & Gardner, F. T. Theory of protozoan habituation. *Nature New Biology*, 1971, **230**, 285–287.

Applewhite, P. B., & Gardner, F. T. Tube-escape behavior of paramecia. *Behavioral Biology*, 1973, **9**, 245–250.

Arai, Y., & Gorski, R. A. Effect of anti-estrogen on steroid induced sexual receptivity in ovariectomized rats. *Physiology and Behavior*, 1968 **3**, 351–353.

Archer, J. Effects of aggressive behavior on the adrenal cortex in male laboratory mice. *Journal of Mammalogy*, 1970, **51**, 327–332.

Archer, J. Tests for emotionality in rats and mice: A review. *Animal Behaviour*, 1973, **21**, 205–235.

Armitage, K. B. Social behaviour of a colony of the yellowbellied marmot *(Marmota flaviventris). Animal Behaviour*, 1962, **10**, 319–331.

Armus, H. L. Conditioning of the sensitive plant, *Mimosa pudica.* In M. R. Denny & S. C. Ratner (Eds.), *Comparative psychology: Research in animal behavior.* (2d ed.) Homewood, Ill.: Dorsey, 1970. Pp. 597–600.

Aronson, L. R., & Cooper, M. L. Desensitization of the glans penis and sexual behavior in cats. In M. Diamond (Ed.), *Reproduction and sexual behavior.* Bloomington, Ind.: Indiana University Press, 1968. Pp. 51–82.

Aschoff, J. Exogenous and endogenous components in circadian rhythms. *Cold Spring Harbor Symposium on Quantitative Biology*, 1960, **25**, 11–26.

Atz, J. W. The application of the idea of homology to behavior. In L. R. Aronson, E. Tobach, D. S. Lehrman, & J. S. Rosenblatt (Eds.), *Development and evolution of behavior.* San Francisco: Freeman, 1970. Pp. 53–74.

Avis, H. H. The neuropharmacology of aggression: A critical review. *Psychological Bulletin,* 1974, **81**, 47–63.

Bacon, W. E., & Stanley, W. C. Reversal learning in neonatal dogs. *Journal of Comparative and Physiological Psychology*, 1970, **70,** 344–350.

Baggerman, B. Some endocrine aspects of fish migration. *General and Comparative Endocrinology*, 1962, Suppl. 1, 188–205.

Baker, R. H. Nutritional strategies of myomorph rodents in North American grasslands. *Journal of Mammalogy*, 1971, **52**, 800–805.

Balaban, M., & Hill, J. Perihatching behavior patterns of chick embryos *(Gallus domesticus). Animal Behaviour*, 1969, **17**, 430–439.

Baldwin, J. D. The ontogeny of social behavior in squirrel monkeys *(Saimiri sciureus)* in a semi-natural environment. *Folia Primatologica*, 1969, **11**, 35–79.

Baldwin, J. D. The social organization of a semifree-ranging troop of squirrel monkeys *(Saimiri sciureus). Folia Primatologica*, 1971, **14**, 23–50.

Bamber, R. T., & Boice, R. The labyrinth method of comparing wild and domestic rats: Origins of animal psychology revisited. *Psychonomic Science*, 1972, **29**, 161–163.

Barash, D. P. The social biology of the Olympic marmot. *Animal Behaviour Monographs*, 1973a, **6**, 171–245.

Barash, D. P. Social variety in the yellow-bellied marmot *(Marmota flaviventris). Animal Behaviour*, 1973b, **21**, 579–584.

Barash, D. P. The evolution of marmot societies: A general theory. *Science*, 1974, **185**, 415–420.

Barber, H. S. North American fireflies of the genus *Photuris. Smithsonian Institution Miscellaneous Collections*, 1951, **117**, 1–58.

Bard, P. The hypothalamus and sexual behavior. *Research Publication of the Association for Research in Nervous and Mental Diseases*, 1940, **20**, 551–579.

Barfield, R. J. Activation of sexual and aggressive behavior by androgen implanted into the male ring dove brain. *Endocrinology*, 1971, **89**, 1470–1476.

Barfield, R. J. & Geyer, L. A. Sexual behavior: Ultrasonic postejaculatory song of the male rat. *Science*, 1972, **176**, 1349–1350.

Barfield, R. J., Wilson, C., & McDonald, P. G. Sexual behavior: Extreme reduction of postejaculatory refractory period by midbrain lesions in male rats. *Science*, 1975, **189**, 147–149.

Barlow, G. W., & Green, R. F. The problems of appeasement and of sexual roles in the courtship behavior of the blackchin mouthbreeder, *Tilapia melanotheron* (Pisces: Cichlidae). *Behaviour*, 1970, **36**, 84–115.

Barnett, S. A. *A study in behaviour.* London: Methuen, 1963.
Barnett, S. A., & Cowan, P. E. Activity, exploration, curiosity, and fear: An ethological study. *Interdisciplinary Science Reviews*, 1976, **1**, 43–62.
Barnett, S. A., & Walker, K. Z. Early stimulation, parental behavior, and the temperature of infant mice. *Developmental Psychobiology*, 1974, **7**, 563–577.
Bateman, A. J. Intra-sexual selection in *Drosophila. Heredity*, 1948, **2**, 349–368.
Bateson, P. P. G. Imprinting. In H. Moltz (Ed.), *The ontogeny of vertebrate behavior.* New York: Academic Press, 1971. Pp. 369–387.
Bayroff, A. G. The experimental social behavior of animals: II. The effect of early isolation of white rats on their competition in swimming. *Journal of Comparative Psychology*, 1940, **29**, 392–306.
Beach, F. A. Effects of injury to the cerebral cortex upon sexually-receptive behavior in the female rat. *Psychosomatic Medicine*, 1944, **6**, 40–55.
Beach, F. A. The snark was a boojum. *American Psychologist*, 1950, **5**, 115–124.
Beach, F. A. The descent of instinct. *Psychological Review*, 1955, **62**, 401–410.
Beach, F. A. Evolutionary aspects of psychoendocrinology. In A. Roe & G. G. Simpson (Eds.), *Behavior and evolution.* New Haven, Conn.: Yale University Press, 1958. Pp. 81–102.
Beach, F. A. Experimental investigations of species-specific behavior. *American Psychologist*, 1960, **15**, 1–18.
Beach, F. A. Ontogeny of "coitus-related" reflexes in the female guinea pig. *Proceedings of the National Academy of Sciences, U.S.A.*, 1966, **56**, 526–533.
Beach, F. A. Cerebral and hormonal control of reflexive mechanisms involved in copulatory behavior. *Physiological Reviews*, 1967, **47**, 289–316.
Beach, F. A. Effects of gonadal hormones on urinary behavior in dogs. *Physiology and Behavior*, 1974, **12**, 1005–1013.
Beach, F. A. Behavioral endocrinology: An emerging discipline. *American Scientist*, 1975, **63**, 178–187.
Beach, F. A. Sexual attractivity, proceptivity, and receptivity in female mammals. *Hormones and Behavior*, 1976, **7**, 105–138.
Beach, F. A., & Jordan, L. Sexual exhaustion and recovery in the male rat. *Quarterly Journal of Experimental Psychology*, 1956, **8**, 121–133.
Beach, F. A., & LeBoeuf, B. J. Coital behavior in dogs: I. Preferential mating in the bitch. *Animal Behaviour*, 1967, **15**, 546–558.
Beck, B. B. A study of problem solving by gibbons. *Behaviour*, 1967, **28**, 95–109.
Beer, C. G. Individual recognition of voice in the social behavior of birds. *Advances in the Study of Behavior*, 1970, **3**, 27–74.
Beer, C. G. Species-typical behavior and ethology. In D. A. Dewsbury & D. A. Rethlingshafer (Eds.), *Comparative psychology: A modern survey.* New York: McGraw-Hill, 1973. Pp. 21–77.
Behrend, E. R., & Bitterman, M. E. Sidman avoidance in the fish. *Journal of the Experimental Analysis of Behavior*, 1963, **6**, 47–52.
Bekoff, M. Social play and play-soliciting by infant canids. *American Zoologist*, 1974a, **14**, 323–340.
Bekoff, M. Social play in coyotes, wolves, and dogs. *Bioscience*, 1974b, **24**, 225–230.
Bekoff, M., & Fox, M. W. Postnatal neural ontogeny: Environment-dependent and/or environment-expectant? *Developmental Psychobiology*, 1972, **5**, 323–341.
Bekoff, M., Hill, H. L., & Mitton, J. B. Behavioral taxonomy in canids by discrimination function analyses. *Science*, 1975, **190**, 1223–1225.

Bennett, D. A., & Francis, D. Learning in *Stentor*. *Journal of Protozoology*, 1972, **19**, 484–487.

Bennett, E. L., & Calvin, M. Failure to train planarians reliably. *Neuroscience Research Program Bulletin*, 1964, **2**, 3–24.

Bennett, M. V. L. Comparative physiology: Electric organs. *Annual Review of Physiology*, 1970, **32**, 471–528.

Bentley, D., & Hoy, R. R. The neurobiology of cricket song. *Scientific American*, 1974, **231**(2), 34–44.

Benzer, S. Genetic dissection of behavior. *Scientific American*, 1973, **229**(6), 24–37.

Bermant, G., & Davidson, J. M. *Biological bases of sexual behavior*. New York: Harper & Row, 1974.

Bermant, G., & Gary, N. E. Discrimination training and reversal in groups of honey bees. *Psychonomic Science*, 1966, **5**, 179–180.

Bernard, B. K. Frog killing (ranacide) in the male rat: Lack of effect of hormonal manipulations. *Physiology and Behavior*, 1974, **12**, 405–408.

Bernstein, I. S. Dominance, aggression, and reproduction in primate societies. *Journal of Theoretical Biology*, 1976, **60**, 459–472.

Bertolini, A., Gessa, G. L., & Ferrari, W. Penile erection and ejaculation: A central effect of ACTH-like peptides in mammals. In M. Sandler & G. L. Gessa (Eds.), *Sexual behavior: Pharmacology and biochemistry*. New York: Raven, 1975. Pp. 247–257.

Bertram, B. The vocal behaviour of the Indian Hill mynah, *Gracula religiosa*. *Animal Behaviour Monographs*, 1970, **3**, 81–192.

Best, J. B. Behavior of planaria in instrumental learning paradigms. *Animal Behaviour*, 1965, Suppl. 1, 69–75.

Bicker, G., & Spatz, H. C. Maze-learning ability of *Drosophila melanogaster*. *Nature*, 1976, **260**, 371.

Bignami, G. Selection for high rates and low rates of conditioning in the rat. *Animal Behaviour*, 1965, **13**, 221–227.

Birdsall, D. A., & Nash, D. Occurrence of successful multiple insemination of females in natural populations of deer mice *(Peromyscus maniculatus)*. *Evolution*, 1973, **27**, 106–110.

Bitterman, M. R. Toward a comparative psychology of learning. *American Psychologist*, 1960, **15**, 704–712.

Bitterman, M. E. Phyletic differences in learning. *American Psychologist*, 1965a, **20**, 396–410.

Bitterman, M. E. The evolution of intelligence. *Scientific American*, 1965b, **212**(1), 92–100.

Bitterman, M. E. Critical commentary. In W. C. Corning, J. A. Dyal, & A. O. D. Willows (Eds.), *Invertebrate learning: Vol. 3. Cephalopods and echinoderms*. New York: Plenum, 1975a. Pp. 139–145.

Bitterman, M. E. The comparative analysis of learning. *Science*, 1975b, **188**, 699–709.

Bitterman, M. E. Flavor aversion studies. *Science*, 1976, **192**, 266–267.

Bitterman, M. E., Wodinsky, J., & Candland, D. K. Some comparative psychology. *American Journal of Psychology*, 1958, **71**, 94–110.

Black-Cleworth, P. The role of electrical discharges in the non-reproductive social behaviour of *Gymnotus carapo* (Gymnotidae, Pisces). *Animal Behaviour Monographs*, 1970, **3**, 1–77.

Blest, A. D. The evolution of protective displays in the Saturnioidea and Sphingidae (Lepidoptera). *Behaviour*, 1957, **11**, 257–310.

Blest, A. D. The concept of ritualisation. In W. H. Thorpe & O. L. Zangwill (Eds.), *Current problems in animal behaviour.* Cambridge: Cambridge University Press, 1961. Pp. 102–124.

Block, R., & McConnell, J. V. Classically conditioned discrimination in the planarian, *Dugesia dorotocephala. Nature*, 1967, **215,** 1465–1466.

Blough, D. S. Method for tracing dark adaptation in the pigeon. *Science*, 1955, **121**, 703–704.

Blough, D. S. Experiments in animal psychophysics. *Scientific American*, 1961, **205**(1), 113–122.

Bock, W. J. The concept of homology. *Annals of the New York Academy of Sciences*, 1969, **167**, 71–73.

Bogert, C. M. The influence of sound on the behavior of amphibians and reptiles. In W. E. Lanyon & W. N. Tavolga (Eds.), *Animal sounds and communication.* Washington, D. C.: American Institute of Biological Sciences, 1960. Pp. 137–320.

Bohus, B., & deWied, D. Pituitary-adrenal system hormones and adaptive behavior. In I. Chester Jones & I. W. Henderson (Eds.), *General, comparative and clinical endocrinology of the adrenal cortex.* Vol. 3. London: Academic Press, 1976, in press.

Boice, R. Avoidance learning in active and passive frogs and toads. *Journal of Comparative and Physiological Psychology*, 1970a, **70**, 154–156.

Boice, R. Effect of domestication on avoidance learning in the Norway rat. *Psychonomic Science*, 1970b, **18**, 13–14.

Boice, R. Laboratizing the wild rat *(Rattus norvegicus). Behavior Research Methods and Instrumentation*, 1971, **3**, 177–182.

Boice, R. Some behavioral tests of domestication in Norway rats. *Behaviour*, 1972, **42**, 198–231.

Boice, R. Domestication. *Psychological Bulletin*, 1973, **80**, 215–230.

Bolles, R. C. Species-specific defense reactions and avoidance learning. *Psychological Review*, 1970, **77**, 32–48.

Bonnell, M. L., & Selander, R. K. Elephant seals: Genetic variation and near extinction. *Science*, 1974, **184**, 908–909.

Borchelt, P. L., Griswold, J. G., & Branchek, R. S. An analysis of sandbathing and grooming in the kangaroo rat *(Dipodomys merriami). Animal Behaviour*, 1976, **24**, 347–353.

Boring, E. G. *A history of experimental psychology.* New York: Appleton-Century-Crofts, 1957.

Borsellino, A., Pierantoni, R., & Schieti-Cavazza, B. Survival in adult mealworm beetles *(Tenebrio molitor)* of learning acquired at the larval stage. *Nature*, 1970, **225**, 963–964.

Bourliere, F. *The natural history of mammals.* (3d ed.) New York: Knopf, 1964.

Bovet, D., Bovet-Nitti, F., & Oliverio, A. Genetic aspects of learning and memory in mice. *Science*, 1969, **163**, 139–149.

Boycott, B. B., & Young, J. Z. The comparative study of learning. *Symposium of the Society for Experimental Biology*, 1950, **4**, 432–453.

Breland, K., & Breland, M. The misbehavior of organisms. *American Psychologist*, 1961, **16**, 681–684.

Broadhurst, P. L. Experiments in psychogenetics: Applications of biometrical genetics to the inheritance of behavior. In H. J. Eysenck (Ed.), *Experiments in personality.* Vol. 1. *Psychogenetics and psychopharmacology.* London: Routledge & Kegan Paul, 1960. Pp. 1–102.

Broadhurst, P. L. An introduction to the diallel cross. In J. Hirsch (Ed.), *Behavior-genetic analysis.* New York: McGraw-Hill, 1967. Pp. 287–304.

Bronson, F. H. Rodent pheromones. *Biology of Reproduction,* 1971, **4**, 344–357.

Brooks, R. J., & Banks, E. M. Behavioural biology of the collared lemming *[Dicrostonyx groenlandicus* (Traill)*]:* An analysis of acoustic communication. *Animal Behaviour Monographs*, 1973, **6**, 1–83.

Brookshire, K. H. Comparative psychology of learning. In M. H. Marx (Ed.), *Learning: Interactions.* New York: Macmillan, 1970. Pp. 291–364.

Brower, L. P. Ecological chemistry. *Scientific American*, 1969, **220**(2), 22–29.

Brower, L. P., & Brower, J. V. Z. Investigations into mimicry. *Natural History*, 1962, **71**, 8–19.

Brower, L. P., Brower, J. V. Z., & Cranston, E. P. Courtship behavior of the queen butterfly, *Danaus gilippus. Zoologica*, 1965, **50**, 1–39.

Brown, F. A., Jr. The "clocks" timing biological rhythms. *American Scientist*, 1972, **60**, 756–766.

Brown, H. M., Dustman, R. E., & Beck, E. C. Sensitization in planaria. *Physiology and Behavior*, 1966, **1**, 305–308.

Brown, J. L. *The evolution of behavior.* New York: Norton, 1975.

Brown, P. L., & Jenkins, H. M. Auto-shaping of the pigeon's key-peck. *Journal of the Experimental Analysis of Behavior*, 1968, **11**, 1–8.

Brown, R. G. B. Species isolation between the herring gull *Larus argentatus* and lesser black-backed gull *L. fuscus. Ibis*, 1967, **109**, 310–317.

Bruce, H. M. A block to pregnancy in the mouse caused by proximity of strange males. *Journal of Reproduction and Fertility*, 1960, **1**, 96–103.

Bruell, J. H. Inheritance of behavioral and physiological characters of mice and the problem of heterosis. *American Zoologist*, 1964, **4**, 125–138.

Bruell, J. H. Behavioral heterosis. In J. Hirsch (Ed.), *Behavior-genetic analysis.* New York: McGraw-Hill, 1967. Pp. 270–286.

Budker, P. *The life of sharks.* New York: Columbia University Press, 1971.

Bunnell, B. N. Mammalian behavior patterns. In D. A. Dewsbury & D. A. Rethlingshafer (Ed.), *Comparative psychology: A modern survey.* New York: McGraw-Hill, 1973.

Bunnell, B. N., Sodetz, F. J., & Shalloway, D. I. Amygdaloid lesions and social behavior in the golden hamster. *Physiology and Behavior*, 1970, **5**, 153–161.

Burghardt, G. M. Chemical-cue preferences of inexperienced snakes: Comparative aspects. *Science*, 1967, **157**, 718–721.

Burghardt, G. M. Instinct and innate behavior: Toward an ethological psychology. In J. A. Nevin (Ed.), *The study of behavior.* Glenview, Ill.: Scott, Foresman, 1973. Pp. 322–400.

Burghardt, G. M., Wilcoxon, H. C., & Czaplicki, J. A. Conditioning in garter snakes: Aversion to palatable prey induced by delayed illness. *Animal Learning and Behavior*, 1973, **1**, 317–320.

Buss, A. H. *The psychology of aggression.* New York: Wiley, 1961.

Butler, R. A. Acquired drives and the curiosity-investigative motives. In R. H. Waters,

D. A. Rethlingshafer & W. E. Caldwell (Eds.), *Principles of comparative psychology*. New York: McGraw-Hill, 1960. Pp. 144–176.

Campbell, B. A. Developmental studies of learning and motivation in infra-primate mammals. In H. W. Stevenson, E. H. Hess, & H. L. Rheingold (Eds.), *Early behavior: Comparative and developmental approaches*. New York: Wiley, 1967. Pp. 43–71.

Campbell, B. A., & Campbell, E. H. Retention and extinction of learned fear in infant and adult rats. *Journal of Comparative and Physiological Psychology*, 1962, **55**, 1–8.

Campbell, B. A., & Jaynes, J. Reinstatement. *Psychological Review*, 1966, **73**, 478–480.

Campbell, B. A., Jaynes, J. R., & Misanin, J. Retention of a light-dark discrimination in rats of different ages. *Journal of Comparative and Physiological Psychology*, 1968, **66**, 467–472.

Campbell, B. A., & Mabry, P. D. Ontogeny of behavioral arousal: A comparative study. *Journal of Comparative and Physiological Psychology*, 1972, **81**, 371–379.

Campbell, B. A., Misanin, J. R., White, B. C., & Lytle, L. D. Species differences in ontogeny of memory: Indirect support for neural maturation as a determinant of forgetting. *Journal of Comparative and Physiological Psychology*, 1974, **87**, 193–202.

Campbell, B. A., & Spear, N. E. Ontogeny of memory. *Psychological Review*, 1972, **79**, 215–236.

Campbell, C. B. G., & Hodos, W. The concept of homology and the evolution of the nervous system. *Brain, Behavior, and Evolution*, 1970, **3**, 353–367.

Campbell, D. T. On the conflicts between biological and social evolution and between psychology and moral tradition. *American Psychologist*, 1975, **30**, 1103–1126.

Candland, D. K. Discriminability of facial regions used by the domestic chicken in maintaining the social dominance order. *Journal of Comparative and Physiological Psychology*, 1969, **69**, 281–285.

Candland, D. K., & Campbell, B. A. Development of fear in the rat as measured by behavior in the open field. *Journal of Comparative and Physiological Psychology*, 1962, **55**, 593–596.

Candland, D. K., & Nagy, Z. M. The open field: Some comparative data. *Annals of the New York Academy of Sciences*, 1969, **159**, 831–851.

Carey, M., & Nolan, V., Jr. Polygyny in indigo buntings: A hypothesis tested. *Science*, 1975, **190**, 1296–1297.

Carlson, N. R., & Thomas, G. J. Maternal behavior of mice with limbic lesions. *Journal of Comparative and Physiological Psychology*, 1968, **66**, 731–737.

Carlsson, S. G., & Larsson, K. Mating in male rats after local anesthetization of the glans penis. *Zeitschrift für Tierpsychologie*, 1964, **21**, 854–856.

Carpenter, C. R. Sexual behavior of free ranging rhesus monkeys *(Macaca mulatta):* 1. Specimens, procedures and behavioral characteristics of estrus. *Journal of Comparative Psychology*, 1942, **33**, 113–142.

Carr, W. J., Loeb, L. S., & Dissinger, M. L. Responses of rats to sex odors. *Journal of Comparative and Physiological Psychology*, 1965, **59**, 370–377.

Carr, W. J., Solberg, B., & Pfaffmann, C. The olfactory threshold for estrous female urine in normal and castrated male rats. *Journal of Comparative and Physiological Psychology*, 1962, **55**, 415–417.

Carran, A. B. Biometrics of reversal learning in mice: II. Diallel cross. *Journal of Comparative and Physiological Psychology*, 1972, **78**, 466–470.

Carran, A. B. Biometrics of reversal learning in mice: III. Complete reciprocated Mendelian analysis. *Journal of Comparative and Physiological Psychology*, 1975, **88**, 878–881.

Caul, W. F., & Barrell, R. J. Shuttle-box versus Y-maze avoidance: Value of multiple response measures in interpreting active-avoidance performance of rats. *Journal of Comparative and Physiological Psychology*, 1973, **84**, 572–578.

Chalmers, N. R. The social behavior of free living mangabeys in Uganda. *Folia Primatologica*, 1968, **8**, 263–281.

Chester, R. V., & Zucker, I. Influence of male copulatory behavior on sperm transport, pregnancy, and pseudopregnancy in female rats. *Physiology and Behavior*, 1970, **5**, 35–43.

Christian, J. J. Social subordination, population density, and mammalian evolution. *Science*, 1970a, **168**, 84–90.

Christian, J. J. Mammalian evolution: Is it due to social subordination? *Science*, 1970a, **168**, 345–346.

Christian, J. J., & Davis, D. E. Endocrines, behavior, and population. *Science*, 1964, **146**, 1550–1560.

Clark, E., Aronson, L. R., & Gordon, M. Mating behavior patterns in two sympatric species of xiphophorin fishes; their inheritance and significance in sexual isolation. *Bulletin of the American Museum of Natural History*, 1954, **103**, 139–225.

Clark, T. K., Caggiula, A. R., McConnell, R. A., & Antelman, S. M. Sexual inhibition is reduced by rostral midbrain lesions in the male rat. *Science*, 1975, **170**, 169–171.

Clemens, L. G. Development and behavior. In D. A. Dewsbury & D. A. Rethlingshafer (Eds.), *Comparative psychology: A modern survey*. New York: McGraw-Hill, 1973. Pp. 238–268.

Clemens, L. G., Wallen, K., & Gorski, R. A. Mating behavior: Facilitation in the female rat after cortical application of potassium chloride. *Science*, 1967, **157**, 1208–1209.

Collins, R. A. Aggression in mice selectively bred for brain weight. *Behavior Genetics*, 1970, **1**, 169–171.

Collins, R. L. Inheritance of avoidance conditioning in mice: A diallel study. *Science*, 1964, **143**, 1188–1190.

Collins, R. L., & Fuller, J. L. Audiogenic seizure prone (asp): A gene affecting behavior in linkage group VIII of the mouse. *Science*, 1968, **162**, 1137–1139.

Conaway, C. H. Ecological adaptation and mammalian reproduction. *Biology of Reproduction*, 1971, **4**, 239–247.

Conner, R. L. Hormones, biogenic amines, and aggression. In S. Levine (Ed.), *Hormones and behavior*. New York: Academic Press, 1972. Pp. 209–233.

Conner, R. L., Stolk, J. M., Barchas, J. D., Dement, W. C., & Levine, S. The effect of parachlorophenylalanine (PCPA) on shock-induced fighting behavior in rats. *Physiology and Behavior*, 1970, **5**, 1221–1224.

Connor, J. L. Genetic mechanisms controlling the domestication of a wild house mouse population *(Mus musculus* L*). Journal of Comparative and Physiological Psychology*, 1975, **89**, 118–130.

Connor, J. L., & Winston, H. Genetic analysis of conditioned emotional responses in the mouse *(Mus musculus* L.*). Journal of Comparative and Physiological Psychology*, 1972, **81**, 37–44.

Corning, W. C., Dyal, J. A., & Willows, A. O. D. (Eds.) *Invertebrate learning*. Vol. 1. *Protozoans through annelids*. New York: Plenum, 1973a.

Corning, W. D., Dyal, J. A., & Willows, A. O. D. *Invertebrate learning.* Vol. 2. *Arthropods and gastropod mollusks.* New York: Plenum, 1973b.

Corning, W. C., Dyal, J. A., & Willows, A. O. D. *Invertebrate learning.* Vol. 3. *Cephalopods and echinoderms.* New York: Plenum, 1975.

Corning, W. C., & Kelly, S. Platyhelminthes: The turbellarians. In W. C. Corning, J. A. Dyal, & A. O. D. Willows (Eds.), *Invertebrate learning.* Vol. 1. *Protozoans through annelids.* New York: Plenum, 1973. Pp. 171–224.

Corning, W. C., & Lahue, R. Invertebrate strategies in comparative learning studies. *American Zoologist,* 1972, **12**, 455–469.

Corning, W. C., & vonBurg, R. Protozoa. In W. C. Corning, J. A. Dyal, & A. O. D. Willows (Eds.), *Invertebrate learning.* Vol. 1. *Protozoans through annelids.* New York: Plenum, 1973. Pp. 49–122.

Cox, C. R., & LeBoeuf, B. J. Female incitation of male competition: A mechanism in sexual selection. *American Naturalist,* 1977, **111**, 317–335.

Crancher, P., King, M. G., Bennett, A., & Montgomery, R. B. Conditioning of a free operant in *Octopus cyaneus. Journal of the Experimental Analysis of Behavior,* 1972, **17**, 359–362.

Crane, J. Basic patterns of display in fiddler crabs (Ocypodidae, Genus *Uca*). *Zoologica,* 1957, **42**, 69–82.

Crawford, M. P. The cooperative solving by chimpanzees of problems requiring serial responses to color cues. *Journal of Social Psychology,* 1941, **13**, 259–280.

Crews, D. Coition-induced inhibition of sexual receptivity in female lizards *(Anolis carolinensis). Physiology and Behavior,* 1973, **11**, 463–468.

Crews, D. Psychobiology of reptilian reproduction. *Science,* 1975, **189**, 1059–1065.

Crook, J. H. Social organization and the environment: Aspects of contemporary social ethology. *Animal Behaviour,* 1970, **18**, 197–209.

Crook, J. H., & Gartlan, J. S. Evolution of primate societies. *Nature,* 1966, **210**, 1200–1203.

Crossley, S. A. Changes in mating behavior produced by selection for ethological isolation between ebony and vestigial mutants of *Drosophila melanoggster. Evolution,* 1974, **28**, 631–647.

Cullen, E. Adaptations in the kittiwake to cliff-nesting. *Ibis,* 1957, **99**, 275–302.

Cullen, J. M. The aerial display of the Arctic tern and other species. *Ardea,* 1960, **48**, 1–39.

Cullen, J. M. Ritualization of animal activities in relation to phylogeny, speciation, and ecology. *Philosophical Transactions of the Royal Society of London,* 1966, **251B**, 363–374.

Daanje, A. On the locomotory movements of birds, and the intention movements derived from them. *Behaviour,* 1950, **3**, 48–98.

Daly, M. Early stimulation of rodents: A critical review of present interpretations. *British Journal of Psychology,* 1973, **64**, 435–460.

Darby, C. L., & Riopelle, A. J. Observational learning in the rhesus monkey. *Journal of Comparative and Physiological Psychology,* 1959, **52**, 94–98.

Darwin, C. *The origin of species.* London: John Murray, 1859.

Darwin, C. *The descent of man and selection in relation to sex.* London: John Murray, 1871.

Darwin, C. *The expression of the emotions in man and animals.* London: Appleton, 1873. (Reprinted: Chicago: University of Chicago Press, 1965.)

Davenport, R. K., & Rogers, C. M. Intellectual performance of differentially reared chimpanzees: I. Delayed response. *American Journal of Mental Deficiency*, 1968, **72**, 674–680.

Davenport, R. K., Rogers, C. M., & Menzel, E. W. Intellectual performance of differentially reared chimpanzees: II. Discrimination-learning set. *American Journal of Mental Deficiency*, 1969, **73**, 963–969.

Davidson, J. R., & Levine, S. Endocrine regulation of behavior. *Annual Review of Physiology*, 1972, **34**, 375–408.

Davis, D. E. The role of density in aggressive behaviour of house mice. *Animal Behaviour*, 1958, **6**, 207–210.

Davis, D. E. The physiological analysis of aggressive behavior. In W. Etkin (Ed.), *Social behavior and organization among vertebrates*. Chicago: University of Chicago Press, 1964. Pp. 53–74.

Davis, H. N., Gray, G. D., & Dewsbury, D. A. Maternal age and male behavior in relation to successful reproduction by female rats *(Rattus norvegicus)*. *Journal of Comparative and Physiological Psychology*, 1977, **91**, 281–289.

Davis, R. E., & Agranoff, B. W. Stages of memory formation in goldfish: Evidence for an environmental trigger. *Proceedings of the National Academy of Sciences, U.S.A.*, 1966, **55**, 555–559.

Dawkins, R. *The selfish gene*. Oxford: Oxford University Press, 1976.

Dawkins, R., & Carlisle, T. R. Parental investment, mate desertion and a fallacy. *Nature*, 1976, **262**, 131–133.

Deag, J. M., & Crook, J. H. Social behavior and "agonistic buffering" in the wild Barbary macaque *Macaca sylvania* L. *Folia Primatologica*, 1971, **15**, 183–200.

DeCoursey, P. J. Phase control of activity in a rodent. *Cold Spring Harbor Symposium on Quantitative Biology*, 1960, **25**, 49–55.

DeFries, J. C. Pleiotropic effects of albinism on open-field behaviour in mice. *Nature*, 1969, **221**, 65–66.

DeFries, J. C., & Hegmann, J. P. Genetic analysis of open-field behavior. In G. Lindzey & D. D. Thiessen (Eds.), *Contributions to behavior-genetic analysis: The mouse as a prototype*. New York: Appleton Century Crofts, 1970. Pp. 23–56.

DeFries, J. C., Hegmann, J. P., & Halcomb, R. A. Response to 20 generations of selection for open-field activity in mice. *Behavioral Biology*, 1974, **11**, 481–495.

DeFries, J. C., & McClearn, G. E. Social dominance and Darwinian fitness in the laboratory mouse. *American Naturalist*, 1970, **104,** 408–411.

DeFries, J. C., Thomas, E. A., Hegmann, J. P., & Weir, M. W. Open-field behavior in mice: Analysis of maternal effects by means of ovarian transplantation. *Psychonomic Science*, 1967, **8**, 207–208.

DeFries, J. C., Weir, M. W., & Hegmann, J. P. Differential effects of prenatal maternal stress on offspring behavior in mice as a function of genotype and stress. *Journal of Comparative and Physiological Psychology*, 1967, **63**, 332–334.

DeFries, J. C., Wilson, J. R., & McClearn, G. E. Open-field behavior in mice: Selection response and situational generality. *Behavior Genetics*, 1970, **1**, 195–211.

Delgado, J. M. R. Cerebral heterostimulation in a monkey colony. *Science*, 1963, **141**, 161–163.

Delgado, J. M. R. Aggressive behavior evoked by radio-stimulation in monkey colonies. *American Zoologist*, 1966, **6**, 669–681.

Delius, J. D. Agonistic behaviour of juvenile gulls, a neuroethological study. *Animal Behaviour,* 1973, **21**, 236–246.

Denenberg, V. H. Critical periods, stimulus input, and emotional reactivity: A theory of infantile stimulation. *Psychological Review,* 1964, **71**, 335–351.

Denenberg, V. H. A consideration of the usefulness of the critical period hypothesis as applied to the stimulation of rodents in infancy. In G. Newton & S. Levine (Eds.), *Early experience and behavior: The psychobiology of development.* Springfield, Ill.: Charles C. Thomas, 1968. Pp. 142–167.

Denenberg, V. H., & Karas, G. C. Interactive effects of infantile and adult experiences upon weight gain and mortality in the rat. *Journal of Comparative and Physiological Psychology,* 1961, **54**, 685–689.

Denenberg, V. H., & Morton, J. R. C. Effects of preweaning and postweaning manipulations upon problem-solving behavior. *Journal of Comparative and Physiological Psychology,* 1962, **55**, 1096–1098.

Denenberg, V. H., Morton, J. R. C., Kline, N. J., & Grota, L. J. Effects of duration of infantile stimulation upon emotionality. *Canadian Journal of Psychology,* 1962, **16**, 72–76.

Denenberg, V. H., & Rosenberg, K. M. Nongenetic transmission of information. *Nature,* 1967, **216**, 549–550.

Denenberg, V. H., & Smith, S. A. Effects of infantile stimulation and age upon behavior. *Journal of Comparative and Physiological Psychology,* 1963, **56**, 307–312.

Denenberg, V. H., & Whimbey, A. E. Behavior of adult rats is modified by the experiences their mothers had as infants. *Science,* 1963, **142**, 1192–1193.

de Ruiter, L. Countershading in caterpillars. *Archives Neerlandaises de Zoologie,* 1956, **11**, 285–341.

Desforges, M. F., & Wood-Gush, D. G. M. A behavioral comparison of domestic and mallard ducks: Spatial relationships in small flocks. *Animal Behaviour,* 1975, **23**, 698–705.

Desjardins, C., Maruniak, J. A., & Bronson, F. H. Social rank in house mice: Differentiation revealed by ultraviolet visualization of urinary marking patterns. *Science,* 1973, **182**, 939–941.

Dethier, V. G. *To know a fly.* San Francisco: Holden-Day, 1962.

Dethier, V. G. The hungry fly. *Psychology Today,* 1967, **1**(2), 64–72.

Dethier, V. G. A surfeit of stimuli: A paucity of receptors. *American Scientist,* 1971, **59**, 706–715.

Dethier, V. G., & Stellar, E. *Animal behavior: Its evolutionary and neurological basis.* Englewood Cliffs, N.J.: Prentice-Hall, 1961.

Deutsch, J. A. The physiological basis of memory. *Annual Review of Psychology,* 1969, **20**, 85–104.

Deutsch, J. A. The cholinergic synapse and the site of memory. *Science,* 1971, **174**, 788–794.

Devine, J. V. Stimulus attributes and training procedures in learning-set formation of rhesus and cebus monkeys. *Journal of Comparative and Physiological Psychology,* 1970, **73**, 62–67.

Devine, M. C. Copulatory plugs in snakes: Enforced chastity. *Science,* 1975, **187**, 844–845.

Dews, P. B. Some observations on an operant in the octopus. *Journal of the Experimental Analysis of Behavior,* 1959, **2**, 57–63.

Dewsbury, D. A. Stimulus-produced changes in the discharge rate of an electric fish and their relation to arousal. *Psychological Record,* 1966, **16**, 495–504.

Dewsbury, D. A. Patterns of copulatory behavior in male mammals. *Quarterly Review of Biology,* 1972, **47**, 1–33.

Dewsbury, D. A. Comparative psychologists and their quest for uniformity. *Annals of the New York Academy of Sciences,* 1973, **223**, 147–167.

Dewsbury, D. A. A diallel cross analysis of genetic determinants of copulatory behavior in rats. *Journal of Comparative and Physiological Psychology,* 1975a, **88**, 713–722.

Dewsbury, D. A. The normal heterosexual pattern of copulatory behavior in male rats: Effects of drugs that alter brain monoamine levels. In M. Sandler & G. L. Gessa (Eds.), *Sexual behavior: Pharmacology and biochemistry.* New York: Raven, 1975b. Pp. 169–179.

Dewsbury, D. A. Diversity and adaptation in rodent copulatory behavior. *Science,* 1975c, **190**, 947–954.

Dewsbury, D. A. The comparative method in studies of reproductive behavior. In T. E. McGill, D. A. Dewsbury, & B. D. Sachs (Eds.) *Sex behavior: Status and prospectus.* New York: Plenum, 1977, in press.

Dewsbury, D. A., & Estep, D. Q. Pregnancy in cactus mice: Effects of prolonged copulation. *Science,* 1975, **187**, 552–553.

Dewsbury, D. A., Lanier, D. L., Estep, D. Q., Oglesby, J. M., & Dudley, D. Prolonged copulatory behavior in rats and hamsters: Protection against subsequent inseminations. 1977, unpublished manuscript.

Dewsbury, D. A., & Lovecky, D. V. Copulatory behavior of old-field mice *(Peromyscus polionotus)* from different natural populations. *Behavior Genetics,* 1974, **4**, 347–355.

Diakow, C. Male-female interactions and the organization of mammalian mating patterns. *Advances in the Study of Behavior,* 1974, **5**, 227–268.

Diamond, M., Diamond, L., & Mast, M. Visual sensitivity and sexual arousal levels during the menstrual cycle. *Journal of Nervous and Mental Disease,* 1972, **155**, 170–176.

Diamond, M., & Yanagimachi, R. Induction of pseudopregnancy in the golden hamster. *Journal of Reproduction and Fertility,* 1968, **17**, 165–168.

Diamond, S. Four hundred years of instinct controversy. *Behavior Genetics,* 1974, **4**, 237–252.

DiGiusto, E. L., Cairncross, K., & King, M. G. Hormonal influences on fear-motivated responses. *Psychological Bulletin,* 1971, **75**, 432–444.

Dilger, W. C. The comparative ethology of the African parrot genus *Agapornis. Zeitschrift für Tierpsychologie,* 1960, **17**, 649–685.

Dilger, W. C. The behavior of lovebirds. *Scientific American,* 1962, **208**(1), 88–98.

Dilly, P. N. Delayed responses in *Octopus. Journal of Experimental Biology,* 1963, **40**, 393–401.

Disterhoft, J. F. Learning in the intact cockroach *(Periplaneta americana)* when placed in a punishment situation. *Journal of Comparative and Physiological Psychology,* 1972, **79**, 1–7.

Donoso, A. O., & Cukier, J. O. Oestrogen as a depressor of noradrenaline concentration in the anterior hypothalamus. *Nature,* 1968, **218**, 969.

Doty, R. L., & Anisko, J. J. Procaine hydrochloride olfactory block eliminates mounting in the male golden hamster. *Physiology and Behavior,* 1973, **10**, 393–397.

Drew, W. G., & Batt, J. A contribution to the evolutionary theory of dreaming: An hypothesis on the role of the pineal in species and specimen protection. *Biological Psychiatry,* 1972, **4**, 131–146.

Drickamer, L. C. Sexual maturation of female house mice: Social inhibition. *Developmental Psychobiology,* 1974, **7**, 257–265.

Dudley, D. Contributions of paternal care to the growth and development of the young in *Peromyscus californicus. Behavioral Biology,* 1974, **11**, 155–166.

Dunlap, J. L., Gerall, A. A., & McLean, L. D. Enhancement of receptivity in neonatally castrated males by prepuberal ovarian transplants. *Physiology and Behavior,* 1973, **10**, 701–705.

Duvall, S. W., Bernstein, I. S., & Gordon, T. P. Paternity and status in a rhesus monkey group. *Journal of Reproduction and Fertility,* 1976, **47**, 25–31.

Dyal, J. A., & Corning, W. C. Invertebrate learning and behavior taxonomies. In W. C. Corning, J. A. Dyal, & A. O. D. Willows (Eds.), *Invertebrate learning.* Vol. 1. *Protozoans through annelids.* New York: Plenum, 1973.

Ehrman, L. The mating advantage of rare males in *Drosophila. Proceedings of the National Academy of Sciences, U.S.A.,* 1970, **65**, 345–348.

Ehrman, L. Genetics and sexual selection. In B. Campbell (Ed.), *Sexual selection and the descent of man 1871–1971.* Chicago: Aldine, 1972.

Eibl-Eibesfeldt, I. *Galapagos.* New York: Doubleday, 1961.

Eibl-Eibesfeldt, I. Angeborenes and Erbworbenes in Verhalten einiger Säuger. *Zeitschrift für Tierpsychologie,* 1963, **20**, 705–754.

Eibl-Eibesfeldt, I. *Ethology: The biology of behavior.* (2d ed.) New York: Holt, 1975.

Eichelman, B. S., & Thoa, N. B. The aggressive monoamines. *Biological Psychiatry,* 1973, **6**, 143–164.

Eisenberg, J. F. The social organizations of mammals. *Handbuch der Zoologie,* 1965, **8**, 1–91.

Eisenberg, J. F., & Kleiman, D. G. Olfactory communication in mammals. *Annual Review of Ecology and Systematics,* 1972, **3**, 1–32.

Eisenberg, J. F., & Leyhausen, P. The phylogenesis of predatory behavior in mammals. *Zeitschrift für Tierpsychologie,* 1972, **30**, 59–93.

Eisenberg, J. F., Muckenhirn, N. A., & Rudran, R. The relation between ecology and social structure in primates. *Science,* 1972, **176**, 863–874.

Eisenstein, E. M. The use of invertebrate systems for studies on the bases of learning and memory. In G. C. Quarton, T. Melnechuk, & F. O. Schmitt (Eds.), *The Neurosciences: A study program.* New York: Rockefeller University Press, 1976. Pp. 653–665.

Emlen, S. T. An experimental analysis of the parameters of bird song eliciting species recognition. *Behaviour,* 1972, **41**, 130–171.

Emlen, S. T. The stellar-orientation system of a migratory bird. *Scientific American,* 1975, **233**(2), 102–111.

Emlen, S. T. Altruism in mountain bluebirds? *Science,* 1976, **191**, 808–809.

Englehardt, F., Woodard, W. T., & Bitterman, M. E. Discrimination reversal in the goldfish as a function of training conditions. *Journal of Comparative and Physiological Psychology,* 1973, **85**, 144–150.

Erickson, C. J. Induction of ovarian activity in female ring doves by androgen treatment of castrated males. *Journal of Comparative and Physiological Psychology,* 1970, **71**, 210–215.

Erickson, C. J., & Martinez-Vargas, C. The hormonal basis of cooperative nest-building. In P. Wright, O. G. Caryl, & D. M. Vowles (Eds.), *Neural and endocrine aspects of behaviour in birds.* Amsterdam: Elsevier, 1975. Pp. 91–109.

Erickson, C. J., & Zenone, P. G. Courtship differences in male ring doves: Avoidance of cuckoldry? *Science,* 1976, **192**, 1353–1354.

Estep, D. Q. Copulatory behavior of *Rattus rattus.* Doctoral dissertation, University of Florida, 1975.

Estep, D. Q., Lanier, D. L., & Dewsbury, D. A. Copulatory behavior and nest building behavior of wild house mice *(Mus musculus). Animal Learning & Behavior,* 1975, **3**, 329–336.

Estes, R. D. Behavior and life history of the wildebeest *(Connochaetes taurinus* Burchell*). Nature,* 1966, **212**, 999–1000.

Estes, R. D. Social organization of the African Bovidae. In V. Geist & F. Walther (Eds.), *The behavior of ungulates and its relation to management.* Morges, Switzerland: International Union for Conservation of Nature and Natural Resources, 1974. Pp. 166–205.

Etkin, W. Types of social organization in birds and mammals. In W. Etkin (Ed.), *Social behavior and organization among vertebrates.* Chicago: University of Chicago Press, 1964. Pp. 256–297.

Evans, H. E. Comparative ethology and the systematics of spider wasps. *Systematic Zoology,* 1953, **2**, 155–172.

Ewer, R. F. *Ethology of mammals.* New York: Plenum, 1968.

Ewer, R. F. Book review. *Animal Behaviour,* 1971, **19**, 802–807.

Fagen, R. Selective and evolutionary aspects of animal play. *American Naturalist,* 1974, **108**, 850–858.

Falconer, D. S. *Introduction to quantitative genetics.* New York: Ronald, 1960.

Fantz, R. L. Ontogeny of perception. In A. M. Schrier, H. F. Harlow, & F. Stollnitz (Eds.), *Behavior of nonhuman primates.* New York: Academic Press, 1965. Pp. 365–403.

Fantz, R. L. Visual perception and experience in early infancy: A look at the hidden side of behavior development. In H. W. Stevenson, E. H. Hess, & H. L. Rheingold (Eds.), *Early behavior: Comparative and developmental approaches.* New York: Wiley, 1967. Pp. 181–224.

Fentress, J. C. Development and patterning of movement sequences in inbred mice. In J. A. Kiger, Jr. (Ed.), *The biology of behavior.* Corvallis, Ore.: Oregon State University Press, 1972. Pp. 83–131.

Fentress, J. C. Development of grooming in mice with amputated forelimbs. *Science,* 1973, **179**, 704–705.

Ferster, C. B., & Skinner, B. F. *Schedules of reinforcement.* New York: Appleton Century Crofts, 1957.

Festing, M. F. W. Water escape learning in mice: I. Strain differences and biometrical considerations. *Behavior Genetics,* 1973a, **3**, 13–24.

Festing, M. F. W. Water escape learning in mice: II. Replicated selection for increased learning speed. *Behavior Genetics,* 1973b, **3**, 25–36.

Festing, M. F. W. Water escape learning in mice: III. A diallel study. *Behavior Genetics,* 1974, **4**, 111–124.

File, S. E., & Scott, E. M. Acquisition and retention of habituation in the preweaning rat. *Developmental Psychobiology,* 1974, **9**, 97–107.

Fischer, G. J., & Campbell, G. L. The development of passive avoidance conditioning in Leghorn chicks. *Animal Behaviour,* 1964, **12**, 268–269.

Fisler, G. F. Mammalian organizational systems. *Los Angeles County Museum Contributions in Science,* 1969, **167**, 1–32.

Fletcher, H. J. The delayed-response problem. In A. M. Schrier, H. F. Harlow, & F. Stollnitz (Eds.), *Behavior of nonhuman primates: Modern research trends.* Vol. 1. New York: Academic Press, 1965. Pp. 129–165.

Flood, N. B., & Overmier, J. B. Effects of telencephalic lesions and olfactory lesions on appetitive learning in goldfish. *Physiology and Behavior,* 1971, **6**, 35–40.

Forgays, D. G., & Read, J. M. Crucial periods for free-environmental experience in the rat. *Journal of Comparative and Physiological Psychology,* 1962, **55**, 816–818.

Fouts, R. S. Acquisition and testing of gestural signs in four young chimpanzees. *Science,* 1973, **180**, 978–980.

Fox, M. W. Reflex-ontogeny and behavioral development of the mouse. *Animal Behaviour,* 1965, **13**, 234–241.

Fox, M. W. Neuro-behavioral ontogeny: A synthesis of ethological and neurophysiological concepts. *Brain Research,* 1966, **2**, 3–20.

Fox, R., Lehmkuhle, S. W., & Westendorf, D. H. Falcon visual acuity. *Science,* 1976, **192**, 263–265.

Frank, L. H., & Meyer, M. E. Social facilitation in the chick: Evidence for facilitation of learning. *Bulletin of the Psychonomic Society,* 1974, **3**, 196–198.

Franklin, W. L. The social behavior of the vicuña. In V. Geist & F. Walther (Eds.), *The behavior of ungulates and its relation to management.* Morges, Switzerland: International Union for Conservation of Nature and Natural Resources, 1974. Pp. 477–487.

French, J. W. Trial and error learning in paramecium. *Journal of Experimental Psychology,* 1940, **26**, 609–613.

Friedmann, J. K. A diallel analysis of the genetic underpinnings of mouse sleep. *Physiology and Behavior,* 1974, **12**, 169–175.

Frith, H. J. *The mallee-fowl.* London: Angus and Robertson, 1962.

Fulker, D. W. Mating speed in male *Drosophila melanogaster:* A psychogenetic analysis. *Science,* 1966, **153**, 203–205.

Fuller, J. L. Genetics and individual differences. In R. H. Waters, D. A. Rethlingshafer, & W. E. Caldwell (Eds.), *Principles of comparative psychology.* New York: McGraw-Hill, 1960. Pp. 325–354.

Fuller, J. L. Experiential deprivation and later behavior. *Science,* 1967, **158**, 1645–1652.

Fuller, J. L., & Thompson, W. R. *Behavior genetics.* New York: Wiley, 1960.

Fuller, J. L., & Wimer, R. E. Behavior genetics. In D. A. Dewsbury & D. A. Rethlingshafer (Eds.), *Comparative psychology: A modern survey.* New York: McGraw-Hill, 1973. Pp. 197–237.

Galef, B. G., Jr. Aggression and timidity: Responses to novelty in feral Norway rats. *Journal of Comparative and Physiological Psychology,* 1970, **70**, 370–381.

Galef, B. G., Jr. Social transmission of acquired behavior: A discussion of tradition and social learning in vertebrates. *Advances in the Study of Behavior,* 1976, **6**, 77–100.

Galef, B. G., Jr., & Clark, M. M. Parent-offspring interactions determine time and place of first ingestion of solid food by wild rat pups. *Psychonomic Science,* 1971, **25**, 15–16.

Galef, B. G., Jr., & Sherry, D. F. Mother's milk: A medium for transmission of cues reflecting the flavor of mother's diet. *Journal of Comparative and Physiological Psychology,* 1973, **83**, 374–378.

Gallup, G. G. Animal hypnosis: Factual status of a fictional concept. *Psychological Bulletin,* 1974, **81**, 836–853.

Gamow, R. I., & Harris, J. F. The infrared receptors of snakes. *Scientific American,* 1973, **228**(5), 94–100.

Garcia, J., Ervin, F. R., & Koelling, R. A. Learning with prolonged delay of reinforcement. *Psychonomic Science,* 1966, **5**, 121–122.

Garcia, J., Hankins, W. G., & Rusiniak, K. W. Behavioral regulation of the milieu interne in man and rat. *Science,* 1974, **185**, 824–831.

Garcia, J., Hankins, W. G., & Rusiniak, K. W. Flavor aversion studies. *Science,* 1976, **192**, 265–266.

Garcia, J., & Koelling, R. A. Relation of cue to consequence in avoidance learning. *Psychonomic Science,* 1966, **4**, 123–124.

Gardner, L. E. Retention and overhabituation of a dual-component response in *Lumbricus terrestris. Journal of Comparative and Physiological Psychology,* 1968, **66**, 315–318.

Gardner, R. A., & Gardner, B. T. Teaching sign language to a chimpanzee. *Science,* 1969, **165**, 664–672.

Garten, C. T., Jr. Relationships between aggressive behavior and genic heterozygosity in the oldfield mouse, *Peromyscus polionotus. Evolution,* 1976, **30**, 59–72.

Gates, M. G., & Allee, W. C. Conditioned behavior of isolated and grouped cockroaches on a simple maze. *Journal of Comparative Psychology,* 1933, **15**, 331–358.

Geist, V. The evolution of horn-like organs. *Behaviour,* 1966, **27**, 175–214.

Geist, V. A consequence of togetherness. *Natural History,* 1967, **76**, 24–30.

Gelber, B. Investigations of the behavior of *Paramecium auratus:* I. Modification of behavior after training with reinforcement. *Journal of Comparative and Physiological Psychology,* 1952, **45**, 58–65.

Gelber, B. Food or training in paramecium? *Science,* 1957, **126**, 1340–1341.

Gelber, B. Studies of the behavior of *Paramecium aurelia. Animal Behaviour,* 1965, Suppl. 1, 21–29.

Gelperin, A. Rapid food-aversion learning by a terrestrial mollusk. *Science,* 1975, **189**, 567–570.

Gerall, A. A., Napoli, A. M., & Cooper, U. C. Daily and hourly estrous running in intact, spayed, and estrone implanted rats. *Physiology and Behavior,* 1973, **10**, 225–229.

Gerall, H. D., Ward, I. L., & Gerall, A. A. Disruption of the male rat's sexual behavior induced by social isolation. *Animal Behaviour,* 1967, **15**, 54–58.

Gerhardt, H. C. The vocalizations of some hybrid treefrogs: Acoustic and behavioral analyses. *Behaviour,* 1974, **49**, 130–151.

Ghiselin, M. T. Darwin and evolutionary psychology. *Science,* 1973, **179**, 964–968.

Ghiselin, M. T. *The economy of nature and the evolution of sex.* Berkeley: University of California Press, 1974.

Gleason, K. K., & Reynierse, J. H. The behavioral significance of pheromones in vertebrates. *Psychological Bulletin,* 1969, **71**, 58–73.

Glickman, S. E. Curiosity has killed more mice than cats. *Psychology Today,* 1971, **5**(5), 55–56, 86.

Glickman, S. E. Responses and reinforcement. In R. A. Hinde & J. Stevenson-Hinde (Eds.), *Constraints on learning: Limitations and predispositions.* New York: Academic Press, 1973. Pp. 207–241.

Glickman, S. E., & Schiff, B. B. A biological theory of reinforcement. *Psychological Review,* 1967, **74**, 81–109.

Glickman, S. E., & Sroges, R. W. Curiosity in zoo animals. *Behaviour,* 1966, **24**, 151–188.

Gold, R. M. Hypothalamic obesity: The myth of the ventromedial nucleus. *Science,* 1973, **182**, 488–490.

Goldfoot, D. A., Kravetz, M. A., Goy, R. W., & Freeman, S. K. Lack of effect of vaginal lavages and aliphatic acids on ejaculatory responses in rhesus monkeys: Behavioral and chemical analyses. *Hormones and Behavior,* 1976, **7**, 1–27.

Gonzalez, R. C., & Bitterman, M. E. Spaced-trials partial reinforcement effect as a function of contrast. *Journal of Comparative and Physiological Psychology,* 1969, **67**, 94–103.

Gonzalez, R. C., Potts, A., Pitcoff, K., & Bitterman, M. E. Runway performance of goldfish as a function of complete and incomplete reduction in amount of reward. *Psychonomic Science,* 1972, **27**, 305–307.

Gordon, T. P., Rose, R. M., & Bernstein, I. S. Seasonal rhythm in plasma testosterone levels in the rhesus monkey *(Macaca mulatta):* A three year study. *Hormones and Behavior,* 1976, **7**, 229–243.

Gossette, R. L. Variation in magnitude of negative transfer on successive discrimination reversal (SDR) tasks across species. *Perceptual and Motor Skills,* 1969, **29**, 803–811.

Gossette, R. L. Note on the calibration of inter-species successive discrimination reversal (SDR) performance differences. Qualitative vs. quantitative scaling. *Perceptual and Motor Skills,* 1970, **31**, 95–104.

Gossette, R. L., & Gossette, M. F. Examination of the reversal index (RI) across fifteen different mammalian and avian species. *Perceptual and Motor Skills,* 1967, **24**, 987–990.

Gossette, R. L., Kraus, G., & Speiss, J. Comparison of successive discrimination reversal (SDR) performances of seven mammalian species on a spatial task. *Psychonomic Science,* 1968, **12**, 193–194.

Gottlieb, G. Prenatal behavior of birds. *Quarterly Review of Biology,* 1968, **43**, 148–174.

Gottlieb, G. Ontogenesis of sensory function in birds and mammals. In E. Tobach, L. R. Aronson, & E. Shaw (Eds.), *The biopsychology of development.* New York: Academic Press, 1971. Pp. 67–128.

Gottlieb, G. Neglected developmental variables in the study of species identification in birds. *Psychological Bulletin,* 1973, **79**, 362–372.

Gottlieb, G. Comparative psychology. *American Psychologist,* 1976, **31**, 295–297.

Gould, J. L. Genetics and molecular ethology. *Zeitschrift für Tierpsychologie,* 1974, **36**, 267–292.

Gould, J. L. Honey bee recruitment: The dance-language controversy. *Science,* 1975, **189**, 685–693.

Gould, J. L. The dance-language controversy. *Quarterly Review of Biology,* 1976, **51**, 211–244.

Gould, S. J. Positive allometry of antlers in the "Irish elk," *Megaloceros giganteus. Nature,* 1973, **244**, 375–376.

Goy, R. W., & Jakway, J. S. The inheritance of patterns of sexual behavior in female guinea pigs. *Animal Behaviour,* 1959, **7**, 142–149.

Goy, R. W., & Jakway, J. S. Role of inheritance in determination of sexual behavior patterns. In E. L. Bliss (Ed.), *Roots of behavior.* New York: Harper, 1962. Pp. 96–112.

Graves, H. B. Early social responses in *Gallus:* A functional analysis. *Science,* 1973, **182**, 937–939.

Graves, H. B., & Siegel, P. B. Bidirectional selection for responses of *Gallus domesticus* chicks to an imprinting situation. *Animal Behaviour,* 1969, **17**, 683–691.

Gray, G. D., Davis, H. N., Kenney, A. McM., & Dewsbury, D. A. Effect of mating on plasma levels of LH and progesterone in montane voles *(Microtus montanus). Journal of Reproduction and Fertility,* 1976, **47**, 89–91.

Gray, G. D., Kenney, A. McM., & Dewsbury, D. A. Adaptive significance of the copulatory behavior pattern of male meadow voles *(Microtus pennsylvanicus)* in relation to the induction of ovulation and implantation in females. *Journal of Comparative and Physiological Psychology,* 1977, in press.

Gray, J. *How animals move.* Cambridge: Cambridge University Press, 1953.

Gray, P. H. The early animal behaviorists, prelegomenon to ethology. *Isis,* 1968–1969, **59**, 372–383.

Greeno, J. G., & Bjork, R. A. Mathematical learning theory and the new "mental forestry." *Annual Review of Psychology,* 1973, **24**, 81–116.

Greenough, W. T. Experiential modification of the developing brain. *American Scientist,* 1975, **63**, 37–46.

Griffin, D. R. Bird navigation. In A. Wolfson (Ed.), *Recent advances in avian biology.* Urbana, Ill.: University of Illinois Press, 1955. Pp. 154–197.

Griffin, D. R. *Listening in the dark.* New Haven, Conn.: Yale University Press, 1958.

Griffin, D. R., & Hopkins, C. D. Sounds audible to migrating birds. *Animal Behaviour,* 1974, **22**, 672–678.

Griffin, D. R., Webster, F. A., & Michael, C. R. The echolocation of flying insects by bats. *Animal Behaviour,* 1960, **8**, 141–154.

Grossman, S. P. Eating or drinking elicited by direct adrenergic or cholinergic stimulation of hypothalamus. *Science,* 1960, **132**, 301–302.

Grossman, S. P. *A textbook of physiological psychology.* New York: Wiley, 1967.

Grossman, S. P. Aggression, avoidance, and reaction to novel environments in female rats with ventromedial hypothalamic lesions. *Journal of Comparative and Physiological Psychology,* 1972, **78**, 274–283.

Groves, P. M., & Thompson, R. F. Habituation: A dual-process theory. *Psychological Review,* 1970, **77**, 419–450.

Gruendel, A. D., & Arnold, W. J. Effects of early social deprivation on reproductive behavior of male rats. *Journal of Comparative and Physiological Psychology,* 1969, **67**, 123–128.

Grunt, J. A., & Young, W. C. Differential reactivity of individuals and the response of the male guinea pig to testosterone propionate. *Endocrinology,* 1952, **51**, 237–248.

Guhl, A. M. The social order of chickens. *Scientific American,* 1956, **194**(2), 42–46.

Gustavson, C. R., Kelly, D. J., Sweeney, M., & Garcia, J. Prey-lithium aversions: I. Coyotes and wolves. *Behavioral Biology,* 1976, **17**, 61–72.

Guthrie, R. D. A new theory of mammalian rump patch evolution. *Behaviour,* 1971, **38**, 132–145.

Hailman, J. P. Cliff-nesting adaptations in the Galapagos swallow-tailed gull. *Wilson Bulletin,* 1965, **77**, 346–362.

Hailman, J. P. The ontogeny of an instinct: The pecking response in chicks of the laughing gull *(Larus atricilla* L.*). Behaviour,* 1967, Suppl. 15, 1–159.

Hailman, J. P. How an instinct is learned. *Scientific American,* 1969, **221**(6), 98–106.

Halas, E. S., James, R. L., & Knutson, C. S. An attempt at classical conditioning in the planarian. *Journal of Comparative and Physiological Psychology,* 1962, **55**, 959–971.

Halcomb, R. A., Hegmann, J. P., & DeFries, J. C. Open-field behavior in mice: A diallel analysis of selected lines. *Behavior Genetics,* 1975, **5,** 217–231.

Hale, E. B. Domestication and the evolution of behavior. In E. S. E. Hafez (Ed.), *The behaviour of domestic animals.* (2d Ed.) Baltimore: Williams & Wilkins, 1969. Pp. 22–42.

Hall, K. R. L., & DeVore, I. Baboon social behavior. In I. DeVore (Ed.), *Primate behavior: Field studies of monkeys and apes.* New York: Holt, 1965. Pp. 53–110.

Hall, K. R. L., & Goswell, M. J. Aspects of social learning in captive patas monkeys. *Primates,* 1964, **5**, 59–70.

Hamburger, V. Some aspects of the embryology of behavior. *Quarterly Review of Biology,* 1963, **38**, 342–365.

Hamburger, V. Development of embryonic motility. In E. Tobach, L. R. Aronson, & E. Shaw (Eds.), *The biopsychology of development.* New York: Academic Press, 1971. Pp. 45–66.

Hamilton, T. C., Thompson, J. M., & Eisenstein, E. M. Quantitative analysis of ciliary and contractile responses during habituation training in *Spirostomum ambiguum. Behavioral Biology,* 1974, **12**, 393–407.

Hamilton, W. D. The genetical theory of social behaviour: I and II. *Journal of Theoretical Biology,* 1964, **7**, 1–52.

Hamilton, W. D. Geometry for the selfish herd. *Journal of Theoretical Biology,* 1971, **31**, 295–311.

Haney, R. E. Classical conditioning of a plant: *Mimosa pudica. Worm Runner's Digest,* 1969, **11**(1), 5–12.

Hanzel, T. E., & Rucker, W. B. Trial and error learning in Paramecium: A replication. *Behavioral Biology,* 1973, **7**, 873–880.

Harlow, H. F. Social facilitation of feeding in the albino rat. *Journal of Genetic Psychology,* 1932, **43**, 211–221.

Harlow, H. F. The formation of learning sets. *Psychological Review,* 1949, **56**, 51–56.

Harlow, H. F. The heterosexual affectional system in monkeys. *American Psychologist,* 1962, **17**, 1–9.

Harlow, H. F. *Learning to love.* New York: Ballantine, 1971.

Harlow, H. F., Harlow, M. K., & Hansen, E. W. The maternal affectional system of rhesus monkeys. In H. L. Rheingold (Ed.), *Maternal behavior in mammals.* New York: Wiley, 1963. Pp. 254–281.

Harlow, H. F., Harlow, M. K., Rueping, R. R., & Mason, W. A. Performance of infant rhesus monkeys on discrimination learning, delayed response, and discrimination learning set. *Journal of Comparative and Physiological Psychology,* 1960, **53**, 113–121.

Harlow, H. F., Harlow, M. K., & Suomi, S. J. From thought to therapy: Lessons from a primate laboratory. *American Scientist,* 1971, **59**, 538–549.

Harlow, H. F., Uehling, H., & Maslow, A. H. Comparative behavior of primates: I. Delayed reaction tests on primates from lemur to the orang-outan. *Journal of Comparative Psychology,* 1932, **13**, 313–343.

Hart, B. L. Sexual reflexes and mating behavior in the male dog. *Journal of Comparative and Physiological Psychology,* 1967, **64**, 388–399.

Hart, B. L. Sexual reflexes and mating behavior in the male rat. *Journal of Comparative and Physiological Psychology,* 1968a, **65**, 452–460.

Hart, B. L. Neonatal castration: Influence on neural organization of sexual reflexes in male rats. *Science,* 1968, **160**, 1135–1136.

Hart, B. L. Gonadal hormones and sexual reflexes in the female rat. *Hormones and Behavior,* 1969, **1**, 65–71.

Hart, B. L. Effects of testosterone propionate and dihydrotestosterone on penile morphology and sexual reflexes of spinal male rats. *Hormones and Behavior,* 1973, **4**, 239–246.

Hart, B. L. Gonadal androgen and sociosexual behavior of male mammals: A comparative analysis. *Psychological Bulletin,* 1974, **81**, 383–400.

Hausfater, G. Dominance and reproduction in baboons *(Papio cynocephalus).* A quantitative analysis. *Contributions to Primatology,* 1975, **7**, 1–150.

Hay, D. A. Strain differences in maze-learning ability of *Drosophila melanogaster. Nature,* 1975, **257**, 44–46.

Hebb, D. O. Heredity and environment in mammalian behavior. *British Journal of Animal Behaviour,* 1953, **1**, 43–47.

Hediger, H. *Wild animals in captivity.* New York: Dover, 1964.

Hediger, H. Environmental factors influencing the reproduction of zoo animals. In F. A. Beach (Ed.), *Sex and behavior.* New York: Wiley, 1965. Pp. 319–354.

Heffner, H. E., Ravizza, R. J., & Masterton, B. Hearing in primitive mammals: IV. Bushbaby *(Galago senegalenis). Journal of Auditory Research,* 1969, **9**, 19–23.

Hegmann, J. P. Physiological function and behavior genetics: I. Genetic variance for peripheral conduction velocity in mice. *Behavior Genetics,* 1972, **2**, 55–67.

Heiligenberg, W. Processes governing behavioral states of readiness. *Advances in the Study of Behavior,* 1974, **5**, 173–200.

Heimer, L., & Larsson, K. Drastic changes in the mating behaviour of male rats following lesions in the junction of diencephalon and mesencephalon. *Experientia,* 1964, **20**, 1–4.

Hein, A., & Held, R. Dissociation of the visual placing response into elicited and guided components. *Science,* 1967, **158**, 390–392.

Heinroth, O., & Heinroth, K. *The birds.* Ann Arbor, Mich.: University of Michigan Press, 1958.

Held, R., & Bauer, J. A., Jr. Visually guided reaching in infant monkeys after restricted rearing. *Science,* 1967, **155**, 718–720.

Held, R., & Hein, A. Movement-produced stimulation in the development of visually guided behavior. *Journal of Comparative and Physiological Psychology,* 1963, **56**, 872–876.

Henderson, N. D. Prior treatment effects on open-field behavior of mice—A genetic analysis. *Animal Behaviour,* 1967, **15**, 364–376.

Henderson, N. D. The confounding effects of genetic variables in early experience research: Can we ignore them? *Developmental Psychobiology,* 1968a, **1**, 146–152.

Henderson, N. D. Genetic analysis of acquisition and retention of a conditioned fear in mice. *Journal of Comparative and Physiological Psychology,* 1968b, **65**, 325–330.

Henderson, N. D. Genetic influences on the behavior of mice can be obscured by laboratory rearing. *Journal of Comparative and Physiological Psychology,* 1970a, **72**, 505–511.

Henderson, N. D. Brain weight increases resulting from environmental enrichment: A directional dominance in mice. *Science,* 1970b, **169**, 776–778.

Henton, W. W., Smith, J. C., & Tucker, D. Odor discrimination in pigeons. *Science,* 1966, **153**, 1138.

Herberg, L. J., Pye, J. G., & Blundell, J. E. Sex differences in the hypothalamic regulation of food hoarding: Hormones versus calories. *Animal Behaviour,* 1972, **20**, 186–191.

Heron, W. T. The inheritance of brightness and dullness in maze learning ability in the rat. *Journal of Genetic Psychology,* 1941, **59**, 41–49.

Hess, E. H. Imprinting: An effect of early experience. *Science,* 1959, **130**, 133–141.

Hess, E. H. "Imprinting" in a natural laboratory. *Scientific American,* 1972, **227**(2), 24–31.

Hess, E. H. Comparative sensory processes. In D. A. Dewsbury & D. A. Rethlingshafer (Eds.), *Comparative psychology: A modern survey.* New York: McGraw-Hill, 1973. Pp. 344–394.

Heston, L. L. The genetics of schizophrenic and schizoid disease. *Science,* 1970, **167**, 249–256.

Hilgard, E. R., & Bower, G. H. *Theories of learning.* (3d ed.) New York: Appleton Century Crofts, 1966.

Hinde, R. A. The behavior of certain cardueline F_1 inter-species hybrids. *Behaviour,* 1956, **9**, 202–213.

Hinde, R. A. Interaction of internal and external factors in integration of canary reproduction. In F. A. Beach (Ed.), *Sex and behavior.* New York: Wiley, 1965. Pp. 381–415.

Hinde, R. A. *Animal behavior: A synthesis of ethology and comparative psychology.* (2d ed.) New York: McGraw-Hill, 1970.

Hinde, R. A. Constraints on learning—An introduction to the problems. In R. A. Hinde & J. Stevenson-Hinde (Eds.), *Constraints on learning: Limitations and predispositions.* New York: Academic Press, 1973. Pp. 1–19.

Hinde, R. A. *Biological bases of human social behaviour.* New York: McGraw-Hill, 1974.

Hinde, R. A., & Spencer-Booth, Y. Effects of brief separation from mother on rhesus monkeys. *Science,* 1971, **173**, 111–118.

Hinde, R. A., & Stevenson-Hinde, J. (Eds.) *Constraints on learning: Limitations and predispositions.* New York: Academic Press, 1973.

Hinde, R. A., & Tinbergen, N. The comparative study of species-specific behavior. In A. Roe & G. G. Simpson (Eds.), *Behavior and evolution.* New Haven: Yale University Press, 1958. Pp. 251–268.

Hirsch, J. Behavior genetics and individuality understood. *Science,* 1963, **142**, 1436–1442.

Hirsch, J. Behavior–genetic analysis. In J. Hirsch (Ed.), *Behavior–genetic analysis.* New York: McGraw-Hill, 1967. Pp. 416–435.

Hirsch, J., & Boudreau, J. C. Studies in experimental behavior genetics: I. The heritability of phototaxis in a population of *Drosophila melanogaster. Journal of Comparative and Physiological Psychology,* 1958, **51**, 647–651.

Hobson, E. S. Feeding patterns among tropical reef fishes. *American Scientist,* 1975, **63**, 382–392.

Hodos, W. Progressive ratio as a measure of reward strength. *Science,* 1961, **134**, 943–944.

Hodos, W. Evolutionary interpretations of neural and behavioral studies of living vertebrates. In F. O. Schmitt (Ed.), *The neurosciences second study program.* New York: Rockefeller University Press, 1970. Pp. 26–39.

Hodos, W., & Campbell, C. B. G. *Scala naturae:* Why there is no theory in comparative psychology. *Psychological Review,* 1969, **76**, 337–350.

Hoffman, H. S., Searle, J. L., Toffey, S., & Kozma, F., Jr. Behavioral control by an imprinted stimulus. *Journal of the Experimental Analysis of Behavior,* 1966, **9**, 177–189.

Hogan, J. A. Fighting and reinformcement in the Siamese fighting fish *(Betta splendens). Journal of Comparative and Physiological Psychology,* 1967, **64**, 356–359.

Hogan, J. A. Development of food recognition in young chicks: I. Maturation and nutrition. *Journal of Comparative and Physiological Psychology,* 1973a, **83**, 355–366.

Hogan, J. A. Development of food recognition in young chicks: II. Learned associations over long delays. *Journal of Comparative and Physiological Psychology,* 1973b, **83**, 367–373.

Hogan, J. A. How young chicks learn to recognize food. In R. A. Hinde & J. Stevenson-Hinde (Eds.), *Constraints on learning: Limitations and predispositions.* New York: Academic Press, 1973c. Pp. 119–139.

Hogan, J. A., Kleist, S., & Hutchings, C. S. L. Display and food as reinforcers in the Siamese fighting fish *(Betta splendens). Journal of Comparative and Physiological Psychology,* 1970, **70**, 351–357.

Hollis, J. H. Habituatory response decrement in pupae of *Tenebrio molitor. Animal Behaviour,* 1963, **11**, 161–163.

Holmes, E., & Gruenberg, G. Learning in plants? *Worm Runner's Digest,* 1965, **7**(1), 9–11.

Holmes, E., & Yost, M. "Behavioral" studies in the sensitive plant. *Worm Runner's Digest,* 1966, **8**(2), 38–40.

Holmes, T. M., Aksel, R., & Royce, J. R. Inheritance of avoidance behavior in *Mus musculus. Behavior Genetics,* 1974, **4**, 357–371.

Honig, W. K. (Ed.) *Operant behavior: Areas of research and application.* New York: Appleton Century Crofts, 1966.

Hoogland, R., Morris, D., & Tinbergen, N. The spines of sticklebacks *(Gasterosteus* and *Pygosteus)* as means of defense against predators *(Perca* and *Esox). Behaviour,* 1957, **10**, 205–236.

Hopkins, C. D. Electric communication in fish. *American Scientist,* 1974, **62**, 426–437.

Horel, J. A. The brain and behavior in phylogenetic perspective. In D. A. Dewsbury & D. A. Rethlingshafer (Eds.), *Comparative psychology: A modern survey.* New York: McGraw-Hill, 1973. Pp. 271–300.

Horn, J. M. Aggression as a component of relative fitness in four inbred strains of mice. *Behavior Genetics,* 1974, **4**, 373–381.

Horridge, G. A. Learning of leg position by the ventral nerve cord in headless insects. *Proceedings of the Royal Society of London,* 1962, **157B**, 33–52.

Horridge, G. A. The electrophysiological approach to learning in isolatable ganglia. *Animal Behaviour,* 1965, **13**, Suppl. 1, 163–182.

House, E. L., & Pansky, B. A. *A functional approach to neuroanatomy.* New York: McGraw-Hill, 1967.

Hrdy, S. B. Male-male competition and infanticide among the langurs *(Presbytis entellus). Folia Primatologica,* 1974, **22**, 19–58.

Hrdy, S. B. Care and exploitation of nonhuman primate infants by conspecifics other than the mother. *Advances in the Study of Behavior,* 1976, **6**, 101–158.

Hubel, D. H., & Wiesel, T. N. Receptive fields of cells in striate cortex of very young, visually inexperienced kittens. *Journal of Neurophysiology,* 1963, **26**, 994–1002.

Huber, F. Untersuchungen über die Function des zentral Nervensystems und insbesondere des Gehirnes bei der Fortbewegung und der Lauterzeugung der Grillen. *Zeitschrift für vergleichende Physiologie,* 1960, **44**, 60–132.

Huber, F. Central nervous control of sound production in crickets and some speculations on its evolution. *Evolution,* 1962, **16**, 429–444.

Huber, J. C., Rucker, W. B., & McDiarmid, C. G. Retention of escape training and activity changes in single paramecia. *Journal of Comparative and Physiological Psychology,* 1974, **86**, 258–266.

Hughes, C. W., Jr. Early experience in domestication. *Journal of Comparative and Physiological Psychology,* 1975, **88**, 407–417.

Hughes, R. N. Social facilitation of locomotion and exploration in rats. *British Journal of Psychology,* 1975, **60**, 385–388.

Hullett, J. W., & Homzie, M. J. Sensitizaton effect in the classical conditioning of *Dugesia dorotocephala. Journal of Comparative and Physiological Psychology,* 1966, **62**, 227–230.

Hulse, S. H., Deese, J., & Egeth, H. *The psychology of learning.* New York: McGraw-Hill, 1975.

Hunsaker, D. Ethological isolating mechanisms in the *Scelopolus torquatus* group of lizards. *Evolution,* 1962, **16**, 62–74.

Hunt, E. Computer simulation: Artificial intelligence studies and their relevance to psychology. *Annual Review of Psychology,* 1968, **19**, 135–168.

Hunter, M. W. III, & Kamil, A. C. Object-discrimination learning set and hypothesis behavior in the northern bluejay *(Cyanocitta cristata). Psychonomic Science,* 1971, **22**, 271–272.

Hunter, M. W. III, & Kamil, A. C. Marginal learning-set formation by the crow *(Corvus brachyrhynchos). Bulletin of the Psychonomic Society,* 1975, **5**, 373–375.

Hutchison, J. B. Hypothalamic mechanisms of sexual behavior, with special references to birds. *Advances in the Study of Behavior,* 1976, **6**, 159–200.

Hutchison, J. B., & Poynton, J. C. A neurological study of the clasp reflex in *Xenopus laevis* (Daudin). *Behaviour,* 1963, **22**, 41–63.

Huxley, J. S. The present standing of the theory of sexual selection. In G. R. deBeer (Ed.), *Evolution: Essays on aspects of evolutionary biology.* Oxford: Oxford University Press, 1938. Pp. 11–42.

Hyde, J. S. Inheritance of learning ability in mice: A diallel-environmental analysis. *Journal of Comparative and Physiological Psychology,* 1974, **86**, 116–123.

Hydén, H., & Lange, P. W. A differentiation in RNA response in neurons early and late during learning. *Proceedings of the National Academy of Sciences, U.S.A.,* 1965, **53**, 946–952.

Immelmann, K. Sexual and other long-term aspects of imprinting in birds and other species. *Advances in the Study of Behavior,* 1972, **4**, 147–174.

Jacobson, A. L., Horowitz, S. D., & Fried, C. Classical conditioning, pseudoconditioning, or sensitization in the planarian. *Journal of Comparative and Physiological Psychology,* 1967, **64**, 73–79.

Jakway, J. S. Inheritance of patterns of mating behavior in the male guinea pig. *Animal Behaviour,* 1959, **7**, 150–162.

Jameson, E. W., Jr. Natural history of the prairie vole. *University of Kansas Publications of the Museum of Natural History,* 1947, **1**, 125–151.

Jaynes, J. The historical origins of "ethology" and "comparative psychology." *Animal Behaviour,* 1969, **17**, 601–606.

Jensen, D. D. Experiments on "learning" in paramecia. *Science,* 1957a, **125**, 191–192.

Jensen, D. D. More on "learning" in paramecia. *Science,* 1957b, **126**, 1341–1342.

Jensen, D. D. Operationism and the question "Is this behavior learned or innate?" *Behaviour,* 1961, **17**, 1–8.

Jensen, D. D. Paramecia, planaria, and pseudo-learning. *Animal Behaviour,* 1965, Suppl. 1, 9–20.

Jensen, D. D. Polythetic biopsychology: An alternative to behaviorism. In J. H. Reynierse (Ed.), *Current issues in animal learning.* Lincoln, Nebr.: University of Nebraska Press, 1970. Pp. 1–31.

Jerison, H. J. *Evolution of the brain and intelligence.* New York: Academic Press, 1973.

Jerison, H. J. Paleoneurology and the evolution of mind. *Scientific American,* 1976, **234**(1), 90–101.

Johnsgard, P. A. The taxonomy of the Anatidae—A behavioral analysis. *Ibis,* 1961, **103,** 71–85.

Johnsgard, P. A. *Animal behavior.* (2d ed.) Dubuque, Iowa: Wm. C. Brown, 1972.

Johnson, D. F., & Phoenix, C. H. Hormonal control of female sexual attractiveness, proceptivity, and receptivity in rhesus monkeys. *Journal of Comparative and Physiological Psychology,* 1976, **90**, 473–483.

Johnson, R. N. *Aggression in man and animals.* Philadelphia: Saunders, 1972.

Johnson, R. N. & Johnson, L. D. Intra- and interspecific social and aggressive behavior in the male Siamese fighting fish, *Betta splendens. Animal Behaviour,* 1973, **21**, 665–672.

Johnston, R. E. Sexual excitation function of hamster vaginal secretion. *Animal Learning & Behavior,* 1975, **3**, 161–166.

Johnston, R. E., & Zahorik, D. M. Taste aversions to sexual attractants. *Science,* 1975, **189**, 893–894.

Jolly, A. *The evolution of primate behavior.* New York: Macmillan, 1972.

Jolly, A. Primate birth hour. *International Zoo Yearbook,* 1973, **13**, 391–397.

Jouvet, M. Biogenic amines and the states of sleep. *Science,* 1969, **163**, 32–41.

Jouvet-Mounier, D., Astic, L., & Lacote, D. Ontogenesis of the states of sleep in rat, cat, and guinea pig during the first postnatal month. *Developmental Psychobiology,* 1969, **2**, 216–239.

Kalat, J. W. Taste-aversion learning in infant guinea pigs. *Developmental Psychobiology,* 1975, **8**, 383–387.

Kamil, A. C., & Hunter, M. W. III. Performance on object-discrimination learning set by the Greater Hill myna *(Gracula religiosa). Journal of Comparative and Physiological Psychology,* 1970, **73**, 68–73.

Karli, P. The Norway rat's killing response to the white mouse. *Behaviour,* 1956, **10**, 81–103.

Katz, M. S., & Deterline, W. A. Apparent learning in the paramecium. *Journal of Comparative and Physiological Psychology,* 1958, **51**, 243–247.

Kaufman, D. W. Differential owl predation on white and agouti *Mus musculus. Auk,* 1974a, **91**, 145–150.

Kaufman, D. W. Differential predation on active and inactive prey by owls. *Auk,* 1974b, **91**, 172–173.

Kaufmann, J. H. A three-year study of mating behavior in a freeranging band of rhesus monkeys. *Ecology,* 1965, **46,** 500–512.

Kaufmann, J. H. Is territory definable? In A. H. Esser (Ed.), *Behavior and Environment: The use of space by animals and men.* New York: Plenum, 1971. Pp. 36–40.

Kaufmann, J. H. The ecology and evolution of social organization in the kangaroo family (Macropodidae). *American Zoologist,* 1974a, **14**, 51–62.

Kaufmann, J. H. Social ethology of the whiptail wallaby, *Macropus parryi,* in northeastern New South Wales. *Animal Behaviour.* 1974b, **22**, 281–369.

Kawamura, S. The process of sub-culture propagation among Japanese macaques. In C. H. Southwick (Ed.), *Primate social behavior: An enduring problem.* Princeton, N.J.: Van Nostrand, 1963. Pp. 82–90.

Keesey, R. E., & Powley, T. L. Hypothalamic regulation of body weight. *American Scientist,* 1975, **63**, 558–565.

Keeton, W. T. The orientational and navigational basis of homing in birds. *Advances in the Study of Behavior,* 1974a, **5**, 47–132.

Keeton, W. T. The mystery of pigeon homing. *Scientific American,* 1974b, **231**(6), 96–107.

Keeton, W. T., & Brown, A. I. Homing behavior of pigeons not disturbed by application of an olfactory stimulus. *Journal of Comparative Physiology,* 1976, **105**, 259–266.

Kellogg, W. N. Echo ranging in the porpoise. *Science,* 1958, **128**, 982–988.

Kellogg, W. N. *Porpoises and sonar.* Chicago: University of Chicago Press, 1961.

Kellogg, W. N. Communication and language in the home-raised chimpanzee. *Science,* 1968, **162**, 423–427.

Kessel, E. L. The mating activities of balloon flies. *Systematic Zoology,* 1955, **4**, 97–104.

Kessler, S. Selection for and against ethological isolation between *Drosophila pseudoobscura* and *Drosophila persimilis. Evolution,* 1966, **20**, 634–645.

Kessler, S., & Kraemer, H. C. Gene substitutions and geotaxis in *Drosophila melanogaster. Journal of Comparative and Physiological Psychology,* 1975, **89**, 274–278.

Kettlewell, H. B. D. Insect survival and selection for pattern. *Science,* 1965, **148,** 1290–1296.

Kevan, P. G. Sir Thomas More on imprinting: Observations from the sixteenth century. *Animal Behaviour,* 1976, **24** 16–17.

Kimble, G. A. *Hilgard and Marquis' conditioning and learning.* New York: Appleton Century Crofts, 1961.

King, H. D. Life processes in gray Norway rats during fourteen years in captivity. *American Anatomical Memoirs,* 1939, **17**, 1–72.

King, H. D., & Donaldson, H. H. Life processes and size of the body and organs of the gray Norway rat during ten generations in captivity. *American Anatomical Memoirs,* 1929, **14**, 1–106.

King, J. A. Social behavior, social organization and population dynamics in a black-tailed prairie dog town in the Black Hills of South Dakota. *Contributions of the Laboratory of Vertebrate Biology of the University of Michigan,* 1955, **67**, 1–123.

King, J. A. Parameters relevant to determining the effect of early experience upon adult behavior in animals. *Psychological Bulletin,* 1958, **55**, 46–58.

King, J. A. The social behavior of prairie dogs. *Scientific American,* 1959, **201**(4), 128–140.

King, J. A. Maternal behavior in *Peromyscus.* In H. L. Rheingold (Ed.), *Maternal behavior in mammals.* New York: Wiley, 1963. Pp. 58–93.

King, J. A. *Biology of* Peromyscus *(Rodentia).* Lawrence, Kans.: American Society of Mammalogists, 1968.

King, J. A. Ecological psychology: An approach to motivation. *Nebraska Symposium on Motivation,* 1970, **18**, 1–33.

King, J. A., & Vestal, B. M. Visual acuity of *Peromyscus. Journal of Mammalogy,* 1974, **55**, 238–243.

Kirby, R. H. Acquisition, extinction, and retention of an avoidance response in rats as a function of age. *Journal of Comparative and Physiological Psychology,* 1963, **56**, 158–162.

Klinghammer, E. Factors influencing choice of mate in altricial birds. In H. W. Stevenson, E. H. Hess, & H. L. Rheingold (Eds.), *Early behavior: Comparative and developmental approaches.* New York: Wiley, 1967. Pp. 5–42.

Klopfer, P. H. Stimulus preferences and imprinting. *Science,* 1967, **156**, 1394–1396.

Klopfer, P. H. Mother love: What turns it on? *American Scientist,* 1971, **59**, 404–407.

Klopfer, P. H. *An introduction to animal behavior: Ethology's first century.* Englewood Cliffs, N.J.: Prentice-Hall, 1974.

Klopfer, P. H., & Hatch, J. J. Experimental considerations. In T. A. Sebeok (Ed.), *Animal communication: Techniques of study and results of research.* Bloomington, Ind.: Indiana University Press, 1968.

Knight, G. R., Robertson, A., & Waddington, C. H. Selection for sexual isolation within a species. *Evolution,* 1956, **10**, 14–22.

Koford, C. B. The vicuña and the puna. *Ecological Monographs,* 1957, **27**, 153–219.

Komisaruk, B. R. Effects of local brain implants of progesterone on reproductive behavior in ring dove. *Journal of Comparative and Physiological Psychology,* 1967, **64**, 219–224.

Komisaruk, B. R. Strategies in neuroendocrine neurophysiology. *American Zoologist,* 1971, **11**, 741–754.

Komisaruk, B. R., Adler, N. T., & Hutchison, J. B. Genital sensory field: Enlargement by estrogen treatment in female rats. *Science,* 1972, **178**, 1295–1298.

Konishi, M. The role of auditory feedback in the vocal behavior of the domestic fowl. *Zeitschrift für Tierpsychologie,* 1963, **20**, 349–367.

Konishi, M. The attributes of instinct. *Behaviour,* 1966, **27**, 316–328.

Konishi, M. How the owl tracks its prey. *American Scientist,* 1973, **61**, 414–424.

Koopman, K. F. Natural selection for reproductive isolation between *Drosophila pseudoobscura* and *Drosophila persimilis. Evolution,* 1950, **4**, 135–148.

Korenbrot, C. C., Schomberg, D. W., & Erickson, C. J. Radioimmunoassay of plasma estradiol during the breeding cycle of ring doves *(Streptopelia risoria). Endocrinology,* 1974, **94**, 1126–1132.

Kostowski, W., Rewerski, W., & Piechocki, T. Effects of some steroids on aggressive behaviour in mice and rats. *Neuroendocrinology,* 1970, **6**, 311–318.

Kow, L. M., Malsbury, C. W., & Pfaff, D. W. Effects of progesterone on female reproductive behavior in rats: Possible modes of action and role in behavioral sex differences. In W. Montagna & W. A. Sadler (Eds.), *Reproductive behavior.* New York: Plenum, 1974. Pp. 179–210.

Krane, R. V., & Wagner, A. R. Taste aversion learning with a delayed shock US: Implications for the "generality of the laws of learning." *Journal of Comparative and Physiological Psychology,* 1975, **88**, 882–889.

Krasne, F. B. Excitation and habituation of the crayfish escape reflex: The depolarizing response in lateral giant fibres of the isolated abdomen. *Journal of Experimental Biology,* 1969, **50**, 29–46.

Krebs, C. J., Gaines, M. S., Keller, B. L., Myers, J. H., & Tamarin, R. H. Population cycles in small rodents. *Science,* 1973, **179**, 35–44.

Kruuk, H. Surplus killing by carnivores. *Journal of Zoology*, 1972, **166**, 233–244.

Kuehn, R. E., & Young, W. C. Mating behavior in adult *Macaca mulatta. American Zoologist,* 1965, **5**, 687–688. (Abstract)

Kummer, H. *Primate societies: Group techniques of ecological adaptation.* Chicago: Aldine-Atherton, 1971.

Kuo, Z. Y. *The dynamics of behavior development.* New York: Random House, 1967.

Kupfermann, I. Neurophysiology of learning. *Annual Review of Psychology,* 1975, **26**, 367–391.

Kuse, A. R., & DeFries, J. C. Social dominance and Darwinian fitness in laboratory mice: An alternative test. *Behavioral Biology,* 1976, **16**, 113–116.

Lack, D. *Darwin's finches.* Cambridge: Cambridge University Press, 1947.

Lack, D. Darwin's finches. *Scientific American,* 1953, **188**(4), 66–72.

Lagerspetz, K. M. J. Aggression and aggressiveness in laboratory mice. In S. Garattini & E. B. Sigg (Eds.), *Aggressive behaviour.* Amsterdam: Excerpta Medica, 1969. Pp. 77–85.

Lagler, K. F., Bardach, J. E., & Miller, R. R. *Ichthyology.* New York: Wiley, 1962.

Lahue, R., Kokkinidis, L., & Corning, W. Telson reflex habituation in *Limulus polyphemus. Journal of Comparative and Physiological Psychology,* 1975, **89**, 1061–1069.

Lamb, M. E. Physiological mechanisms in the control of maternal behavior in rats: A review. *Psychological Bulletin,* 1975, **82**, 104–119.

Land, R. B., & McGill, T. E. The effects of the mating pattern of the mouse on the formation of corpora lutea. *Journal of Reproduction and Fertility,* 1967, **13**, 121–125.

Landry, S. The rodentia as omnivores. *Quarterly Review of Biology,* 1970, **45**, 351–372.

Lanier, D. L., & Dewsbury, D. A. Studies of copulatory behavior in northern grasshopper mice *(Onychomys leucogaster). Animal Behaviour*, 1977, **25**, 185–192.

Lanier, D. L., Estep, D. Q., & Dewsbury, D. A. Copulatory behavior of golden hamsters: Effects on pregnancy. *Physiology and Behavior,* 1975, **15**, 209–212.

Larsson, K. *Conditioning and sexual behavior in the male albino rat.* Stockholm: Almqvist & Wiksell, 1956.

Larsson, K. Mating behavior in male rats after cerebral cortex ablation: I. Effects of lesions in the dorsolateral and median cortex. *Journal of Experimental Zoology,* 1962, **151**, 167–176.

Larsson, K. Mating behavior in male rats after cerebral cortex ablation: II. Effects of lesions in the frontal lobes compared to lesions in the posterior half of the hemispheres. *Journal of Experimental Zoology,* 1964, **155**, 203–214.

Larsson, K. Individual differences in reactivity to androgen in male rats. *Physiology and Behavior,* 1966, **1**, 255–258.

Lashley, K. S. In search of the engram. *Society of Experimental Biology symposium No. 4: Physiological mechanisms in animal behavior.* Cambridge: Cambridge University Press, 1950.

Latane, B., Joy, V., Meltzer, J., Lubell, B., & Cappell, H. Stimulus determinants of social attraction in rats. *Journal of Comparative and Physiological Psychology,* 1972, **79**, 13–21.

Layne, J. N. Nest-building behavior in three species of deer mice, *Peromyscus. Behaviour,* 1969, **35**, 288–303.

Layne, J. N. Climbing behavior of *Peromyscus floridanus* and *Peromyscus gossypinus. Journal of Mammalogy,* 1970, **51**, 580–591.

Layne, J. N. Activity responses of two species of *Peromyscus* (Rodentia, Muridae) to varying light cycles. *Oecologia,* 1971, **7**, 223–241.

Layne, J. N., & Ehrhart, L. M. Digging behavior of four species of deer mice *(Peromyscus). American Museum Novitates,* 1970, **2429**, 1–16.

Lazarus, J., & Crook, J. H. The effects of luteinizing hormone, oestrogen and ovariectomy on the agonistic behaviour of female *Quelea quelea. Animal Behaviour,* 1973, **21**, 49–60.

LeBoeuf, B. J. Interindividual associations in dogs. *Behaviour,* 1967, **29**, 268–295.

LeBoeuf, B. J. Male-male competition and reproductive success in elephant seals. *American Zoologist,* 1974, **14**, 163–176.

LeBoeuf, B. J. Sex and evolution. In T. E. McGill, D. A. Dewsbury, & B. D. Sachs (Eds.), *Sex and behavior: Status and prospectus.* New York: Plenum, 1977, in press.

LeBoeuf, B. J., & Peterson, R. S. Social status and mating activity in elephant seals. *Science,* 1969, **163**, 91–93.

Lee, C. T., & Griffo, W. Early androgenization and aggression pheromone in inbred mice. *Hormones and Behavior,* 1973, **4**, 181–189.

Lee, M. H. S., & Williams, D. I. Changes in licking behaviour of rat mother following handling of young. *Animal Behaviour,* 1974, **22**, 679–681.

Lee, R. Conditioning of a free operant response in planaria. *Science,* 1963, **139**, 1048–1049.

Lehrman, D. S. A critique of Konrad Lorenz's theory of instinctive behavior. *Quarterly Review of Biology,* 1953, **28**, 337–363.

Lehrman, D. S. The reproductive behavior of ring doves. *Scientific American,* 1964, **211**(5), 48–54.

Lehrman, D. S. Interaction between internal and external environments in the regulation of the reproductive cycle of the ring dove. In F. A. Beach (Ed.), *Sex and behavior.* New York: Wiley, 1965. Pp. 355–380.

Lehrman, D. S. Semantic and conceptual issues in the nature-nurture problem. In L. R. Aronson, E. Tobach, D. S. Lehrman, & J. S. Rosenblatt (Eds.), *Development and evolution of behavior.* San Francisco: Freeman, 1970. Pp. 17–52.

Leon, M. Maternal pheromone. *Physiology and Behavior,* 1974, **13**, 441–453.

Leshner, A. I. The adrenals and the regulatory nature of running wheel activity. *Physiology and Behavior,* 1971, **6**, 551–558.

Leshner, A. I. A model of hormones and agonistic behavior. *Physiology and Behavior,* 1975, **15**, 225–235.

Leshner, A. I., & Candland, D. K. Endocrine effects of grouping and dominance rank in squirrel monkeys. *Physiology and Behavior,* 1972, **8**, 441–445.

Lettvin, J. Y., Maturana, H. R., McCulloch, W. S., & Pitts, W. H. What the frog's eye tells the frog's brain. *Proceedings of the Institute of Radio Engineers,* 1959, **47**, 1940–1951.

Levine, B. A. Effects of drive and incentive magnitude on serial discrimination reversal learning in pigeons and chickens. *Journal of Comparative and Physiological Psychology,* 1974, **86**, 730–735.

Levine, L. Studies on sexual selection in mice: I. Reproductive competition between albino and black-agouti males. *American Naturalist,* 1958, **92**, 21–26.

Levine, S. A further study of infantile handling and adult avoidance learning. *Journal of Personality,* 1956, **25**, 70–80.

Levine, S. Hormones and conditioning. *Nebraska Symposium on Motivation,* 1968, **16**, 85–101.

Levine, S. Introduction and basic concepts. In S. Levine (Ed.), *Hormones and behavior.* New York: Academic Press, 1972. Pp. 1–9.

Levine, S., & Mullins, R. F., Jr. Hormonal influences on brain organization in infant rats. *Science,* 1966, **152**, 1585–1592.

Levine, S., & Otis, L. S. The effects of handling during pre- and post-weaning on the resistance of the albino rat in adulthood. *Canadian Journal of Psychology,* 1958, **12**, 103–108.

Levine, S., & Wetzel, A. Infantile experiences, strain differences and avoidance learning. *Journal of Comparative and Physiological Psychology,* 1963, **56**, 879–881.

Levy, E., Allen, A., Caton, W., & Holmes, E. An attempt to condition the sensitive plant: *Mimosa pudica. Journal of Biological Psychology,* 1970, **12**(1), 86–87.

Lewis, C. I. *Mind and the world order.* New York: Dover, 1929.

Leyhausen, P. The communal organization of solitary mammals. *Symposium of the Zoological Society of London,* 1965, **14**, 249–263.

Lin, N., & Michener, C. D. Evolution of sociality in insects. *Quarterly Review of Biology,* 1972, **47**, 131–159.

Lincoln, G. A., Guinness, F., & Short, R. V. The way in which testosterone controls the social and sexual behavior of the red deer stag *(Cervus elaphus). Hormones and Behavior,* 1972, **3**, 375–396.

Lindauer, M. *Communication among social bees.* Cambridge, Mass.: Harvard University Press, 1961.

Lints, C. E., & Harvey, J. A. Altered sensitivity to footshock and decreased brain content of serotonin following brain lesions in the rat. *Journal of Comparative and Physiological Psychology*, 1969, **67**, 23–31.

Lindzey, G., & Thiessen, D. D. Introduction: *Mus musculus,* mechanisms and man. In G. Lindzey & D. D. Thiessen (Eds.), *Contributions to behavior-genetic analysis: The mouse as a prototype.* New York: Appleton Century Crofts, 1970. Pp. ix–xviii.

Lisk, R. D. Inhibitory centers in sexual behavior in the male rat. *Science,* 1966, **152**, 669–670.

Lisk, R. D. Oestrogen and progesterone synergism and elicitation of maternal nest-building in the mouse *(Mus musculus). Animal Behaviour,* 1971, **19**, 606–610.

Lissmann, H. W., & Machin, K. E. The mechanism of object location in *Gymnarchus niloticus* and similar fish. *Journal of Experimental Biology,* 1958, **35**, 451–486.

Lloyd, J. E. Studies on the flash communication system in *Photinus* fireflies. *Miscellaneous Publications of the Museum of Zoology, University of Michigan,* 1966, **130**, 1–95.

Lloyd, J. E. Flashes of *Photuris* fireflies: Their value and use in recognizing species. *Florida Entomologist,* 1969, **52**, 29–35.

Lloyd, J. E. Aggressive mimicry in *Photuris* fireflies: Signal repertoires by *Femmes fatales. Science,* 1975, **187**, 452–453.

Lockard, R. B. The albino rat: A defensible choice or a bad habit. *American Psychologist,* 1968, **23**, 734–742.

Lockard, R. B. Reflections on the fall of comparative psychology: Is there a message for us all? *American Psychologist,* 1971, **26**, 168–179.

Lockard, R. B. Experimental inhibition of activity of kangaroo rats in the natural habitat by an artificial moon. *Journal of Comparative and Physiological Psychology,* 1975, **89**, 263–266.

Logan, C. A. Topographic changes in responding during habituation to water-stream stimulation in sea anemones *(Anthopleura elegantissima). Journal of Comparative and Physiological Psychology,* 1975, **89**, 105–117.

Loizos, C. Play in mammals. In P. A. Jewell & C. Loizos (Eds.), *Play, exploration and territory in mammals.* London: Academic Press, 1966. Pp. 1–9

Lorenz, K. Z. The comparion in the bird's world. *Auk,* 1937a, **54**, 245–273.

Lorenz, K. Z. Über die Bildung des Instinktbegriffes. *Die Naturwissenschaften,* 1937b, **25**, 289–300. (Trans. R. Martin)

Lorenz, K. Z. Vergleichende Bewegungsstudien an Anatinen. *Journal für Ornithologie Suppl.,* 1941, **79**, 194–294. (Trans. R. Martin).

Lorenz, K. The comparative method in studying innate behaviour patterns. *Symposium of the Society for Experimental Biology,* 1950, **4**, 221–268.

Lorenz, K. Z. The evolution of behavior. *Scientific American,* 1958, **199**(6), 67–78.

Lorenz, K. Z. *Evolution and modification of behavior.* Chicago: University of Chicago Press, 1965.

Lorenz, K. Z. The fashionable fallacy of dispensing with description. *Naturwissenschaften,* 1973, **60**, 1–9.

Lorenz, K. Z. Analogy as a source of knowledge. *Science,* 1974, **185**, 229–234.

Lorenz, K. Z., & Tinbergen, N. Taxis und Instinkthandlung in der Eirollbewegung der Graugans. *Zeitschrift für Tierpsychologie,* 1938, **2**, 1–29.

Lott, D. F., & Brody, P. N. Support of ovulation in the ring dove by auditory and visual stimuli. *Journal of Comparative and Physiological Psychology,* 1966, **62**, 311–313.

Lott, D., Scholz, S. D., & Lehrman, D. S. Exteroceptive stimulation of the reproductive system of the female ring dove *(Streptopelia risoria)* by the mate and by the colony milieu. *Animal Behaviour,* 1967, **15**, 433–437.

Lowes, G., & Bitterman, M. E. Reward and learning in the goldfish. *Science,* 1967, **157**, 455–457.

Loy, J. Estrous behavior of free-ranging rhesus monkeys *(Macaca mulatta). Primates,* 1971, **12**, 1–31.

Luttge, W. G. Effects of anti-estrogens on testosterone stimulated male sexual behavior and peripheral target tissues in the castrate male rat. *Physiology and Behavior,* 1975, **14**, 839–846.

Luttge, W. G., & Hall, N. R. Differential effectiveness of testosterone and its metabolites in the induction of male sexual behavior in two strains of albino mice. *Hormones and Behavior,* 1973, **4**, 31–43.

Lynch, C. B., & Hegmann, J. P. Genetic differences influencing behavioral temperature regulation in small mammals: I. Nesting by *Mus musculus. Behavior Genetics,* 1972, **2**, 43–53.

MacArthur, R. H., & Wilson, E. O. *The theory of island biogeography.* Princeton, N.J.: Princeton University Press, 1967.

Mackintosh, N. J. *The psychology of animal learning.* London: Academic Press, 1974.

Mackintosh, N. J., & Mackintosh, J. Performance of *Octopus* over a series of reversals of a simultaneous discrimination. *Animal Behaviour,* 1964, **12**, 321–324.

Maier, N. R. F., & Schneirla, T. C. *Principles of animal psychology.* New York: McGraw-Hill, 1935.

Maier, R. A., & Maier, B. M. *Comparative animal behavior.* Belmont, Calif.: Brooks/Cole, 1970.

Malsbury, C. W., & Pfaff, D. W. Neural and hormonal determinants of mating behavior in adult male rats. A review. In L. V. DiCara (Ed.), *Limbic and autonomic nervous systems research.* New York: Plenum, 1973. Pp. 85–136.

Manning, A. The sexual behavior of two sibling *Drosophila* species. *Behaviour,* 1959, **15**, 123–145.

Manning, A. Evolutionary changes and behavior genetics. *Genetics Today,* 1964, *Proceedings XI Congress of Genetics,* 807–813.

Manning, A. *An introduction to animal behavior.* (1st ed.) Reading, Mass.: Addison-Wesley, 1967.

Manning, A. Evolution of behavior. In J. L. McGaugh (Ed.), *Psychobiology: Behavior from a biological perspective.* New York: Academic Press, 1971. Pp. 1–52.

Marler, P. Studies of fighting in chaffinches. 3. Proximity as a cause of aggression. *British Journal of Animal Behaviour,* 1956, **4**, 23–30.

Marler, P. Development in the study of animal communication. In P. R. Bell (Ed.), *Darwin's biological work.* Cambridge: Cambridge University Press, 1959. Pp. 150–206.

Marler, P. The filtering of external stimuli during instinctive behavior. In W. H. Thorpe & O. L. Zangwill (Eds.), *Current problems in animal behaviour.* Cambridge: Cambridge University Press, 1963. Pp. 150–166.

Marler, P. A comparative approach to vocal learning: Song development in white-crowned sparrows. *Journal of Comparative Psychology Monograph,* 1970, **71**(2), 1–25.

Marler, P., & Hamilton, W. J. III. *Mechanisms of animal behavior.* New York: Wiley, 1966.

Martan, J., & Shepherd, B. A. Role of the copulatory plug in reproduction of the guinea pig. *Journal of Experimental Zoology,* 1976, **196**, 79–84.

Mason, W. A. The effects of social restriction on the behavior of rhesus monkeys: I. Free social behavior. *Journal of Comparative and Physiological Psychology,* 1960, **53**, 582–589.

Mason, W. A. The effects of social restriction on the behavior of rhesus monkeys: II. Tests of gregariousness. *Journal of Comparative and Physiological Psychology,* 1961, **54**, 287–290.

Mason, W. A., & Lott, D. F. Ethology and comparative psychology. *Annual Review of Psychology,* 1976, **27**, 129–154.

Masterton, B., & Skeen, L. C. Origins of anthropoid intelligence: Prefrontal system and delayed alternation in hedgehog, tree shrew, and bush baby. *Journal of Comparative and Physiological Psychology,* 1972, **81**, 423–433.

Masterton, R. B., Bitterman, M. E., Campbell, C. B. G., & Hotton, N. (Eds.) *Evolution of brain and behavior in vertebrates.* Hillsdale, N.J.: Erlbaum, 1976.

Masterton, R. B., Hodos, W., & Jerison, H. (Eds.) *Evolution, brain, and behavior. Persistent problems.* Hillsdale, N.J.: Erlbaum, 1976.

Matthews, G. V. T. *Bird navigation.* Cambridge: Cambridge University Press, 1955.

Maturana, H. R., Lettvin, J. Y., McCulloch, W. S., & Pitts, W. H. Anatomy and physiology of vision in the frog *(Rana pipiens). Journal of General Physiology Supplement,* 1960, **43**(2), 129–175.

Maynard Smith, J. Evolution and the theory of games. *American Scientist,* 1976, **64**, 41–45.

Maynard Smith, J., & Parker, G. A. The logic of asymmetric contests. *Animal Behaviour,* 1976, **24**, 159–175.

Mayr, E. Behavior and systematics. In A. Roe & G. G. Simpson (Eds.), *Behavior and evolution.* New Haven Conn.: Yale University Press, 1958. Pp. 341–362.

Mayr, E. *Animal species and evolution.* Cambridge, Mass.: Harvard University Press, 1963.

Mayr, E. Sexual selection and natural selection. In B. G. Campbell (Ed.), *Sexual selection and the descent of man.* Chicago: Aldine, 1972. Pp. 87–104.

McBride, G. The study of social organizations. *Behaviour,* 1976, **59**, 96–115.

McClearn, G. E. Strain differences in activity in mice: Influences of illumination. *Journal of Comparative and Physiological Psychology,* 1960, **53**, 142–143.

McClearn, G. E., & DeFries, J. C. *Introduction to behavioral genetics.* San Francisco: Freeman, 1973.

McClintock, M. K. Menstrual synchrony and suppression. *Nature*, 1971, **229**, 244–245.

McClintock, M. K. Sociobiology of reproduction in the Norway rat *(Rattus norvegicus):* Estrous synchrony and the role of the female rat in copulatory behavior. Doctoral dissertation, University of Pennsylvania, 1974.

McCollom, R. E., Siegel, P. B., & VanKrey, H. P. Responses to androgen in lines of chickens selected for mating behavior. *Hormones and Behavior,* 1971, **2**, 31–42.

McConnell, J. V. Cannibals, chemicals, and contiguity. *Animal Behaviour,* 1965, Suppl. 1, 61–66.

McConnell, J. V. Comparative physiology: Learning in invertebrates. *Annual Review of Physiology,* 1966, **28**, 107–136.

McConnell, J. V. Specific factors influencing planarian behavior. In W. C. Corning & S. C. Ratner (Eds.), *Chemistry of learning: Invertebrate research.* New York: Plenum, 1967. Pp. 217–233.

McConnell, J. V., & Mpitsos, G. Effects of the presence or absence of slime on classical conditioning in planarians. *American Zoologist,* 1965, **5**, 122. (Abstract)

McEwen, B. S. Interactions between hormones and nerve tissue. *Scientific American,* 1976, **235**(1), 48–58.

McFarland, D. J. (Ed.) *Motivational control systems analysis.* New York: Academic Press, 1974.

McGaugh, J. L. Time-dependent processes in memory storage. *Science,* 1966, **153**, 1351–1358.

McGaugh, J. L., & Herz, M. J. *Memory consolidation.* San Francisco: Albion, 1972.

McGill, T. E. Sexual behavior in three inbred strains of mice. *Behaviour,* 1962, **19**, 341–350.

McGill, T. E. An enlarged study of genotype and recovery of sex drive in male mice. *Psychonomic Science,* 1969, **15**, 250–251.

McGill, T. E. Induction of luteal activity in female house mice. *Hormones and Behavior,* 1970a, **1**, 211–222.

McGill, T. E. Genetic analysis of male sexual behavior. In G. Lindzey & D. D. Thiessen (Eds.), *Contributions to behavior-genetic analysis: The mouse as a prototype.* New York: Appleton Century Crofts, 1970b. Pp. 57–88.

McGill, T. E. *Readings in animal behavior.* (1st Ed.) New York: Holt, 1965.

McGill, T. E., & Blight, W. C. The sexual behavior of hybrid male mice compared with the sexual behavior of males of the inbred parent strains. *Animal Behaviour,* 1963a, **11**, 480–483.

McGill, T. E., & Blight, W. C. Effects of genotype on the recovery of sex drive in the male mouse. *Journal of Comparative and Physiological Psychology,* 1963b, **56**, 887–888.

McGill, T. E., & Haynes, C. M. Heterozygosity and retention of ejaculatory reflex after castration in male mice. *Journal of Comparative and Physiologial Psychology,* 1973, **84**, 423–429.

McGill, T. E., & Manning, A. Genotype and retention of the ejaculatory reflex in castrated male mice. *Animal Behavior,* 1976, **24**, 507–518.

McGill, T. E., & Ransom, T. W. The effects of genotypic change on conclusions regarding modes of inheritance of elements of behavior. *Animal Behaviour,* 1968, **16**, 88–91.

McGrew, W. C., & Tutin, C. E. G. Chimpanzee tool use in dental grooming. *Nature,* 1973, **241**, 477–478.

McKinney, F. The comfort movements of Anatidae. *Behaviour,* 1965, **25**, 120–220.

McKinney, F. The behavior of ducks. In E. S. E. Hafez (Ed.), *The behaviour of domestic animals.* Baltimore: Williams & Wilkins, 1969. Pp. 593–626.

Meddis, R. On the function of sleep. *Animal Behaviour,* 1975, **23**, 676–691.

Medewar, P. B. Some immunological and endocrinological problems raised by the evolution of viviparity in vertebrates. *Symposium of the Society for Experimental Biology,* 1953, **7**, 320–338.

Melzack, R. Early experience: A neuropsychological approach to heredity-environment interactions. In G. Newton & S. Levine (Eds.), *Early experience and behavior: The psychobiology of development.* Springfield, Ill.: Charles C. Thomas, 1968. Pp. 65–82.

Messeri, P., Oliverio, A., & Bovet, D. Relations between avoidance and activity: A diallel study in mice. *Behavioral Biology,* 1972, **7**, 733–742.

Metzgar, L. H. An experimental comparison of screech owl predation on resident and transient white-footed mice *(Peromyscus leucopus). Journal of Mammalogy,* 1967, **48**, 387–391.

Meyer, M. E. Discriminative basis for astronavigation in birds. *Journal of Comparative and Physiological Psychology,* 1964, **58**, 403–406.

Meyerriecks, A. J. Comparative breeding behavior of four species of North American herons. *Nuttall Ornithological Club Publications,* 1960, **2**, 1–158.

Meyerson, B. J. Central nervous monoamines and hormone induced estrus behavior in the spayed rat. *Acta Physiologica Scandanavica,* 1964, **63**, Suppl. 241, 1–32.

Michael, R. P., & Keverne, E. B. Pheromones in the communication of sexual status in primates. *Nature,* 1968, **218,** 746–749.

Michael, R. P., & Keverne, E. B. An annual rhythm in the sexual activity of the male rhesus monkey, *Macaca mulatta,* in the laboratory. *Journal of Reproduction and Fertility,* 1971, **25**, 95–98.

Michael, R. P., & Saayman, G. S. Individual differences in the sexual behaviour of male rhesus monkeys *(Macaca mulatta)* under laboratory conditions. *Animal Behaviour,* 1967, **15**, 460–466.

Michael, R. P., Saayman, G. S., & Zumpe, D. Inhibition of sexual receptivity by progesterone in rhesus monkeys. *Journal of Endocrinology,* 1967, **39**, 309–310.

Miles, R. C. Discrimination-learning sets. In A. M. Schrier, H. F. Harlow, & F. Stollnitz (Eds.), *Behavior of nonhuman primates: Modern research trends.* Vol. I. New York: Academic Press, 1965. Pp. 51–95.

Millar, R. D. Free-operant comparisons of wild and domestic Norway rats. *Journal of Comparative and Physiological Psychology,* 1975, **89**, 913–922.

Miller, N. E. Certain factors of learning relevant to the search for its physical basis. In

G. C. Quarton, T. Melnechuk, & F. O. Schmitt (Eds.), *The neurosciences: A study program.* New York: Rockefeller University Press, 1967. Pp. 643–652.

Milligan, W. L., Powell, D. A., & Borasio, G. Sexual variables and shock-elicited aggression. *Journal of Comparative and Physiological Psychology,* 1973, **83**, 441–450.

Misanin, J. R., Nagy, Z. M., Keiser, E. F., & Bowen, W. Emergence of long-term memory in the neonatal rat. *Journal of Comparative and Physiological Psychology,* 1971, **77**, 188–199.

Mitchell, D. Experiments on neophobia in wild and laboratory rats: A reevaluation. *Journal of Comparative and Physiological Psychology,* 1976, **90**, 190–197.

Mitchell, G. Paternalistic behavior in primates. *Psychological Bulletin,* 1969, **71**, 399–417.

Moltz, H. Contemporary instinct theory and the fixed action pattern. *Psychological Review,* 1965, **72**, 27–47.

Moltz, H., Lubin, M., Leon, M., & Numan, M. Hormonal induction of maternal behavior in the ovariectomized nulliparous rat. *Physiology and Behavior,* 1970, **5**, 1373–1377.

Moltz, H., Rowland, D., Steele, M., & Halaris, A. Hypothalamic norepinephrine: Concentration and metabolism during pregnancy and lactation in the rat. *Neuroendocrinology,* 1975, **19**, 252–258.

Money, J., & Ehrhardt, A. A. *Man & woman, boy & girl.* Baltimore: Johns Hopkins University Press, 1972.

Montagna, W., & Sadler, W. A. *Reproductive behavior.* New York: Plenum, 1974.

Moorcroft, W. H., Lytle, L. D., & Campbell, B. A. Ontogeny of starvation-induced behavioral arousal in the rat. *Journal of Comparative and Physiological Psychology,* 1971, **75**, 59–67.

Morgan, C. L. *Introduction to comparative psychology.* New York: Scribner, 1894.

Morris, D. The reproductive behavior of the zebra finch *(Poephila guttata)* with special reference to pseudo-female behavior and displacement activities. *Behaviour,* 1954, **6**, 271–322.

Morris, D. The feather postures of birds and the problem of the origin of social signals. *Behaviour,* 1956, **9**, 75–113.

Moss, F. A. (Ed.) *Comparative psychology.* (2d Ed.) New York: Prentice-Hall, 1942.

Moss, R. L., & McCann, S. M. Induction of mating behavior in rats by luteinizing hormone-releasing factor. *Science,* 1973, **181**, 177–179.

Mountjoy, P. T., Bos, J. H., Duncan, M. O., & Verplank, R. B. Falconry: Neglected aspect of the history of psychology. *Journal of the History of the Behavioral Sciences,* 1969, **5**, 59–67.

Moyer, K. E. Kinds of aggression and their physiological basis. *Communications in Behavioral Biology,* 1968, **2**, 65–87.

Moyer, K. E. *The psychobiology of aggression.* New York: Harper & Row, 1976.

Mpitsos, G. J., & Collins, S. D. Learning: Rapid aversive conditioning in the gastropod mollusk *Pleurobranchaea. Science,* 1975, **188**, 954–957.

Mpitsos, G. J., & Davis, W. J. Learning: Classical and avoidance conditioning in the mollusk *Pleurobranchaea. Science,* 1973, **180**, 317–320.

Mueller, H. C. Hawks select odd prey. *Science,* 1975, **188**, 953–954.

Mulligan, J. A. Singing behavior and its development in the song sparrow, *Melospiza melodia. University of California Publications in Zoology,* 1966, **81**, 1–76.

Murphey, R. M. Instrumental conditioning in the fruit fly, *Drosophila melanogaster. Animal Behaviour,* 1967, **15**, 153–161.

Murphy, M. R., & Schneider, G. E. Olfactory bulb removal eliminates mating behaviour in the male golden hamster. *Science,* 1970, **167**, 302–304.

Myers, J. H., & Krebs, C. J. Population cycles in rodents. *Scientific American,* 1974, **230**(6), 38–46.

Myrberg, A. A., Jr., & Thresher, R. E. Interspecific aggression and its relevance to the concept of territoriality in reef fishes. *American Zoologist,* 1974, **14**, 81–96.

Nadler, R. D. Intrahypothalamic locus for induction of androgen sterilization in neonatal female rats. *Neuroendocrinology,* 1972, **9**, 349–357.

Narayanan, C. H., Fox, M. W., & Hamburger, V. Prenatal development of spontaneous and evoked activity in the rat *(Rattus norvegicus albinus). Behaviour,* 1971, **41**, 100–134.

Narins, P. M., & Capranica, R. R. Sexual differences in the auditory system of the tree frog *Eleutherodactylus coqui. Science,* 1976, **192**, 378–380.

Nelson, J. B. Colonial and cliff nesting in the gannet. *Ardea,* 1967, **55**, 60–90.

Nelson, M. C. Classical conditioning in the blowfly *(Phormia regina):* Associative and excitatory factors. *Journal of Comparative and Physiological Psychology,* 1971, **77**, 353–368.

Nice, M. M. The role of territory in bird life. *American Midland Naturalist,* 1941, **26**, 441–487.

Nigrosh, B. J., Slotnick, B. M., & Nevin, J. A. Olfactory discrimination, reversal learning, and stimulus control in rats. *Journal of Comparative and Physiological Psychology,* 1975, **89**, 285–294.

Nisbet, I. C. T. Courtship-feeding, egg-size and breeding success in common terns. *Nature,* 1973, **241**, 141–142.

Noble, G. K. Dominance in the life of birds. *Auk,* 1939, **56**, 263–273.

Noirot, E. Ultrasounds and maternal behavior in small rodents. *Developmental Psychobiology,* 1972, **5**, 371–387.

Nottebohm, F. Ontogeny of bird song. *Science,* 1970, **167**, 950–956.

Nottebohm, F. The origins of vocal learning. *American Naturalist,* 1972, **106**, 116–140.

Nottebohm, F., & Nottebohm, M. Vocalizations and breeding behavior of surgically deafened ring doves *Streptopelia risoria. Animal Behaviour,* 1971, **19**, 313–328.

Numan, M. Medial preoptic area and maternal behavior in the female rat. *Journal of Comparative and Physiological Psychology,* 1974, **87**, 746–759.

Nyby, J., Wallace, P., Owen, K., & Thiessen, D. D. An influence of hormones on hoarding behavior in the Mongolian gerbil *(Meriones unguiculatus). Hormones and Behavior,* 1973, **4**, 283–288.

O'Boyle, M. Rats and mice together: The predatory nature of the rat's mouse-killing response. *Psychological Bulletin,* 1974, **81**, 261–269.

Olds, J., Disterhoft, J. F., Segal, M., Kornblith, C. L., & Hirsh, R. Learning centers of rat brain mapped by measuring latencies of conditioned unit responses. *Journal of Neurophysiology,* 1972, **35**, 202–219.

Oliverio, A., & Messeri, P. An analysis of single-gene effects on avoidance, maze, wheel running, and exploratory behavior in the mouse. *Behavioral Biology,* 1973, **8**, 771–783.

Oppenheim, R. W. Color preferences in the pecking response of newly hatched ducks *(Ana platyrhynchos). Journal of Comparative and Physiological Psychology,* 1968, **66**, Suppl., 1–17.

Oppenheim, R. W. The ontogeny of behavior in the chick embryo. *Advances in the Study of Behavior,* 1974, **5**, 133–172.

Orians, G. H. On the evolution of mating systems in birds and mammals. *American Naturalist,* 1969, **103**, 589–603.

Osborne, G. L., Caul, W. F., & Vanstrum, R. Avoidance differences between rats and gerbils. *Animal Learning and Behavior,* 1976, **4**, 151–159.

Papi, F., Fiore, L., Fiaschi, V., & Benvenuti, S. An experiment for testing the hypothesis of olfactory navigation of homing pigeons. *Journal of Comparative Physiology,* 1973, **83**, 93–102.

Parker, G. A. Sperm competition and its evolutionary consequences in the insects. *Biological Reviews,* 1970, **45**, 525–567.

Parsons, P. A. Male mating speed as a component of fitness in *Drosophila. Behavior Genetics,* 1974, **4**, 395–404.

Patterson, D. J. Habituation in a protozoan *Vorticella convallaria. Behaviour,* 1973, **45**, 304–311.

Patterson, I. J. Timing and spacing of broods in the black-headed gull *Larus ridibundus. Ibis,* 1965, **107**, 433–459.

Pavlov, I. P. Conditioned reflexes: An investigation of the physiological activity of the cerebral cortex. Oxford: Oxford University Press, 1927. (Dover edition, New York, 1960, Trans. G. V. Anrep.)

Payne, R. S. & McVay, S. Songs of humpback whales. *Science,* 1971, **173**, 587–597.

Peeke, H. V. S., Herz, M. J., & Wyers, E. J. Forward conditioning, backward conditioning, and pseudoconditioning sensitization in the earthworm *(Lumbricus terrestris). Journal of Comparative and Physiological Psychology,* 1967, **64**, 534–536.

Peters, J. J., & Isaacson, R. L. Acquisition of active and passive responses in two breeds of chickens. *Journal of Comparative and Physiological Psychology,* 1963, **56**, 793–796.

Petersen, B., Lundgren, L., & Wilson, L. The development of flight capacity in a butterfly. *Behaviour,* 1956, **10**, 324–339.

Petersen, M. R., & Lyon, D. D. An application of the species-specific defense reaction hypothesis. *Psychological Record,* 1975, **25**, 21–37.

Pfaff, D. Nature of sex hormone effects on rat sex behavior: Specificity of effects and individual patterns of response. *Journal of Comparative and Physiological Psychology,* 1970, **73**, 349–358.

Pfaff, D., Lewis, C., Diakow, C., & Keiner, M. Neurophysiological analysis of mating behavior responses as hormone-sensitive reflexes. *Progress in Physiological Psychology,* 1973, **5**, 253–297.

Phillips, R. E., & Youngren, O. M. Brain stimulation and species-typical behaviour: Activities evoked by electrical stimulation of the brains of chickens *(Gallus gallus). Animal Behaviour,* 1971, **19**, 757–779.

Phoenix, C. H., Copenhaver, K. H., & Brenner, R. M. Scanning electron microscopy of penile papillae in intact and castrated male rats. *Hormones and Behavior,* 1976, **7**, 217–222.

Phoenix, C. H., Goy, R. W., & Resko, J. A. Psychosexual differentiation as a function of androgenic stimulation. In M. Diamond (Ed.) *Reproduction and sexual behavior.* Bloomington, Ind.: Indiana University Press, 1968. Pp. 33–49.

Pianka, E. R. On r- and K-selection. *American Naturalist,* 1970, **104**, 592–597.

Pietras, R. J., & Moulton, D. G. Hormonal influences on odor detection in rats: Changes associated with the estrous cycle, pseudopregnancy, ovariectomy, and administration of testosterone propionate. *Physiology and Behavior,* 1974, **12**, 475–491.

Pittendrigh, C. Adaptation, natural selection, and behavior. In A. Roe & G. G. Simpson (Eds.), *Behavior and evolution.* New Haven, Conn.: Yale University Press, 1958. Pp. 390–416.

Porter, K. P. *Herpetology.* Philadelphia: Saunders, 1972.

Powell, R. W. Operant responding in the common crow *(Corvus brachyrhynchos). Bulletin of the Psychonomic Society,* 1973, **1**, 401–403.

Power, H. W. Mountain bluebirds: Experimental evidence against altruism. *Science,* 1975, **189**, 142–143.

Power, H. W. Altruism in mountain bluebirds? *Science,* 1976, **191**, 809–810.

Powers, J. B., & Winans, S. S. Sexual behavior in peripherally anosmic male hamsters. *Physiology and Behavior,* 1973, **10**, 361–368.

Powers, J. B., & Winans, S. S. Vomeronasal organ: Critical role in mediating sexual behavior of the male hamster. *Science,* 1975, **187**, 961–963.

Premack, D. Language in chimpanzee? *Science,* 1971, **172**, 808–822.

Price, E. The effect of reproductive performance on the domestication of the prairie deermouse, *Peromyscus maniculatus bairdii. Evolution,* 1967, **211**, 762–770.

Price, E. Differential reactivity of wild and semi-domestic deermice *(Peromyscus maniculatus). Animal Behaviour,* 1970, **18**, 747–752.

Price, E. O. Novelty-induced self-food deprivation in wild and semi-domestic deermice *(Peromyscus maniculatus bairdii). Behaviour,* 1972, **41**, 91–104.

Price, E. O. Some behavioral differences between wild and domestic Norway rats: Gnawing and platform jumping. *Animal Learning and Behavior,* 1973, **1**, 312–316.

Price, E. O., & Huck, U. W. Open-field behavior of wild and domestic Norway rats. *Animal Learning and Behavior,* 1976, **4**, 125–130.

Price, E. O., & Loomis, S. Maternal influence on the response of wild and domestic Norway rats to a novel environment. *Developmental Psychobiology,* 1973, **6**, 203–208.

Priestnall, R. Effects of handling on maternal behavior in the mouse *(Mus musculus). Animal Behaviour,* 1973, **21**, 383–386.

Pryor, K. The porpoise caper. *Psychology Today,* 1969, **3**(7), 46–49, 64.

Pryor, K. W., Haag, R., & O'Reilly, J. The creative porpoise: Training for novel behavior. *Journal of the Experimental Analysis of Behavior,* 1969, **12**, 653–661.

Puente, A. E., & Dewsbury, D. A. Courtship and copulatory behavior of bottlenosed dolphins *(Tursiops truncatus). Cetology,* 1976, **21**, 1–9.

Quadagno, D. M., & Rockwell, J. The effect of gonadal hormones in infancy on maternal behavior in the adult rat. *Hormones and Behavior,* 1972, **3**, 55–62.

Quadagno, D. M., Shryne, J., Anderson, C., & Gorski, R. A. Influence of gonadal hormones on social, sexual, emergence, and open field behaviour in the rat *(Rattus norvegicus). Animal Behaviour,* 1972, **20**, 732–740.

Quinn, W. G., Harris, W. A., & Benzer, S. Conditioned behavior in *Drosophila melanogaster. Proceedings of the National Academy of Sciences, U.S.A.,* 1974, **71**, 708–712.

Rabb, G. B. Evolutionary aspects of the reproductive behavior of frogs. In J. L. Vial (Ed.), *Evolutionary biology of the Anurans.* Columbia, Mo.: University of Missouri Press, 1973. Pp. 213–227.

Rachlin, H. *Behavior and learning.* San Francisco: Freeman, 1976.

Raisman, G., & Field, P. M. Sexual dimorphism in the neuropil of the preoptic area of the rat and its dependence on neonatal androgen. *Brain Research,* 1973, **54**, 1–29.

Rajecki, D. W. Imprinting in precocial birds: Interpretation, evidence, and evaluation. *Psychological Bulletin,* 1973, **79**, 48–58.

Ralls, K. Auditory sensitivity in mice: *Peromyscus* and *Mus musculus. Animal Behaviour,* 1967, **15**, 123–128.

Randall, P. K., & Campbell, B. A. Ontogeny of behavioral arousal in rats: Effect of maternal and sibling presence. *Journal of Comparative and Physiological Psychology,* 1976, **90**, 453–459.

Ransom, T. W., & Ransom, B. S. Adult male-infant relations among baboons *(Papio anubis). Folia Primatologica,* 1971, **16**, 179–195.

Ratner, S. C., & Gilpin, A. R. Habituation and retention of habituation of responses to air puff of normal and decerebrate earthworms. *Journal of Comparative and Physiological Psychology,* 1974, **86**, 911–918.

Ratner, S. C., & Miller, K. R. Classical conditioning in earthworms, *Lumbricus terrestris. Journal of Comparative and Physiological Psychology,* 1959, **52**, 102–105.

Redican, W. K., & Mitchell, G. A longitudinal study of paternal behavior in adult rhesus monkeys. *Developmental Psychology,* 1973, **8**, 135–136.

Rensch, B. Increase of learning capability with increase of brain-size. *American Naturalist,* 1956, **90**, 81–95.

Rensch, B. The intelligence of elephants. *Scientific American,* 1957, **196**(2), 44–49.

Rescorla, R. A. Pavlovian conditioning and its proper control procedures. *Psychological Review,* 1967, **74**, 71–80.

Resko, J. A., Feder, H. H., & Goy, R. W. Androgen concentrations in plasma and testis of developing rats. *Journal of Endocrinology,* 1968, **40**, 485–491.

Richards, M. P. M. Infantile handling in rodents: A reassessment in the light of recent studies of maternal behavior. *Animal Behaviour,* 1966, **14**, 582.

Richter, C. P. The use of the Norway rat for psychiatric research. *Journal of Nervous and Mental Disorders,* 1949, **110**, 370–386.

Richter, C. P. The effects of domestication and selection on the behavior of the Norway rat. *Journal of the National Cancer Institute,* 1954, **15**, 727–738.

Riesen, A. H. Perception in comparative perspective. In D. A. Dewsbury & D. A. Rethlingshafer (Eds.), *Comparative psychology: A modern survey.* New York: McGraw-Hill, 1973. Pp. 395–426.

Roberts, W. W., Steinberg, M. L., & Means, L. W. Hypothalamic mechanisms for sexual, aggressive, and other motivational behaviors in the opossum, *Didelphis virginiana. Journal of Comparative and Physiological Psychology,* 1967, **64**, 1–15.

Robertson, D. R. Social control of sex reversal in a coral-reef fish. *Science,* 1972, **177**, 1007–1009.

Roeder, K. D. *Nerve cells and insect behavior.* (Rev. Ed.) Cambridge, Mass.: Harvard University Press, 1967.

Roeder, K. D. Episodes in insect brains. *American Scientist,* 1970, **58**, 378–389.

Roeder, K. D., & Treat, A. E. The detection and evasion of bats by moths. *American Scientist,* 1961, **49**, 135–148.

Romer, A. S. *The vertebrate body.* (3d ed.) Philadelphia: Saunders, 1962.

Rose, R. M., Holaday, J. W., & Bernstein, I. S. Plasma testosterone, dominance rank and aggressive behavior in male rhesus monkeys. *Nature,* 1971, **231**, 366–368.

Rosenblatt, J. S. Effects of experience on sexual behavior in male cats. In F. A. Beach (Ed.), *Sex and behavior.* New York: Wiley, 1965. Pp. 416–439.

Rosenblatt, J. S. Nonhormonal basis of maternal behavior in the rat. *Science,* 1967, **156**, 1512–1514.

Rosenblatt, J. S., & Lehrman, D. S. Maternal behavior in the laboratory rat. In H. L. Rheingold (Ed.), *Maternal behavior in mammals.* New York: Wiley, 1963. Pp. 8–57.

Rosenkoetter, J. S., & Boice, R. Earthworm pheromones and T-maze performance. *Journal of Comparative and Physiological Psychology,* 1975, **88**, 904–910.

Rosenzweig, M. R., Krech, D., Bennett, E. L., & Diamond, M. C. Modifying brain chemistry and anatomy by enrichment of impoverishment of experience. In G. Newton & S. Levine (Eds.), *Early experience and behavior: The psychobiology of development.* Springfield, Ill.: Charles C Thomas, 1968. Pp. 258–298.

Rosenzweig, M. R., & Leiman, A. L. Brain functions. *Annual Review of Psychology,* 1968, **19**, 55–98.

Ross, S., Sawin, P. B., Zarrow, M. X., & Denenberg, V. H. Maternal behavior in the rabbit. In H. L. Rheingold (Ed.), *Maternal behavior in mammals.* New York: Wiley, 1963. Pp. 94–121.

Rothenbuhler, W. C. Genetic and evolutionary considerations of social behavior of honey bees and some related insects. In J. Hirsch (Ed.), *Behavior-genetic analysis.* New York: McGraw-Hill, 1967. Pp. 61–106.

Rovner, J. S. Copulatory pattern supports generic placement of *Scizocosa avida* (Walckenaer) (Araneae: Lycosidae). *Psyche,* 1973, **80**, 245–248.

Rowell, T. *The social behaviour of monkeys.* Middlesex, England: Penguin, 1972.

Rozin, P., & Kalat, J. W. Specific hungers and poison avoidance as adaptive specializations of learning. *Psychological Review,* 1971, **78**, 459–486.

Rudegeair, T. J., Jr. The reproductive behavior and ecology of the white ibis *(Eudocimus albus).* Doctoral dissertation, University of Florida, 1975.

Rumbaugh, D. M. The learning and sensory capacities of the squirrel monkey in phylogenetic perspective. In L. A. Rosenblum & R. W. Cooper (Eds.), *The squirrel monkey.* New York: Academic Press, 1968. Pp. 255–317.

Rumbaugh, D. M., & Gill, T. V. Language and the acquisition of language-type skills by a chimpanzee *(Pan). Annals of the New York Academy of Sciences,* 1976, **270**, 90–123.

Rumbaugh, D. M., & Steinmetz, G. T. Discrimination reversal skills of the lowland gorilla *(Gorilla g. gorilla). Folia Primatalogica,* 1971, **16**, 144–152.

Rusak, B., & Zucker, I. Biological rhythms and animal behavior. *Annual Review of Psychology,* 1975, **26**, 137–171.

Russell, E. M., & Pearce, G. A. Exploration of novel objects by marsupials. *Behaviour,* 1971, **40**, 312–322.

Russell, P. A. "Infantile stimulation" in rodents: A consideration of possible mechanisms. *Psychological Bulletin,* 1971, **75**, 192–202.

Russo, F., & Caldwell, W. E. Biomagnetic phenomena: Some implications for the behavioral and neurophysiological sciences. *Genetic Psychology Monographs,* 1971, **84**, 177–243.

Sachs, B. D. Photoperiodic control of reproductive behavior and physiology of the male Japanese quail *(Coturnix coturnix japonica). Hormones and Behavior,* 1969, **1**, 7–24.

Sadleir, R. M. F. S. *The ecology of reproduction in wild and domestic mammals.* London: Methuen, 1969.

Sales, G. D. Ultrasound and mating behavior in rodents with some observations on other behavioral situations. *Journal of Zoology,* 1972, **168**, 149–164.

Salis, P. J., & Dewsbury, D. A. *p*-Chlorophenylalanine facilitates copulatory behavior in male rats. *Nature,* 1971, **232**, 400–401.

Sanders, G. D. The cephalopods. In W. C. Corning, J. A. Dyal, & A. O. D. Willows (Eds.), *Invertebrate learning: Vol. 3. Cephalopods and echinoderms.* New York: Plenum, 1975. Pp. 1–101.

Sauer, E. G. Die Sternorientierung nächtlich ziehender Grasmücken *(Sylvia atricapilla borin* und *curruca). Zeitschrift für Tierpsychologie,* 1957, **14**, 29–70.

Schaefer, T. Some methodological implications of the research on "early handling" in the rat. In G. Newton & S. Levine (Eds.), *Early experience and behavior: The psychobiology of development.* Springfield, Ill.: Charles C Thomas, 1968. Pp. 102–141.

Schein, M. W. On the irreversibility of imprinting. *Zeitschrift für Tierpsychologie,* 1963, **20**, 462–467.

Schlesinger, K., Boggan, W. O., & Freedman, D. X. Genetics of audiogenic seizures: III. Time response relationships between drug administration and seizure susceptibility. *Life Sciences,* 1970, **9**, 721–729.

Schmidt-Koenig, K. Current problems in bird orientation. *Advances in the Study of Behavior,* 1965, **1**, 217–278.

Schneider, D. The sex-attractant receptor of moths. *Scientific American,* 1974, **231**(1), 28–35.

Schneiderman, N., Fuentes, I., & Gormezano, I. Acquisition and extinction of the classically conditioned eyelid response in the albino rabbit. *Science,* 1962, **136**, 650–652.

Schneiderman, N., & Gormezano, I. Conditioning of the nictitating membrane of the rabbit as a function of CS-US interval. *Journal of Comparative and Physiological Psychology,* 1964, **57**, 188–195.

Schneirla, T. C. Ant learning as a problem in comparative psychology. In P. L. Harriman (Ed.), *Twentieth century psychology.* New York: Philosophical Library, 1946. Pp. 276–305.

Schneirla, T. C. Interrelationships of the "innate" and the "acquired" in instinctive behavior. In P. P. Grassé (Ed.), *L'Instinct dans le comportement des animaux et de l'homme.* Paris: Masson, 1956. Pp. 387–452.

Schneirla, T. C. Theoretical considerations of cyclic processes in doryline ants. *Proceedings of the American Philosophical Society,* 1957, **101**, 106–133.

Schneirla, T. C., Rosenblatt, J. S., & Tobach, E. Maternal behavior in the cat. In H. L. Rheingold (Ed.), *Maternal behavior in mammals.* New York: Wiley, 1963. Pp. 122–168.

Scholz, A. T., Horrall, R. M., Cooper, J. C., & Hasler, A. D. Imprinting to chemical cues: The basis for home stream selection in salmon. *Science,* 1976, **192**, 1247–1249.

Schusterman, R. J., & Balliet, R. F. Conditioned vocalizations as a technique for determining visual acuity thresholds in sea lions. *Science,* 1970, **169**, 498–501.

Scott, J. P. The organization of comparative psychology. *Annals of the New York Academy of Sciences,* 1973, **223,** 7–40.

Scott, J. P., & Fuller, J. L. *Genetics and the social behavior of the dog.* Chicago: University of Chicago Press, 1965.

Scott, J. P., & McCray, C. Allelomimetic behavior in dogs: Negative effects of competition on social facilitation. *Journal of Comparative and Physiological Psychology,* 1967, **63**, 316–319.

Scudder, G. G. E. Species concepts and speciation. *Canadian Journal of Zoology,* 1974, **52**, 1121–1134.

Searle, L. V. The organization of hereditary maze-brightness and maze-dullness. *Genetic Psychology Monographs,* 1949, **39**, 279–325.

Sebeok, T. A. Animal communication. *Science,* 1965, **147**, 1006–1014.

Seghers, B. H. Schooling behavior in the guppy *(Poecilia reticulata):* An evolutionary response to predation. *Evolution,* 1974, **28**, 486–489.

Selander, R. K. Sexual selection and dimorphism in birds. In B. G. Campbell (Ed.), *Sexual selection and the descent of man, 1871–1971.* Chicago: Aldine, 1972. Pp. 180–230.

Selander, R. K., & Kaufman, D. W. Genic variability and strategies of adaptation in animals. *Proceedings of the National Academy of Sciences, U.S.A.,* 1973, **70**, 1875–1877.

Selander, R. K., Smith, M. H., Yang, S. Y., Johnson, W. E., & Gentry, J. B. Biochemical polymorphism and systematics in the genus *Peromyscus:* I. Variation in the old-field mouse *(Peromyscus polionotus). Studies in Genetics,* 1971, **6**, 49–90.

Seligman, M. E. P. Control group and conditioning: A comment on operationism. *Psychological Review,* 1969, **76**, 484–491.

Seligman, M. E. P. On the generality of the laws of learning. *Psychological Review,* 1970, **77**, 406–418.

Seligman, M. E. P. Reply to Malone. *Psychological Review,* 1973, **80**, 306.

Seligman, M. E. P., & Hager, J. L. (Eds.) *Biological boundaries of learning.* New York: Appleton Century Crofts, 1972.

Sevenster, P. Motivation and learning in sticklebacks. In D. Ingle (Ed.), *The central nervous system and fish behavior.* Chicago: University of Chicago Press, 1968. Pp. 233–245.

Shaw, E. Schooling in fishes: Critique and review. In L. R. Aronson, E. Tobach, D. S. Lehrman, & J. S. Rosenblatt (Eds.), *Development and evolution of behavior.* San Francisco: Freeman, 1970.

Sheen, F. J. *Peace of soul.* New York: McGraw-Hill, 1949.

Shettleworth, S. J. Constraints on learning. *Advances in the Study of Behavior,* 1972, **4**, 1–68.

Shettleworth, S. J. Reinforcement and the organization of behavior in golden hamsters: Hunger, environment, and food reinforcement. *Journal of Experimental Psychology, Animal Behavior Processes,* 1975, **104**, 56–87.

Shumake, S. A., & Caudill, C. J. Operant conditioning of licking in vampire bats, *Desmodus rotundus. Behavior Research Methods and Instrumentation,* 1974, **6**, 464–470.

Sidman, M. Avoidance conditioning with brief shock and no exteroceptive warning signal. *Science,* 1953, **118**, 157–158.

Siegel, R. K., & Jarvik, M. E. Learning in the land snail (*Helix aspersa* Muller). *Bulletin of the Psychonomic Society,* 1974, **4**, 476–478.

Silver, R., Reboulleau, C., Lehrman, D. S., & Feder, H. H. Radioimmunoassay of plasma progesterone during the reproductive cycle of male and female ring doves *(Streptopelia risoria). Endocrinology,* 1974, **94**, 1547–1554.

Simmons, J. A., Howell, D. J., & Suga, N. Information content of bat sonar echoes. *American Scientist,* 1975, **63**, 204–215.

Simpson, K. B., & May, D. Some effects of the oestrus cycle in the female rat. *Journal of the Institute of Animal Technicians,* 1973, **24**, 25–29.

Singh, S. D. Urban monkeys. *Scientific American,* 1969, **221**(1), 108–115.

Skinner, B. F. *The behavior of organisms: An experimental analysis.* New York: Appleton Century Crofts, 1938.

Skinner, B. F. Are theories of learning necessary? *Psychological Review,* 1950, **57**, 193–216.

Skinner, B. F. *Science and human behavior.* New York: Macmillan, 1953.

Skinner, B. F. A case history in the scientific method. *American Psychologist,* 1956, **11**, 221–233.

Sloan, L. R., & Latane, B,. Social deprivation and stimulus satiation in the albino rat. *Journal of Comparative and Physiological Psychology,* 1974, **87**, 1148–1156.

Slotnick, B. M., & Katz, H. M. Olfactory learning-set formation in rats. *Science,* 1974, **185**, 796–798.

Sluckin, W. Imprinting in guinea-pigs. *Nature,* 1968, **220**, 1148.

Smith, E. A. Adoptive suckling in the grey seal. *Nature,* 1968, **217**, 762–763.

Smith, L. T. The interanimal transfer phenomenon: A review. *Psychological Bulletin,* 1974, **81**, 1078–1095.

Smith, N. Effects of interpolated learning on the retention of an escape response in rats as a function of age. *Journal of Comparative and Physiological Psychology,* 1968, **65**, 422–426.

Smith, R. H. Wildness and domestication in *Mus musculus:* A behavioral analysis. *Journal of Comparative and Physiological Psychology,* 1972, **79**, 22–29.

Smith, R. H., & Connor, J. L. The inheritance of behavioral wildness in house mice *(Mus musculus L.). Animal Learning and Behavior,* 1974, **2**, 249–258.

Smith, W. J. Messages of vertebrate communication. *Science,* 1969, **165**, 145–150.

Smith, W. J., Smith, S. L., Oppenheimer, E. C., de Villa, J. G., & Ulmer, F. A. Behavior of a captive population of black-tailed prairie dogs: annual cycle of social behavior. *Behaviour,* 1973, **46**, 189–220.

Snodgrass, R. E. *Principles of insect morphology.* New York: McGraw-Hill, 1935.

Snyder, F. Toward an evolutionary theory of dreaming. *American Journal of Psychiatry,* 1966, **123**, 121–136.

Snyder, L. H. The inheritance of taste deficiency in man. *Ohio Journal of Science,* 1932, **32**, 436–440.

Sodetz, F. J., & Bunnell, B. N. Septal ablation and the social behavior of the golden hamster. *Physiology and Behavior,* 1970, **5**, 79–88.

Southern, H. N. Mimicry in cuckoos' eggs. In J. Huxley, A. C. Hardy, & E. B. Ford (Eds.), *Evolution as a process.* New York: Macmillan, 1954. (Collier edition, 1963, 257–270.)

Southwick, C. H. Effects of maternal environment on aggressive behavior of inbred mice. *Communications in Behavioral Biology,* 1968, **1**, 129–132.

Southwick, C. H., Beg, M. A., & Siddiqi, M. F. Rhesus monkeys in North India. In I. DeVore (Ed.), *Primate behavior: Field studies of monkeys and apes.* New York: Holt, 1965. Pp. 111–159.

Southwick, C. H., & Clark, L. H. Interstrain differences in aggressive behavior and exploratory activity of inbred mice. *Communications in Behavioral Biology,* 1968, **1**, 49–59.

Southwick, C. H., & Siddiqi, M. F. Contrasts in primate social behavior. *Bioscience,* 1974, **24**, 398–406.

Spalding, D. A. Instinct, with original observations on young animals. *Macmillan's Magazine,* 1873, **27**, 283–293. (Reprinted in *British Journal of Animal Behaviour,* 1954, **2**, 2–11.)

Spence, K. W. *Behavior theory and conditioning.* New Haven: Yale University Press, 1956.

Sperry, R. W. How a developing brain gets itself properly wired for adaptive function. In E. Tobach, L. R. Aronson & E. Shaw (Eds.), *The biopsychology of development.* New York: Academic Press, 1971. Pp. 27–44.

Spiegel, R., Price, E., & Huck, U. W. Differential vulnerability of wild, domestic, and hybrid Norway rats to predation by great-horned owls. *Journal of Mammalogy,* 1974, **55**, 336–392.

Spieth, H. T. Courtship behavior in *Drosophila. Annual Review of Entomology,* 1974, **19**, 385–405.

Sprott, R. L., & Staats, J. Behavioral studies using genetically defined mice—a bibliography. *Behavior Genetics,* 1975, **5**, 27–82.

Staats, J. The laboratory mouse. In E. L. Green (Ed.), *Biology of the laboratory mouse.* New York: McGraw-Hill, 1966.

Staddon, J. E. R., & Simmelhag, V. L. The "superstition" experiment: A reexamination of its implications for the principles of adaptive behavior. *Psychological Review,* 1971, **78**, 3–43.

Stanley, S. M. Clades versus clones in evolution: Why we have sex. *Science,* 1975, **190**, 382–383.

Stanley, W. C. Feeding behavior and learning in neonatal dogs. In J. F. Bosma (Ed.), *Second Symposium on oral sensation and perception.* Springfield, Ill.: Charles C Thomas, 1970. Pp. 242–290.

Stanley, W. C., Bacon, W. E., & Fehr, C. Discriminated instrumental learning in neonatal dogs. *Journal of Comparative and Physiological Psychology,* 1970, **70**, 335–343.

Stasik, J. H. Inheritance of T-maze learning in mice. *Journal of Comparative and Physiological Psychology,* 1970, **71**, 251–257.

Stehn, R. A., & Richmond, M. E. Male-induced pregnancy termination in the prairie vole, *Microtus ochrogaster. Science,* 1975, **187**, 1211–1213.

Stephan, H., & Andy, O. J. Quantitative comparative neuroanatomy of primates: An attempt at phylogenetic interpretation. *Annals of the New York Academy of Sciences,* 1969, **167**, 370–387.

Stephens, G. J., & McGaugh, J. L. Biological factors related to learning in the land snail *Helix aspersa* Muller. *Animal Behaviour,* 1972, **20**, 309–315.

Stern, J. J., & Murphy, M. The effects of thryroxine and estradiol benzoate on wheel running activity in female rats. *Physiology and Behaviour,* 1972, **9**, 79–82.

Stewart, J., Skvarenina, A., & Pottier, J. Effects of neonatal androgens on open-field behavior and maze learning in the prepubescent and adult rat. *Physiology and Behavior,* 1975, **14**, 291–295.

Stone, C. P. (Ed.) *Comparative psychology.* (3d Ed.) New York: Prentice-Hall, 1951.

Sutherland, N. S. Visual discrimination of orientation and shape by the octopus. *Nature,* 1957, **179**, 11–13.

Sutherland, N. S. Discrimination of horizontal and vertical events by *Octopus. Journal of Comparative and Physiological Psychology,* 1961, **54**, 43–48.

Sutherland, N. S. Visual discrimination of shape by *Octopus:* Squares and crosses. *Journal of Comparative and Physiological Psychology,* 1962, **55**, 939–943.

Swanson, H. H. Alteration of sex-typical behaviour of hamsters in open field and emergence tests by neo-natal administration of androgen or oestrogen. *Animal Behaviour,* 1967, **15**, 209–216.

Tagliamonte, A., Tagliamonte, P., Gessa, G. L., & Brodie, B. B. Compulsive sexual activity induced by *p*-chlorophenylalanine in normal and pinealectomized male rats. *Science,* 1969, **166**, 1433–1435.

Tapp, J. T., & Markowitz, H. Infant handling: Effects on avoidance learning, brain weight, and cholinesterase activity. *Science,* 1963, **140**, 486–487.

Tardif, G. N., & Murnik, M. R. Frequency-dependent sexual selection among wild-type strains of *Drosophila melanogaster. Behavior Genetics,* 1975, **5**, 373–379.

Tarpy, R. M. *Basic principles of learning.* Glenview, Ill.: Scott, Foresman, 1975.

Teitelbaum, P., & Epstein, A. N. The lateral hypothalamic syndrome: Recovery of feeding and drinking after lateral hypothalamic lesions. *Psychological Review,* 1962, **69**, 74–90.

Teleki, G. The omnivorous chimpanzee. *Scientific American,* 1973a, **228**(1), 32–42.

Teleki, G. *The predatory behavior of wild chimpanzees.* Lewisburg, Pa.: Bucknell University Press, 1973b.

Terkel, J., & Rosenblatt, J. S. Aspects of nonhormonal maternal behavior in the rat. *Hormones and Behavior,* 1971, **2**, 161–171.

Terkel, J., & Rosenblatt, J. S. Humoral factors underlying maternal behavior at paturition: Cross transfusion between freely moving rats. *Journal of Comparative and Physiological Psychology,* 1972, **80**, 365–371.

Thiessen, D. D. Reply to Wilcock on gene action and behavior. *Psychological Bulletin,* 1971, **75**, 103–105.

Thiessen, D. D. *Gene organization and behavior.* New York: Random House, 1972.

Thiessen, D. D. Footholds for survival. *American Scientist,* 1973, **61**, 346–351.

Thiessen, D. D., Lindzey, G., Blum, S., Tucker, A., & Friend, H. C. Visual behavior of the Mongolian gerbil *(Meriones unguiculatus). Psychonomic Science,* 1968, **11**, 23–24.

Thiessen, D. D., & Rice, M. Mammalian scent gland marking and social behavior. *Psychological Bulletin,* 1976, **83**, 505–539.

Thiessen, D. D., & Rodgers, D. A. Population density and endocrine functions. *Psychological Bulletin,* 1961, **58**, 441–451.

Thoman, E. B., & Arnold, W. J. Effects of incubator rearing with social deprivation on maternal behavior in rats. *Journal of Comparative and Physiological Psychology,* 1968, **65**, 441–446.

Thoman, E. B., & Levine, S. Hormonal and behavioral changes in the rat mother as a function of early experience treatments of the offspring. *Physiology and Behavior,* 1970, **5**, 1417–1421.

Thompson, M. L., & Edwards, D. A. Experiential and strain determinants of the estrogen-progesterone induction of sexual receptivity in spayed female mice. *Hormones and Behavior,* 1971, **2**, 299–305.

Thompson, R., & McConnell, J. V. Classical conditioning in the planarian, *Dugesia dorotocephala. Journal of Comparative and Physiological Psychology,* 1955, **48**, 65–68.

Thompson, R. F. The search for the engram. *American Psychologist*, 1976, **31**, 209–227.

Thompson, R. F., Patterson, M. M., & Teyler, T. J. The neurophysiology of learning. *Annual Review of Psychology,* 1972, **23**, 73–104.

Thompson, R. F., & Spencer, W. A. Habituation: A model phenomenon for the study of neuronal substrates of behavior. *Psychological Review,* 1966, **73**, 16–43.

Thompson, T. Operant and classically-conditioned aggressive behavior in Siamese fighting fish. *American Zoologist,* 1966, **6**, 629–641.

Thompson, W. R. The inheritance of behavior: Behavioral differences in fifteen mouse strains. *Canadian Journal of Psychology,* 1953, **7**, 145–155.

Thompson, W. R. The inheritance and development of intelligence. *Proceedings of the Association for Research on Nervous and Mental Diseases,* 1954, **33**, 209–231.

Thompson, W. R. Influence of prenatal maternal anxiety on emotionality in young rats. *Science,* 1957, **125**, 698–699.

Thompson, W. R., & Heron, W. The effects of restricting early experience on the problem-solving capacity of dogs. *Canadian Journal of Psychology*, 1954, **8**, 17–31.

Thorndike, E. L. *Animal intelligence.* New York: Macmillan, 1911.

Thornhill, R. Sexual selection and paternal investment in insects. *American Naturalist,* 1976, **110**, 153–163.

Thorpe, W. H. *Learning and instinct in animals.* (2d ed.) Cambridge, Mass.: Harvard University Press, 1963.

Tinbergen, L. The natural control of insects in pinewoods: I. Factors influencing the intensity of predation by songbirds. *Archieves Neerlandaises de Zoologie,* 1960, **13**, 265–343.

Tinbergen, N. *The study of instinct.* Oxford: Oxford University Press, 1951.

Tinbergen, N. The curious behavior of the stickleback. *Scientific American,* 1952a, **187**(12), 22–26.

Tinbergen, N. "Derived" activities; their causation, biological significance, origin, and emancipation during evolution. *Quarterly Review of Biology,* 1952b, **27**, 1–32.

Tinbergen, N. Comparative studies of the behaviour of gulls *(Laridae)*: A progress report. *Behaviour,* 1959, **15**, 1–70.

Tinbergen, N. On aims and methods of ethology. *Zeitschrift für Tierpsychologie,* 1963a, **20**, 410–429.

Tinbergen, N. The shell menace. *Natural History,* 1963b, **72**, 28–35.

Tinbergen, N. *Animal behavior.* New York: Time, Inc., 1965a.

Tinbergen, N. Some recent studies of the evolution of sexual behavior. In F. A. Beach (Ed.), *Sex and behavior.* New York: Wiley, 1965b. Pp. 1–33.

Tinbergen, N., Broekhuysen, F., Feekes, F., Houghton, C. W., Kruuk, H., & Szulc, E. Egg shell removal by the black-headed gull, *Larus ridibundus* L.: A behaviour component of camouflage. *Behaviour,* 1963, **19**, 74–117.

Tinbergen, N., Impekoven, M., & Franck, D. An experiment on spacing-out as a defence against predation. *Behaviour,* 1967, **28**, 307–321.

Tinbergen, N., Kruuk, H., Paillette, M., & Stamm, R. How do black-headed gulls distinguish between eggs and egg shells? *British Birds,* 1962, **55**, 120–129.

Tinbergen, N., & Perdeck, A. C. On the stimulus situation releasing the begging response in the newly hatched herring gull chick *(Larus argentatus argentatus* Pont). *Behaviour,* 1950, **3**, 1–39.

Tolman, C. W., & Wilson, G. F. Social feeding in domestic chicks. *Animal Behaviour,* 1965, **13**, 134–142.

Tolman, E. C. The inheritance of maze-learning ability in rats. *Journal of Comparative Psychology,* 1924, **4**, 1–18.

Trivers, R. L. The evolution of reciprocal altruism. *Quarterly Review of Biology,* 1971, **46**, 35–57.
Trivers, R. L. Parental investment and sexual selection. In B. G. Campbell (Ed.), *Sexual selection and the descent of man, 1871–1891.* Chicago: Aldine, 1972. Pp. 136–179.
Trivers, R. L. Parent-offspring conflict. *American Zoologist,* 1974, **14**, 249–264.
Trivers, R. L., & Hare, H. Haplodiploidy and the evolution of the social insects. *Science,* 1976, **191**, 249–263.
Trivers, R. L., & Willard, D. E. Natural selection of parental ability to vary the sex ratio of offspring. *Science,* 1973, **179**, 90–92.
Tryon, R. C. Genetic differences in maze-learning ability in rats. *Thirty-ninth yearbook of the National Society for the Study of Education.* Bloomington, Ill.: Public School Publishing, 1940. Pp. 111–119.
Turner, C. D., & Bagnara, J. T. *General endocrinology.* (5th ed.) Philadelphia: Saunders, 1971.
Tyler, P. A., & McClearn, G. E. A quantitative genetic analysis of runway learning in mice. *Behavior Genetics,* 1970, **1**, 57–69.
Udry, J. R., & Morris, N. M. Distribution of coitus in the menstrual cycle. *Nature,* 1968, **220**, 593–596.
Udvary, M. D. F. Mammalian evolution: Is it due to social subordination? *Science,* 1970, **170**, 344–345.
Ulagaraj, S. M., & Walker, T. J. Phonotaxis of crickets in flight: Attraction of male and female crickets to male calling songs. *Science,* 1973, **182**, 1278–1279.
Ungar, G., Desiderio, D. M., & Parr, W. Isolation, identification and synthesis of a specific-behavior-inducing brain peptide. *Nature,* 1972, **238**, 198–202.
Vale, J. R., & Ray, D. A diallel analysis of male mouse sex behavior. *Behavior Genetics,* 1972, **2**, 199–209.
Vale, J. R., Ray, D., & Vale, C. A. The interaction of genotype and exogenous neonatal androgen: Agonistic behavior in female mice. *Behavioral Biology,* 1972, **7**, 321–334.
Vale, J. R., & Vale, C. A. Individual differences and general laws in psychology: A reconciliation. *American Psychologist,* 1969, **24**, 1093–1108.
Valenstein, E. S., Riss, W., & Young, W. C. Experiential and genetic factors in the organization of sexual behavior in male guinea pigs. *Journal of Comparative and Physiological Psychology,* 1955, **48**, 397–403.
van Bergeijk, W. A. Anticipatory feeding behavior in the bullfrog *(Rana catesbiana). Animal Behaviour,* 1967, **15**, 231–238.
Vandenbergh, J. G. Male odor accelerates female sexual maturation in mice. *Endocrinology,* 1969, **84**, 658–660.
Vandenbergh, J. G. The influence of the social environment on sexual maturation in male mice. *Journal of Reproduction and Fertility,* 1971a, **24**, 383–390.
Vandenbergh, J. G. The effects of gonadal hormones on the aggressive behaviour of adult golden hamsters *(Mesocricetus auratus). Animal Behaviour,* 1971b, **19**, 589–594.
Vandenbergh, J. G., & Drickamer, L. C. Reproductive coordination among free-ranging rhesus monkeys. *Physiology and Behavior,* 1974, **13**, 373–376.
van der Lee, S. & Boot, L. M. Spontaneous pseudopregnancy in mice. *Acta Physiologica et Pharmacologica Neerlandica,* 1955, **4**, 442–444.
van Kreveld, D. A. Selective review of dominance-subordination relations in animals. *Genetic Psychology Monographs,* 1970, **81**, 143–173.

van Lawick-Goodall, J. *My friends the wild chimpanzees.* Washington: National Geographic Society, 1967.

van Oortmerssen, G. A. Biological significance, genetics and evolutionary origin of variability in behavior within and between inbred strains of mice *(Mus musculus):* A behavior genetic study. *Behaviour,* 1970, **38**, 1–92.

Verner, J., & Willson, M. F. The influence of habitats on the mating systems of North American passerine birds. *Ecology,* 1966, **47**, 143–147.

Verplanck, W. S. A glossary of some terms used in the objective study of behavior. *Psychological Review,* 1957, **64** (Suppl.), 1–42.

Vestal, B. M., & King, J. A. Relationship of age at eye opening to first optokinetic response in deermice *(Peromyscus). Developmental Psychobiology,* 1968, **1**, 30–34.

von Frisch, K. *Bees: Their vision, chemical senses and language.* (1st ed.) Ithaca, N.Y.: Cornell University Press, 1950.

von Frisch, K. *Bees: Their vision, chemical senses, and language.* (2d ed.) Ithaca, N.Y.: Cornell University Press, 1971.

von Holst, E., & von St. Paul, U. Electrically controlled behavior. *Scientific American,* 1962, **206**(3), 50–59.

von Holst, E., & von St. Paul, U. On the functional organization of drives. *Animal Behaviour,* 1963, **11**, 1–20.

von Uexküll, J. *A stroll through the world of animals and men: A picture book of invisible worlds.* Berlin: Springer, 1934. (Trans. C. H. Schiller in C. H. Schiller, ed., *Instinctive Behavior: The development of a modern concept.* New York: International Universities Press, 1957.)

Voronin, L. G. Some results of comparative-physiological investigations of higher nervous activity. *Psychological Bulletin,* 1962, **59**, 161–195.

Wade, G. N. Sex hormones, regulatory behaviors, and body weight. *Advances in the Study of Behavior,* 1976, **6**, 201–279.

Wahlsten, D. Genetic experiments with animal learning: A critical review. *Behavioral Biology,* 1972, **7**, 143–182.

Walcott, C. The homing of pigeons. *American Scientist,* 1974, **62**, 542–552.

Waldbillig, R. J. Attack, eating, drinking, and gnawing elicited by electrical stimulation of rat mesencephalon and pons. *Journal of Comparative and Physiological Psychology,* 1975, **89**, 200–212.

Walk, R. D. The study of visual depth and distance perception in animals. *Advances in the Study of Behavior,* 1965, **1**, 99–154.

Walker, E. L. Psychological complexity as a basis for a theory of motivation and choice. *Nebraska Symposium on Motivation,* 1964, **7**, 47–95.

Walsh, R. N., & Cummins, R. A. The open-field test: A critical review. *Psychological Bulletin,* 1976, **83**, 482–504.

Wang, G. H. Relationship between "spontaneous" activity and estrous cycle in the white rat. *Comparative Psychology Monographs,* 1923, **2**(6), 1–27.

Ward, I. L. Prenatal stress feminizes and demasculinizes the behavior of males. *Science,* 1972, **175**, 82–84.

Warden, C. J. The historical development of comparative psychology. *Psychological Review,* 1927, **34**, 57–85, 135–168.

Warden, C. J., Jenkins, T. N., & Warner, L. H. *Introduction to comparative psychology.* New York: Ronald, 1934.

Warner, R. R., Robertson, D. R., & Leigh, E. G., Jr. Sex change and sexual selection. *Science,* 1975, **190**, 633–638.

Warren, J. M. Solution of object and positional discriminations by rhesus monkeys. *Journal of Comparative and Physiological Psychology,* 1959, **52**, 92–93.

Warren, J. M. Primate learning in comparative perspective. In A. M. Schrier, H. F. Harlow, & F. Stollnitz (Eds.), *Behavior of nonhuman primates: Modern research trends.* Vol. 1. New York: Academic Press, 1965a. Pp. 249–281.

Warren, J. M. Comparative psychology of learning. *Annual Review of Psychology,* 1965b, **16**, 95–118.

Warren, J. M. Evolution, behavior and the prefrontal cortex. *Acta Neurobiologica Experimental,* 1972, **32**, 581–593.

Warren, J. M. Learning in vertebrates. In D. A. Dewsbury & D. A. Rethlingshafer (Eds.), *Comparative psychology: A modern survey.* New York: McGraw-Hill, 1973. Pp. 471–509.

Wasman, M., & Flynn, J. P. Directed attack elicited from hypothalamus. *Archives of Neurology,* 1962, **6**, 220–227.

Waters, R. H. Historical background of comparative psychology. In C. P. Stone (Ed.), *Comparative psychology.* (3d ed.) New York: Prentice-Hall, 1951.

Watson, J. B. *Behaviorism.* (2d ed.) Chicago: University of Chicago Press, 1930.

Watson, J. B., & Lashley, K. S. An historical and experimental study of homing. *Carnegie Institution Publications,* 1915, **7**(211), 61–83.

Wayner, M. J., & Carey, R. J. Basic drives. *Annual Review of Psychology,* 1973, **24**, 53–80.

Webb, W. B. *Sleep: An experimental approach.* New York: Macmillan, 1968.

Webb, W. B. Sleep as an adaptive response. *Perceptual and Motor Skills,* 1974a, **38**, 1023–1027.

Webb, W. B. The adaptive functions of sleep patterns. In P. Levin & W. P. Koella (Eds.), *Sleep 1974.* Basel: S. Karger, 1974b. Pp. 14–19.

Wecker, S. C. Habitat selection. *Scientific American,* 1964, **211**(4), 109–116.

Weisz, J., & Gunsalus, P. Estrogen levels in immature female rats: True or spurious—ovarian or adrenal? *Endocrinology,* 1973, **93**, 1057–1065.

Welch, B. L., & Welch, A. S. Some aspects of brain biochemistry correlated with general nervous reactivity and aggressiveness. In C. H. Southwick (Ed.), *Animal aggression: Selected readings.* New York: Van Nostrand-Reinhold, 1970.

Wells, M. J. What the octopus makes of it: Our world from another point of view. *American Scientist,* 1961, **49**, 215–227.

Wells, M. J. Learning and movement in octopuses. *Animal Behaviour,* 1965, Suppl. 1, 115–128.

Wells, P. H., & Wenner, A. M. Do honey bees have a language? *Nature,* 1973, **241**, 171–175.

Wenner, A. M. Sound production during the waggle dance of the honeybee. *Animal Behaviour,* 1962, **10**, 79–95.

Wenner, A. M. Information transfer in honey bees: A population approach. In L. Krames, P. Pliner, & T. Alloway, *Nonverbal communication.* New York: Plenum, 1974. Pp. 133–169.

Wenner, A. M., & Johnson, D. L. Simple conditioning in honey bees. *Animal Behaviour,* 1966, **14**, 149–155.

West Eberhard, M. J. The evolution of social behavior by kin selection. *Quarterly Review of Biology,* 1975, **50**, 1–33.

Westerman, R. A. Somatic inheritance of habituation of responses to light in planarians. *Science,* 1963, **140**, 676–677.

Whalen, R. E. The concept of instinct. In J. L. McGaugh (Ed.), *Psychobiology.* New York: Academic Press, 1971. Pp. 53–72.

Whalen, R. E., & Edwards, D. A. Hormonal determinants of the development of masculine and feminine behavior in male and female rats. *Anatomical Record,* 1967, **157**, 173–180.

Whalen, R. E., & Luttge, W. G. Testosterone, androstenedione and dihydrotestosterone: Effects on mating behavior of male rats. *Hormones and Behavior,* 1971, **2**, 117–125.

Whitney, G. D. Vocalization of mice: A single gene unit effect. *Journal of Heredity,* 1969, **60**, 337–340.

Whitney, G. D. Vocalization of mice influenced by a single gene in a heterogeneous population. *Behavior Genetics,* 1973, **3**, 57–64.

Whitten, W. K. Modification of the estrous cycle of the mouse by external stimuli associated with the male. *Journal of Endocrinology,* 1956, **13**, 399–404.

Wickler, W. *The sexual code: The social behavior of animals and men.* Garden City, N.Y.: Doubleday, 1972.

Wickler, W. Ethological analysis of convergent adaptation. *Annals of the New York Academy of Sciences,* 1973, **223**, 65–69.

Wilcock, J. Gene action and behavior: An evaluation of major gene pleiotropism. *Psychological Bulletin,* 1969, **72**, 1–29.

Wilcoxon, H. C., Dragoin, W. B., & Kral, P. A. Illness induced aversions in rat and quail: Relative salience of visual and gustatory cues. *Science,* 1971, **171**, 826–828.

Wiley, R. H. Territoriality and non-random mating in sage grouse, *Centrocercus urophasianus. Animal Behaviour Monographs,* 1973, **6**, 85–169.

Williams, D. R., & Williams, H. Auto-maintenance in the pigeon: Sustained pecking despite contingent non-reinforcement. *Journal of the Experimental Analysis of Behavior,* 1969, **12**, 511–520.

Williams, G. C. *Adaptation and natural selection: A critique of some current evolutionary thought.* Princeton, N.J.: Princeton University Press, 1966.

Williams, G. C. (Ed.) *Group selection.* Chicago: Aldine-Atherton, 1971.

Williams, H. L., Holloway, F. A., & Griffiths, W. J. Physiological psychology: Sleep. *Annual Review of Psychology,* 1973, **24**, 279–316.

Wilson, D. M. Central nervous mechanisms for the generation of rhythmic behavior in arthropods. *Symposium of the Society for Experimental Biology,* 1966, **20**, 199–228.

Wilson, D. M. The flight-control system of the locust. *Scientific American,* 1968, **218**(5), 83–90.

Wilson, E. O. Chemical communication in the social insects. *Science,* 1965, **149**, 1064–1071.

Wilson, E. O. *The insect societies.* Cambridge, Mass.: Harvard University Press, 1971.

Wilson, E. O. On the queerness of social evolution. *Bulletin of the Entomological Society of America,* 1972, **19**, 20–22.

Wilson, E. O. *Sociobiology: The new synthesis.* Cambridge, Mass.: Harvard University Press, 1975.

Wilson, J. R., Adler, N., & LeBoeuf, B. The effects of intromission frequency on successful pregnancy in the female rat. *Proceedings of the National Academy of Sciences, U.S.A.,* 1965, **53**, 1392–1395.

Wilson, P. D., & Riesen, A. H. Visual development in rhesus monkeys neonatally deprived of patterned light. *Journal of Comparative and Physiological Psychology,* 1966, **61**, 87–95.

Wilson, R. C., Vacek, T., Lanier, D. L., & Dewsbury, D. A. Open-field behavior in muroid rodents. *Behavioral Biology,* 1976, **17**, 495–506.

Wimer, R. E., Wimer, C. C., & Roderick, T. H. Genetic variability in forebrain structures between inbred strains of mice. *Brain Research,* 1969, **16**, 257–264.

Winston, H., Lindzey, G., & Connor, J. Albinism and avoidance learning in mice. *Journal of Comparative and Physiological Psychology,* 1967, **63**, 77–81.

Wolff, P. H. Sucking patterns of infant mammals. *Brain, Behavior and Evolution,* 1968, **1**, 354–367.

Wood, D. C. Stimulus specific habituation in a protozoan. *Physiology and Behavior,* 1973, **11**, 349–354.

Woodard, W. T., & Bitterman, M. E. Further experiments on probability learning in goldfish. *Animal Learning and Behavior,* 1973, **1**, 25–28.

Woody, C. D., & Black-Cleworth, P. Differences in excitability of cortical neurons as a function of motor projection in conditioned cats. *Journal of Neurophysiology,* 1973, **36**, 1104–1116.

Woolfenden, G. E. Nesting and survival in a population of Florida scrub jays. *Living Bird,* 1973, **12**, 25–49.

Woolfenden, G. E. Florida scrub jay helpers at the nest. *Auk,* 1975, **92**, 1–15.

Wright, J. W., & Meyer, M. E. A new technique for measuring thermoregulatory behavior in the rat. *Journal of Experimental Analysis of Behavior,* 1969, **12**, 999–1002.

Wynne-Edwards, V. C. *Animal dispersion in relation to social behaviour.* New York: Hafner, 1962.

Yahr, P., & Thiessen, D. D. Steroid regulation of territorial scent marking in the Mongolian gerbil *(Meriones unguiculatus). Hormones and Behavior,* 1972, **3**, 359–368.

Yanai, J., & McClearn, G. E. Assortative mating in mice: 1. Female mating preference. *Behavior Genetics,* 1972, **2**, 173–183.

Yaremko, R. M., Jette, J., & Utter, W. Further study of avoidance conditioning in toads. *Bulletin of the Psychonomic Society,* 1974, **3**, 340–342.

Yeatman, F. R., & Hirsch, J. Attempted replication of, and selective breeding for, instrumental conditioning of *Drosophila melanogaster. Animal Behaviour,* 1971, **19**, 454–462.

Yerkes, R. M. The intelligence of earthworms. *Journal of Animal Behavior,* 1912, **2**, 332–352.

Young, J. Z. *The life of the vertebrates.* Oxford: Oxford University Press, 1950.

Young, J. Z. Learning and discrimination in the octopus. *Biological Reviews,* 1961, **36**, 32–96.

Zajonc, R. B. Social facilitation. *Science,* 1965, **149**, 269–274.

Zajonc, R. B. *Animal social psychology: A reader of experimental studies.* New York: Wiley, 1969.

Zajonc, R. B., Heingartner, A., & Herman, E. M. Social enhancement and impairment of performance in the cockroach. *Journal of Personality and Social Psychology,* 1969, **13**, 83–92.

Zarrow, M. X., Gandelman, R., & Denenberg, V. H. Prolactin: Is it an essential hormone for maternal behavior in the mammal? *Hormones and Behavior,* 1972, **2**, 343–354.

Zeigler, H. P. Feeding behavior of the pigeon. *Advances in the Study of Behavior,* 1976, **7**, 285–389.

Zemlan, F. P., Ward, I. L., Crowley, W. R., & Margules, D. L. Activation of lordotic responding in female rats by suppression of serotonergic activity. *Science,* 1973, **179**, 1010–1011.

Zener, K. The significance of behavior accompanying conditioned salivary secretion for theories of the conditioned response. *American Journal of Psychology,* 1937, **50**, 384–403.

Zepelin, H., & Rechtschaffen, A. Mammalian sleep, longevity, and energy metabolism. *Brain, Behavior and Evolution,* 1974, **10**, 425–470.

Zigmond, R. E., Nottebohm, F., & Pfaff, D. W. Androgen-concentrating cells in the midbrain of a songbird. *Science,* 1973, **179**, 1005–1007.

Zimmermann, R. R., & Torrey, C. C. Ontogeny of learning. In A. M. Schrier, H. F. Harlow, & F. Stollnitz (Eds.), *Behavior of nonhuman primates: Modern research trends*. Vol. 2. New York: Academic Press, 1965. Pp. 405–447.

Name Index

Page numbers in *italic* indicate names appearing in Reference section.

Species Index

Subject Index